Forests in Revolutionary France

This book investigates the economic, strategic, and political importance of forests in early modern and modern Europe and shows how struggles over this vital natural resource both shaped and reflected the ideologies and outcomes of France's long revolutionary period. Until the mid-nineteenth century, wood was the principal fuel for cooking and heating, the primary material for manufacturing worldwide, and the basis for nearly every element of industrial, domestic, military, and maritime activity. Forests also provided essential pasturage. These multifaceted values made forests the subject of ongoing battles for control between the crown, landowning elites, and peasantry, for whom liberty meant preserving their rights to woodland commons.

Focusing on Franche-Comté, an eastern province of France, the book explores the fiercely contested development of state-centered conservation and management from 1669 to 1848. In emphasizing the environmental underpinnings of France's seismic sociopolitical upheavals, it appeals to readers interested in revolution, rural life, and common-pool-resource governance.

Kieko Matteson is Assistant Professor of History at the University of Hawai'i at Mānoa. Her dissertation received the American Society for Environmental History's Rachel Carson Prize and Yale University's Henry A. Turner Prize for outstanding work in European history.

Studies in Environment and History

Editors

J. R. McNeill, *Georgetown University*
Edmund P. Russell, *University of Kansas*

Editors Emeritus

Alfred W. Crosby, *University of Texas at Austin*
Donald Worster, *University of Kansas*

Other Books in the Series

George Colpitts	*Pemmican Empire: Food, Trade, and the Last Bison Hunts in the North American Plains, 1780–1882*
Micah Muscolino	*The Ecology of War in China: Henan Province, the Yellow River, and Beyond, 1938–1950*
John Brooke	*Climate Change and the Course of Global History: A Rough Journey*
Emmanuel Kreike	*Environmental Infrastructure in African History: Examining the Myth of Natural Resource Management*
Gregory T. Cushman	*Guano and the Opening of the Pacific World: A Global Ecological History*
Sam White	*Climate of Rebellion in the Early Modern Ottoman Empire*
Alan Mikhail	*Nature and Empire in Ottoman Egypt: An Environmental History*
Edmund Russell	*Evolutionary History: Uniting History and Biology to Understand Life on Earth*
Richard W. Judd	*The Untilled Garden: Natural History and the Spirit of Conservation in America, 1740–1840*
James L. A. Webb Jr.	*Humanity's Burden: A Global History of Malaria*
Frank Uekoetter	*The Green and the Brown: A History of Conservation in Nazi Germany*
Myrna I. Santiago	*The Ecology of Oil: Environment, Labor, and the Mexican Revolution, 1900–1938*
Matthew D. Evenden	*Fish versus Power: An Environmental History of the Fraser River*
Nancy J. Jacobs	*Environment, Power, and Injustice: A South African History*
Adam Rome	*The Bulldozer in the Countryside: Suburban Sprawl and the Rise of American Environmentalism*
Judith Shapiro	*Mao's War Against Nature: Politics and the Environment in Revolutionary China*
Edmund Russell	*War and Nature: Fighting Humans and Insects with Chemicals from World War I to Silent Spring*
Andrew Isenberg	*The Destruction of the Bison: An Environmental History*
Thomas Dunlap	*Nature and the English Diaspora*
Robert B. Marks	*Tigers, Rice, Silk, and Silt: Environment and Economy in Late Imperial South China*

Mark Elvin Tsui'jung Liu	*Sediments of Time: Environment and Society in Chinese History*
Richard H. Grove	*Green Imperialism: Colonial Expansion, Tropical Island Edens and the Origins of Environmentalism, 1600–1860*
Elinor G. K. Melville	*A Plague of Sheep: Environmental Consequences of the Conquest of Mexico*
J. R. McNeill	*The Mountains of the Mediterranean World: An Environmental History*
Theodore Steinberg	*Nature Incorporated: Industrialization and the Waters of New England*
Timothy Silver	*A New Face on the Countryside: Indians, Colonists, and Slaves in the South Atlantic Forests, 1500–1800*
Michael Williams	*Americans and Their Forests: A Historical Geography*
Donald Worster	*The Ends of the Earth: Perspectives on Modern Environmental History*
Samuel P. Hays	*Beauty, Health, and Permanence: Environmental Politics in the United States, 1955–1985*
Warren Dean	*Brazil and the Struggle for Rubber: A Study in Environmental History*
Robert Harms	*Games Against Nature: An Eco-Cultural History of the Nunu of Equatorial Africa*
Arthur F. McEvoy	*The Fisherman's Problem: Ecology and Law in the California Fisheries, 1850–1980*
Alfred W. Crosby	*Ecological Imperialism: The Biological Expansion of Europe, 900–1900, Second Edition*
Kenneth F. Kiple	*The Caribbean Slave: A Biological History*
Donald Worster	*Nature's Economy: A History of Ecological Ideas, Second Edition*

Forests in Revolutionary France

Conservation, Community, and Conflict, 1669–1848

KIEKO MATTESON
University of Hawai'i at Mānoa

CAMBRIDGE
UNIVERSITY PRESS

Rachel Carson Center
Environment and Society

CAMBRIDGE
UNIVERSITY PRESS

32 Avenue of the Americas, New York, NY 10013-2473, USA

Cambridge University Press is part of the University of Cambridge.

It furthers the University's mission by disseminating knowledge in the pursuit of education, learning, and research at the highest international levels of excellence.

www.cambridge.org
Information on this title: www.cambridge.org/9781107043343

First published 2015

Printed and bound in Great Britain by Clays Ltd, St Ives plc

A catalog record for this publication is available from the British Library.

Library of Congress Cataloging in Publication Data
Matteson, Kieko.
Forests in revolutionary France : conservation, community, and conflict 1669–1848 /
Kieko Matteson, University of Hawai at Manoa.
pages cm. – (Studies in environment and history)
Includes bibliographical references.
ISBN 978-1-107-04334-3
1. Forests and forestry–France–History–17th century. 2. Forests and forestry–France–History–18th century. 3. Forests and forestry–France–History–19th century.
4. Forests and forestry–Political aspects–France–History. I. Title.
SD59.M38 2015
634.90944–dc23 2014035053

ISBN 978-1-107-04334-3 Hardback

Contents

Illustrations

Maps

Abbreviations

1669 Ordinance in Isambert, vol. 18	*Édit portant règlement général pour les Eaux et Forêts, Saint-Germain-en-Laye, 31 août 1669*. Edict no. 571; reprinted in François-André Isambert, Athanase-Jean-Léger Jourdan (Decrusy), and Alphonse-Honoré Taillandier, eds., *Recueil général des anciennes lois françaises, depuis l'an 420 jusqu'à la Révolution de 1789*. 29 vols. (Paris, Belin-Leprieur, 1821–33), 18:219–311.
AD Doubs	Archives départementales du Doubs, Besançon
ADHS	Archives départementales de la Haute-Saône, Vesoul
ADJ	Archives départementales du Jura, Montmorot
AN	Archives nationales, Paris
AP 1ère série, *AP 2ème série*	*Archives parlementaires de 1787 à 1860: Recueil complet des débats législatifs et politiques des chambres françaises, imprimé par ordre du corps législatif sous la direction de mm. J. Mavidal et E. Laurent. Première série (1787 à 1799)*, 82 vols.; *Deuxième série (1800 à 1860)*, 127 vols. (Paris, 1867–1913).
Baudrillart, *Recueil chronologique*, tome 1	Jacques-Joseph Baudrillart, *Traité général des eaux et forêts, chasses et pêches*, part 1, tome 1, *Recueil chronologique des réglemens forestiers: contenant les ordonnances, édits et déclarations des rois de France; les arrêts du conseil et des cours souveraines; les lois, arrêtés du gouvernement, décrets, ordonnances du roi, arrêts de la Cour de Cassation,*

décisions ministérielles, circulaires et instructions administratives (Paris: Imprimerie de Madame Huzard, 1821). (Note: Baudrillart initially intended the *Recueil* as one volume of the planned four-volume *Traité général des eaux et forêts, chasses et pêches*, but the work evolved into a five-volume work unto itself. To avoid confusion with the volume numbering of the *Traité*, I use "tome" to refer to the *Recueil* volumes.)

Baudrillart, *Recueil chronologique*, tome 3 — Jacques-Joseph Baudrillart, *Traité général des eaux et forêts, chasses et pêches*, part 1, tome 3, *Recueil chronologique des réglemens sur les forêts, chasses et pêches, contenant les lois, ordonnances royales, arrêts de la Cour de Cassation, décisions ministérielles, et les circulaires et instructions administratives* (Paris: Imprimerie de Madame Huzard, 1824). (See note on preceding item.)

Duvergier, *Collection complète des lois* — J. B. Duvergier, ed., *Collection complète des lois, décrets, ordonnances, réglemens, avis du Conseil-d'Etat, publiée sur les éditions officielles du Louvre, de l'Imprimerie nationale, par Baudouin; et du Bulletin des lois; de 1788 à 1830 inclusivement, par ordre chronologique*, 2nd ed., 30 vols. (Paris: Chez A. Guyot et Scribe, 1834–38).

JHS — *Journal de la Haute-Saône*, conservative newspaper of the Haute-Saône, 1848, 1852.

Preface and Acknowledgments

> Before I built a wall I'd ask to know
> What I was walling in or walling out,
> And to whom I was like to give offence.
>
> Robert Frost, "Mending Wall," 1914[1]

This book has its origins in Vermont, where for generations my family owned an eighteenth-century farmhouse and several hundred acres of surrounding fields and forest. By the time I appeared on the scene, it was no longer a working enterprise – it housed no animals, other than the woodchucks and chipmunks that lived under the outbuildings, and it grew no commercial crops. Instead, my parents and sisters and I traveled the ten miles from town every weekend to stay overnight, trundle about in the woods, and, in summer months, coax vegetables from the stony soil of our enormous garden.

It was there that I gained an early appreciation for the concept of customary usage and the idea of land as a resource that cannot, and ought not, be possessed exclusively. In the fall, around the time the sugar maples' reds and golds had faded to brown and the "leaf-peepers" had headed back to points south, men dressed in orange vests and camouflage pants would begin appearing at the edge of our meadow, pausing for a cursory wave before trudging, rifles in hand, up the logging road and into the woods. Their cars, large sedans with out-of-state plates, often parked in ways that blocked our egress, underscored their outsider status. My sisters and I would suspend whatever we were doing – digging

[1] Robert Frost, "Mending Wall," in *The Poetry of Robert Frost* (New York: Macmillan, 1979), 34.

potatoes or hauling the last of the winter squash – and tepidly return the men's salute. For us, hunting season meant six weeks of avoiding the woods and dressing in bright orange, lest we be mistaken for a whitetail deer. We had heard the stories – a girl in Maine had been shot in her own yard because she was wearing white mittens. What we resented as perilous trespass, however, was to our father a time-honored tradition. Each time we demanded to know why these intruders were allowed on our property, he would reply in august tones, "our land is not posted" – referring to the small white notices that dotted the property lines of other farms around the county and warned hunters, fishers, and trappers to stay away. Posting, he asserted, was inconsistent with the "neighborly use of private land" that had long characterized the region. "As long as they're not hurting anything or making any messes up there," he would add, gesturing to the mountain behind the farm, "they can come."

No doubt the deer that we saw browsing the meadow's edge in the frosty mornings would have begged to differ with my father's concept of "not hurting anything." And I, for one, remained skeptical of his benevolence toward these strangers, who littered the woods with beer cans and snuff tins and kept us from our rambles during the best season of the year. Still, the message took hold. Legal formalities aside, every property is subject to overlapping interests and competing claims, to assertions expressed through a wall or a wave, and to needs that fluctuate with the passage of time. Just as surely as we accepted that by planting a garden we played host to pollinators and pests, so too our status as the farm's titular landowner in no way invalidated its appeal to others. Instead, we learned to raise our hand in greeting to the hunters, hikers and dirt bikers, in the hopes that when it was our turn to cross their properties, they would do the same. In this same spirit of quid pro quo, we welcomed (albeit more enthusiastically) the maple sugar man in March, who tapped our trees in exchange for two gallons of syrup, and the mowing-machine guys in August, who kept our meadows open in return for hay they cut.

My youthful awareness that the land was not ours alone to possess was eventually rendered manifest in my father's decision to sell it – woodlot by woodlot, meadow by meadow – to pay off property taxes and accumulated debts. Today, he is the uninvited guest on the farm, tottering in his nonagenarian dotage across Vermont's last remaining unposted parcels in search of half-recollected memories and places from long ago.

Though personally painful, the loss of the farm was not entirely unexpected. The fleeting nature of our and our ancestors' presence had been apparent wherever I looked – in the decrepit stone walls surrounding the stagheaded old sugar bush; in former pastures overgrown with beech and hemlock; and in the ancient, caved-in sheep dip on the hillside, where unwilling ovines were once plunged in a chemical cocktail to rid their fleece of ticks and which had since become home to a horde of ground-dwelling critters. These traces of past possessors, and evidence of their replacement and succession, made clear to me in a way no textbook ever could that ecosystems are dynamic and our place within them transient. At the same time, they raised lingering questions about how people adapt and transform the land to serve their needs, how different stakeholders accommodate each other or remain at variance, and how these interests play out in policy and practice. Those questions formed the seedbed of my later research.

This path from recreational avocation to intellectual vocation may seem a winding one. Vermont is a far cry geographically and temporally from France's sanguinary struggles over natural resources in the eighteenth and nineteenth centuries. Nonetheless, for me, the history of forest conservation and customary rights in the revolutionary era that I first became interested in as an undergraduate resonates deeply with the larger questions of access, exploitation, and transformation that were piqued long ago on our family farm. As you will see in the story that follows, peasant communities and practitioners of customary rights, denounced as malefactors by state, industrial, and private landowner interests, succeeded in the long run through tenacity, wiliness, and sheer violence in retaining significant control over their forests – though not before it had ceased to be the most critical element of their everyday survival. Like the moss-covered stone walls that I clambered over in my childhood – barriers delineating practices that no longer took place – communal possession and customary rights in France persist today in a substantially altered ecological and economic context. Understanding this struggle for control, its stakes and evolution – as well as, indirectly, its implications for a future filled with looming and increasingly dire natural resource challenges – is the aim of this book.

In the course of working on this absurdly protracted project – a *longue durée* effort in every sense of the term – I have accumulated an enormous debt of gratitude to many people and institutions. I am grateful for the opportunity to finally be able to thank them. First and foremost, I salute the Rachel Carson Center for Environment and Society at the Ludwig Maximilian University of Munich, especially Christof Mauch and

Helmuth Trischler, whose support enabled me to focus on the manuscript and explore its implications in the context of a warmly collegial and intellectually vibrant community of environmental humanities scholars. The Rachel Carson Center also generously contributed to the book's production costs.

At different stages of the project, I received critical financial support for archival research and instructive field visits to forests throughout Franche-Comté from the University of Hawai'i History Department, Dean's Office, and University Research Council, as well as the Yale Program in Agrarian Studies, the Yale Council on International and Area Studies, and the Mellon Foundation. My research was enabled by the staff of the Archives nationales and Bibliothèque nationale in Paris; the archives départmentales of the Haute-Saône, Doubs, and Jura; Yale's Sterling, Beinecke, Forestry and Environmental Studies, and Lillian Goldman Law Libraries; and the outstandingly efficient University of Hawai'i interlibrary loan department. For helping to create the maps that appear in this book, I am particularly grateful to Janet Dombrowski in the University of Hawai'i's Geospatial Information Services Department and David Olsen in the University of Hawai'i Cartography Lab.

Many people read and commented on the manuscript at different stages. Starting midalphabet and scattering forward, then backward, in the same way that I read novels, I thank for their invaluable insights Peter McPhee, John Merriman, Jeremy Popkin, Naoko Shibusawa, Mari Yoshihara, Laura Lyons, Cindy Franklin, Robert Harms, Monisha Das Gupta, William Beik, David A. Bell, and Ellen F. Arnold. For their patience and faith in the project, I thank my editors, Donald Worster, John McNeill, Edmund Russell, and Deborah Gershenowitz. For inimitable support and reflection on concerns related to the book and otherwise, I am inexpressibly grateful to Frank Zelko, Linda Lierheimer, Matthew Lauzon, Marcus Daniel, Vina Lanzona, Demetria Koninis-Washburn, Andrée Collier, Yuma Totani, Jacqueline Wah, and Winston Welch. For introducing me to France in indelible travels in their Volkswagen camper, I thank Brenda, Ed, and Ursula Owre. For inspiring me to think about ecology and conservation, I thank Mollie Matteson. For their companionship in work and play, I thank all of my sisters, especially Ruth Banchik. For my education, I thank my parents, Mary and Bob Matteson. Angela Kaufman and Xioayan Zuo helped me carve out writing time while my daughters were young; now my daughters are old enough to appreciate a dedication. Amika and Riya, this book is for you.

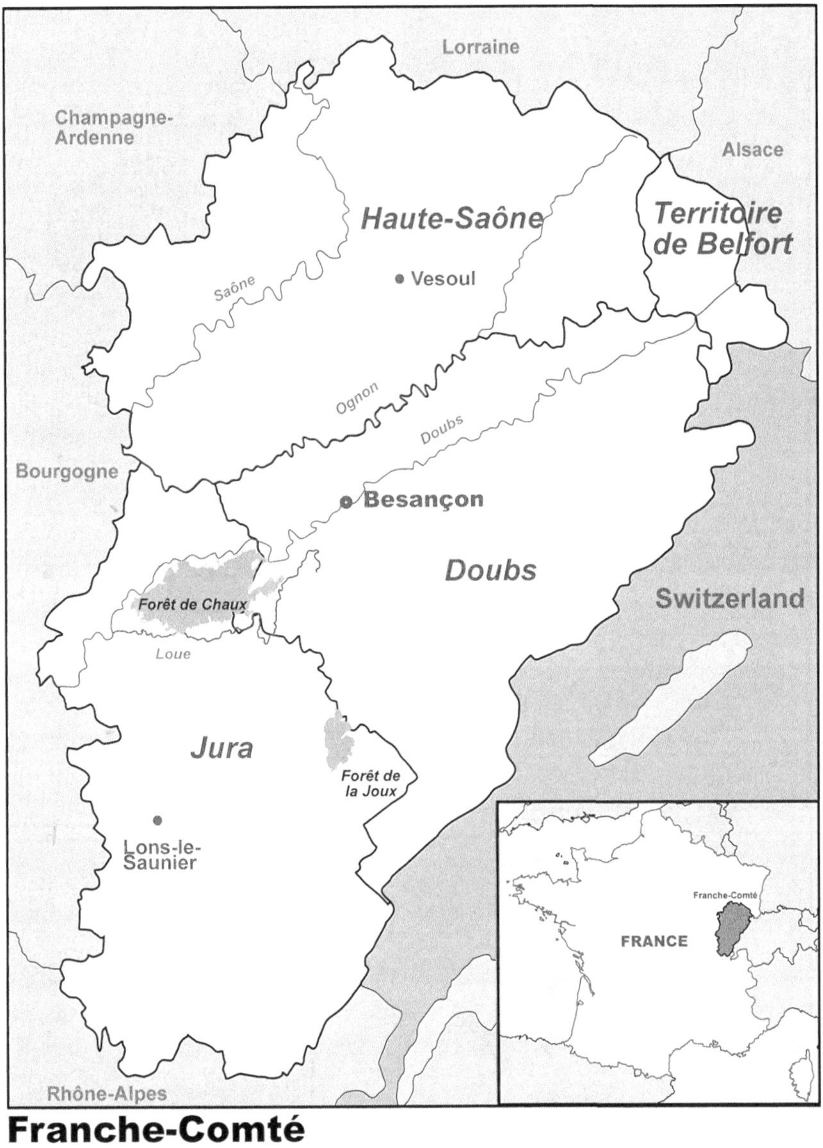

1 Franche-Comté: Political
Source: Map prepared by David Olsen, University of Hawai'i Cartography Lab ©. Printed with permission.

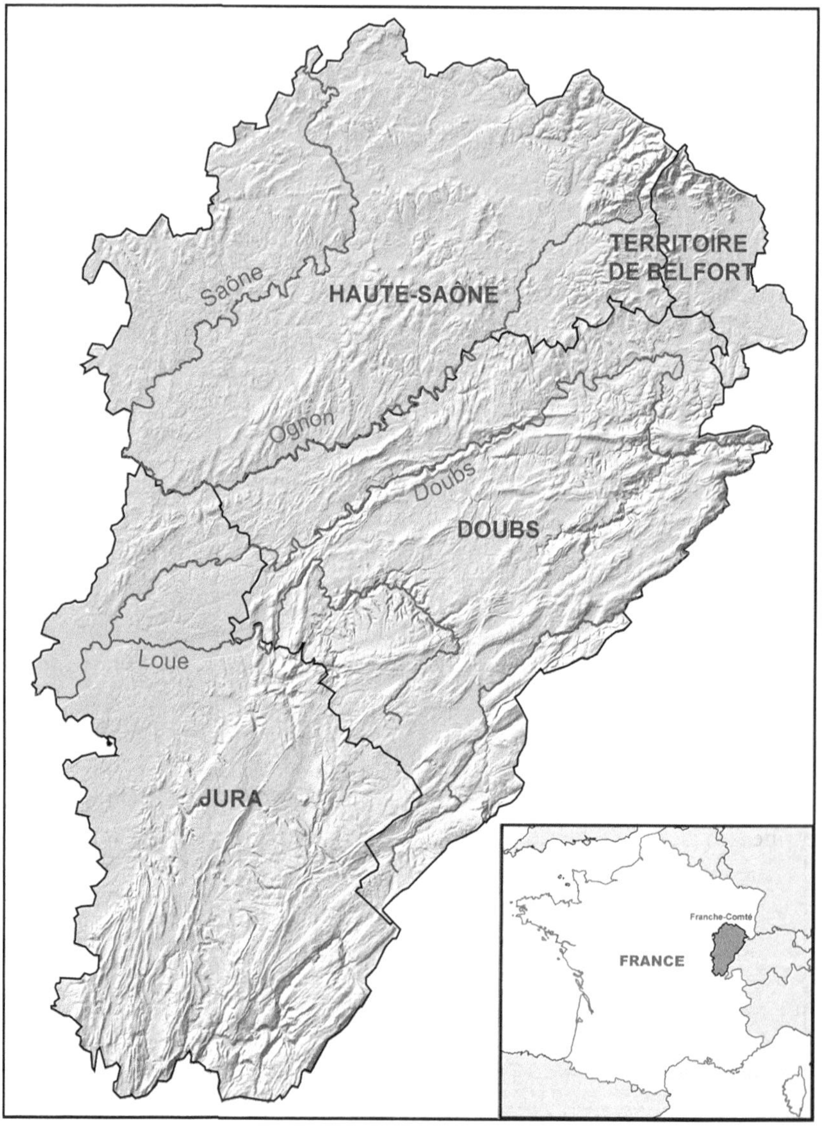

2 **Franche-Comté: Physical**
Source: Map prepared by David Olsen, University of Hawai'i Cartography Lab ©. Printed with permission.

Introduction

In March 1789, amid the deep snow of a lingering winter, five dozen men from the tiny village of Charquemont in the eastern French province of Franche-Comté gathered to discuss their hopes and expectations for the upcoming meeting of the Estates General, a gathering of representatives from across France which was to convene that spring for the first time in 175 years. Like thousands of other communities writing their own petitions, known as *cahiers de doléances*, at that very moment, the Charquemont villagers had much to say to the impending convention.[1] Tax reform, the standardization of weights and measures, freedom of commerce – issues typically associated with the political upheaval that was about to unfold – ranked high among the demands in their fifteen-page petition.[2] But the Charquemontais were equally passionate about another subject: their forests. Still smarting from the imposition three decades earlier of alien and exacting French woodland regulations, Charquemont's petitioners used their *cahier de doléances* as an opportunity to demand an end to the crown's forest oversight. "The community alone [should] have supervision of

[1] Forty thousand parishes drafted *cahiers de doléances*, of which about twenty-five thousand remain in printed and manuscript form. The First and Second Estates – the clergy and nobility, respectively – drafted their *cahiers* separately. Gilbert Shapiro and John Markoff, introduction to *Revolutionary Demands: A Content Analysis of the Cahiers de Doléances of 1789* (Stanford, CA: Stanford University Press, 1998), 3, 114, 233.

[2] On the dominant grievance categories, see Shapiro and Markoff, "What Were the Grievances of France in 1789?" chap. 14 in *Revolutionary Demands*, 253–79.

its woods," they insisted, and "state and private hunting and forest guards [should] be suppressed."[3]

Charquemont's petitioners were far from unique in expressing concern about their woodland resources. Communities from every corner of France registered forest-related grievances in 1789, including complaints regarding the price and availability of wood, outrage about the loss of grazing and gathering rights, and recriminations against the *Eaux et Forêts*, the royal agency in charge of overseeing the country's woods and inland waters.[4] Though diverse in their particulars, the remonstrances reflected France's escalating forest crisis and widening economic predicaments in the eighteenth century. Decades of demographic growth, particularly in the northeast, had catalyzed clearing for agriculture, intensified industrial fuel consumption, and driven up prices for essential goods like grain and firewood, while wages for labor stagnated from midcentury forward.[5] In response, landowners, of whom the crown and seigneurs (feudal landlords) comprised the vast majority, increasingly sought to suppress traditional forest rights and expand their proprietary authority. To profit from rising wood prices and to feed their industrial ventures, proprietors clamped down on centuries-old access rights in their holdings, expanded their use of guards, and appointed agents to oversee exploitation. At the same time, they inflated their claims on communal possessions and instituted forms of management that favored naval timber and charcoal production. The result was not only privation for communities reliant on woodland resources but also a marked transformation of the rural environment. Spaces long shaped by pasturing and wood gathering – with understories scoured by nibbling herds, canopies dominated by stagheaded *baliveaux* (tall trees retained to reseed the forest), and

[3] *Cahier de doléances*, Charquemont, 19 March 1789, preamble and articles 22–23, in Robert Jouvenot, "Les cahiers de doléances en Franche-Comté," in *Questions d'histoire sur la Franche-Comté, à l'occasion du 99e congrès national des Sociétés savantes*, ed. Comité des travaux historiques et scientifiques (CTHS) Orientations de recherche (Besançon: CTHS Bulletin de la section d'histoire moderne et contemporaine, 1974), 245–49. All translations in the book are mine unless otherwise noted.

[4] Shapiro and Markoff, *Revolutionary Demands*, 419–25.

[5] On the historiography of the economic crisis, see Peter Jones, *The Peasantry in the French Revolution* (Cambridge: Cambridge University Press, 1988), 30–33. On demographic growth, see Jacques Dupâquier, ed., *Histoire de la population française* (Paris: Presses Universitaires de France, 1988), 2:75–78. Population increase was most marked in upland communities like Charquemont. André Fel, "Petite Culture 1750–1850," in *Themes in the Historical Geography of France*, ed. Hugh D. Clout (London: Academic Press, 1977), 226; and Jean Boichard, ed., *Le Jura* (Toulouse: Privat, 1986), 100–104.

irregular patches of scrubby *friches* – were converted into closely guarded plots of even-aged *haute futaie* (timber trees) or were cut down and replanted as high-volume coppices, felled on regular rotations, and oxidized on-site into charcoal fuel for iron forges.

The significance of these changes is hard to grasp in an age overrun by plastics and dependent on decomposed fossils. Well into the nineteenth century, France, like most of Europe, remained acutely reliant on wood as its principal energy source and as the essential building block of everyday life.[6] Drawn from a vast array of species prized for their distinct attributes, wood comprised the core of nearly every tool and fitting of agricultural, viticultural, and industrial production, from rakes, stakes, lathes, and crates to baskets, binding cord, footwear, and furniture. It enabled communications and commercial exchange via the bins and boxes, carts and carriages that carried people and commodities to their destinations. It facilitated France's maritime and military objectives in the form of ships, ramparts, and charcoal-smelted artillery. It formed the walls and frames of homes, barns, and outbuildings. Most of all, it served as the country's primary domestic and industrial fuel, even in areas where alternative combustibles like coal or peat were available. Forests themselves provided incalculable resources, including leaf litter for stable bedding; bark for tanning leather; and supplemental edibles and medicinals like mushrooms, sap, roots, and berries.

Seen in this light, the Charquemont petitioners' woodland complaints, together with the comparable objections in thousands of other *cahiers de doléances* in 1789, highlight the forest's fundamental position in France's economy in the early modern period and signal its importance in shaping the rural context of the Revolution. At a time of critical sylvan dependence, material concerns entwined inextricably with political ideals, influencing allegiances and molding expectations. Woodland stakeholders like the Charquemontais hoped the winds of political change would bring them greater freedom and ensure their right to manage and exploit the resource most crucial to their survival. Instead, the regimes that successively emerged from the Revolution would disenfranchise them still further, leading to economic suffering, outrage, and intensifying violence. By the middle of the nineteenth century, communal relations with the

[6] On wood's strategic importance, see Michael Williams, *Deforesting the Earth: From Prehistory to Global Crisis* (Chicago: University of Chicago Press, 2003), 145–67; and F. T. Evans, "Wood since the Industrial Revolution: A Strategic Retreat?" in *History of Technology*, ed. A. Rupert Hall and Norman Smith (London: Mansell, 1982), 46–48.

forest administration would be at their nadir, even as the nation's groundbreaking woodland policies expanded their sway beyond France's borders and into its colonial regimes.

This book investigates the history of forest politics in France over a span of nearly two centuries, from the pinnacle of Louis XIV's reign in the 1670s to the cusp of the Second Empire in 1848. Encompassing some of the country's most transformative moments, this *longue durée* perspective poses some challenges, not least of them the scope of the research required. But forests are long-lived entities, and understanding the origins, evolution, and impact of the interests that shaped them requires an extended frame of analysis.

Focused on the intersection of environmental anxieties – the expansion of arable soil, mounting firewood and timber shortages, and other concerns – with the aims of an increasingly bureaucratic, fiscally desperate French state, I trace the emergence of an autocratic, centralized approach to conservation, the suppression of customary and communal rights, and the rise of proprietary individualism amid forests decimated by economic pressure and political tumult. At a time when woodland access was considered "a necessity as vital as bread," as another *cahier de doléances* put it, and firewood was as strategically vital a commodity as oil is today, the determination of who would control France's forests and how they would benefit was a matter of life-or-death importance.[7] To make sense of this struggle and its consequences, I probe periods of exceptional upheaval as well as the relatively overlooked interstices when new woodland policies took hold.

Geographically, the book is grounded in Franche-Comté, a former Habsburg territory in eastern France that was conquered by Louis XIV in 1674. Comprising the departments of the Doubs, the Jura, and the Haute-Saône, Franche-Comté has long been one of France's most densely wooded and remote areas.[8] Far from the nation's major arteries, coiled hard against the Swiss border, it is regarded as something of a backwater by French cosmopolitans, who start humming the Jacques Brel tune "Vesoul" when I name my region of study. In Brel's rendering, the

[7] Quote from the *cahier de doléances* (grievance petition) of Bourbévelle (Haute-Saône), 17 March 1789, in Charles Godard and Léon Abensour, eds., *Cahiers de doléances du bailliage d'Amont*, vol. 1 (Besançon, 1927), 218.

[8] The tiny Territoire de Belfort in Franche-Comté's northeastern corner is not included in this study because of its geographic isolation and historical dissimilarities.

Haute-Saône capital stands for the absurd lengths one goes to for love. Yet in the realm of forest conservation and debates over environmental policymaking on the national scale, Franche-Comté and its inhabitants played a central role. A site of strict state initiatives and fierce resistance until well into the nineteenth century, Franche-Comté's bitterly fought forest battles – sometimes waged against trees themselves – enduringly influenced conservation initiatives both within France and abroad.[9]

Given their dominance in Franche-Comté, forests unsurprisingly figure strongly in the region's historiography.[10] Studies of woodland geography, exploitation, and conflict constitute a sizeable subgenre of Comtois history unto themselves.[11] Despite their indispensable findings, however, these works for the most part focus on the local picture, to the exclusion of wider political and economic developments in France. In the hopes of speaking to a readership unfamiliar with Franche-Comté and even those unschooled in France's turbulent revolutions, this book attempts a more ambitious, compound approach: keeping the experience of the Francs-Comtois at the center of the frame, I scrutinize the national reform efforts and ideological agendas of lawmakers, landowners, rural stakeholders, and self-styled "improvers," to explore larger questions about the failures and successes of natural resource allocation, environmental conservation, state formation, and revolutions broadly speaking. Why do conservation policies forged in the name of the public good engender resistance?

[9] On France's colonial forestry, see Mathieu Guérin, *Paysans de la forêt à l'époque coloniale: La pacification des aborigènes des hautes terres du Cambodge, 1863–1940*, Bibliothèque d'histoire rurale (Caen: Association d'histoire des sociétés rurales, 2008); and Diana K. Davis, *Resurrecting the Granary of Rome: Environmental History and French Colonial Expansion in North Africa* (Athens: Ohio University Press, 2007). See also Peter Vandergeest and Nancy Lee Peluso, "Empires of Forestry: Professional Forestry and State Power in Southeast Asia, Parts 1 and 2," *Environment and History* 12, nos. 1 and 4 (2006): 31–64, 359–93.

[10] See, for example, Antonio Gonzales and Pierre Gresser, *Nouvelle histoire de la Franche-Comté et des francs-comtois*, (Pontarlier: Éditions du Belvédère, 2014). Jean-Luc Mayaud, *Les secondes républiques du Doubs*, Annales littéraires de l'université de Besançon 338 (Paris: Les Belles Lettres, 1986); R. Locatelli et al., *La Franche-Comté: À la recherche de son histoire (1800–1914)*, Cahiers d'études comtoises 31 (Paris: Les Belles Lettres, 1982); and Maurice Gresset, *La Franche-Comté à la veille de la Révolution*, Collection du bicentenaire de la Révolution française 8 (Paris: Les Belles Lettres, 1988).

[11] In particular, see the prodigious œuvre of François Vion-Delphin, Franche-Comté's foremost forest historian, including François Vion-Delphin et al., eds., *Les hommes et la forêt en Franche-Comté* (Paris: Editions Bonneton, 1990); François Vion-Delphin, "Le braconnage en Franche-Comté: Une pratique populaire au XVIIIe siècle," in *Forêt et chasse*, ed. Andrée Corvol (Paris: L'Harmattan, 2004), 201–11.

How did the ideology of private property come to supplant customary rights and collective usage of common pool resources? How did Comtois communities manage to sustain their traditional practices, despite persistent repression and unfavorable odds, throughout the long revolutionary period and into the twenty-first century? And why does France today, despite its pioneering efforts at forest regulation, have some of the most fragmented and, from a commercial standpoint, underutilized woodlands in Europe?[12]

To address these questions, I draw on extensive archival research and documentary evidence, from grievance petitions, policy proposals, and police reports to legislative debates and lawsuits. I also look to the land and its utilization. Traveling amid Franche-Comté's tiny hamlets, especially in the upland Doubs and Jura, I have been struck time and again by their physical isolation and the challenges inhabitants must have faced in surviving. Burrowed in the depths of a valley, wedged into a river gorge, or huddled atop a rocky plateau, these communities had to make the most of the soil and woods around them. They developed multifaceted and collective modes of exploitation, including silvopastoralism, selective timber felling, and *fruitières* – cheese-making cooperatives – that reflected the constraints of their environment as well as prevailing hierarchies of distribution and status among local denizens. Socially, ecologically, and legally complex, and specific to each village, these arrangements perpetually confounded state and seigneurial attempts to impose a more uniform system of management.

Whether state, communal, or private holdings, forests were defined and governed in terms of use rights, rather than terms of exclusion. The uses, exercised communally or in cooperation with other clearly defined groups or individuals, fell into three categories: rights to wood (firewood, timber, and small wood for tools and crafts); rights of pasturage (in woodland clearings, on understory grasses, and on acorn, beech,

[12] Private ownership accounts for more than 75 percent (12.4 million hectares) of France's forests. Public holdings (state, regional, departmental, and communal) constitute the remaining 4 million hectares. In Franche-Comté, by contrast, private forests comprise 47 percent of the region's forest cover and average less than 2 hectares in size. For data, see Institut national de l'information géographique et forestière, Inventaire forestier, *Le mémento: La forêt en chiffres et en carte*, 2013, http://inventaire-forestier.ign.fr/spip/spip.php?article583, 8–9; and Schéma Régional de Gestion Sylvicole, "Forêt et sylviculture en Franche-Comté," April 2006, 25. On the underexploitation of Franche-Comté forests, see Service régional de l'information statistique et économique, Direction Régionale de l'Alimentation, de l'Agriculture et de la Forêt, Franche-Comté, "Fiche thématique – forêt," May 2010, 3.

and other nut mast), and "secondary rights," including rights to gather sod and fruits. Each of these general categories, in turn, comprised a vast diversity of activities that differed in nature and nomenclature across Franche-Comté and more variably still across France.[13]

Despite these rights and the communal husbanding of resources and labor that made them function, rural life remained precarious, particularly in the windswept and stony uplands. State and private efforts in the nineteenth century to establish manufactures of glasses, clocks, and tobacco pipes in the Doubs and the Jura helped, but the success of this supplemental "winter work" was still not enough to supplant the silvopastoral economy nor reduce the intensity of forest dependence. Forests – or, rather, access to forests' vital benefits – represented the difference between feasibility and despair.

As the following chapters will show, it was this urgent reliance on their forests that rendered Francs-Comtois so hostile to interference and turned them adamantly against external rule. After Louis XIV forcibly annexed the territory in 1674, Franche-Comté's woodlands were slowly and painstakingly reorganized by France. Implementation in the province of France's 1669 *Ordonnance des eaux et forêts* – a landmark of comprehensive, top-down timber controls and woodland management – took nearly five decades, by which point new and sometimes contradictory regulatory schemes were coming to the fore, each bearing the imprint of particular political exigencies, economic theories, and material anxieties. The middle years of the eighteenth century saw the propagation of physiocratic decrees to encourage clearing of forests for agriculture,

[13] The right and practice of feeding of pigs on nut mast, for example, was known variously as *panage*, *porcage*, *glandage*, *glanage*, and *glandée*, each of which had a different, specific meaning. Other terms expressed a complex series of steps in pursuit of a particular benefit, as in the *droit d'écobuage*: the right to clear woods and obtain sod by carving up, drying, and burning sod on-site. On the common rights of Comtois forest users prior to 1789, see Georges Plaisance, "Les droits d'usage forestiers et leur vocabulaire," in *Colloque sur la forêt*, Annales littéraires de l'Université de Besançon 88, Cahiers d'études comtoises 12 (Besançon: Les Belles Lettres, 1966), 209–18; Michel Vernus, *La vie comtoise au temps de l'Ancien Régime*, vol. 1 (Lons-le-Saunier: Éditions Marque-Maillard, 1988), 70–79; and Suzanne Monniot, "Le role de la forêt dans la vie des populations franc-comtoises de la conquête française à la révolution, 1674–1789," *Revue d'histoire moderne* 12, nos. 29/30, n.s. 6 (1937): 449–68.

For examples from specific communities, see Archives nationales (hereafter AN) Q/1/181 6(b) Forêt de l'hôpital du Gros Bois [Doubs], Procès verbaux de reformation de la dite forêt, 3 November 1727; and *Cahier de doléances*, Buffard, 17 March 1789, in François Vion-Delphin and François Lassus, eds., *Le bailliage de Quingey en 1789: Les cahiers de doléances* (Paris: Les Belles Lettres, 1989), 179.

initiatives to increase supplies of naval timber, and innumerable surveys to assess the country's fuel crisis and compute the wood consumption of saltworks and ironworks.[14]

When the Revolution began, legislators aspired to streamline these initiatives and issue an updated version of the 1669 Ordinance that would address France's current and pressing forest concerns. However, their ambitions were thwarted by the nearly impenetrable web of rights in nearly every woodland, large or small, as well as by the rapidly shifting and increasingly discordant political landscape. A 1791 decree decentralizing forest oversight – a popular but poorly conceived solution to the otherwise impossible legislative morass that forest reform represented – only muddled the picture, as did revolutionary lawmakers' subsequent issuance of dozens of other forest decrees, and the nationalization and sale of ecclesiastical and émigré noble properties from 1790 onward. Taking the helm in 1801 after ten years of relative forest free-for-all, Napoleon revived Old Regime controls and reinstalled a powerful woodland bureaucracy. Nonetheless, no substantive replacement for the 1669 Ordinance was attempted until the 1820s, when deepening wood shortages and a resurgent state bureaucracy finally gave rise to a new Forest Code (*Code forestier*). Rigorously enforced from its promulgation in 1827 forward, the Code remains in effect in amended form today.

Each of these efforts, particularly the 1827 Forest Code, were justified in the name of the *bien public* – the public good. But Franche-Comté's woodland inhabitants rarely saw it that way. Rather, to a remarkable extent, they resisted nearly every endeavor to regulate externally the resources they regarded as their own. Whereas policymakers and political theorists insisted that eliminating customary rights and replacing them with state and private ownership was crucial to France's moral and material regeneration, the Francs-Comtois sought to preserve what they had and restore what they had lost. Their tactics ranged from lawsuits, petitions, poaching, and pilfering to arson, assassination, and outright rebellion. Though Franche-Comté communities were divided over how the forest's benefits should be allocated and who

[14] On the surveys, see Bertrand Gille, *Les sources statistiques de l'histoire de France: Des enquêtes au XVIIe siècle à 1870* (Geneva: Libraire Droz, 1964); Bertrand Gille, "L'enquête sur les bois de 1783" (paper presented at the Actes du 88ème congrès national des sociétés savantes, section d'histoire moderne et contemporaine, Clermont-Ferrand, 1963), 627–46; and Bertrand Gille, "Un recensement des usines comtoise en 1744," *Revue d'histoire de la sidérurgie* 2 (1961): 257–76.

should arbitrate them, most agreed that an abstract, centralized model of administration was not desirable.

Over the course of the "long revolution" from 1789 through 1848, rural inhabitants' hopes of achieving liberty, property, and equality on their own terms at first surged, then soured. Residents in the tiny Doubs hamlet of Rouhe, for example, petitioned the Estates General in March 1789 to say that "the *maîtrises* [the crown's provincial forest bureaus] are onerous to all communities," and that they "eagerly desire the suppression of these jurisdictions."[15] Fifty-nine years later, having been subjected to an overall *extension* of woodland restrictions in the intervening decades, Rouhe's aggrieved denizens took a more confrontational approach. Seizing the opportunity afforded by the political chaos of 1848, they took two forest guards hostage and demanded their resignations, asserting that "they were free; they no longer wanted guards, they would administer their woods according to their own will" – in short, that they were the "masters of their woods."[16]

As in so many other kindred episodes, this revolt yielded mixed results. In the short term, Rouhe's citizens had the satisfaction of turning the tables on the state's hated woodland agents and seizing control for themselves. In the long run, the outcome was more ambiguous. The exuberance of 1848 was quickly quashed by a return to the status quo, the restrictions imposed by the 1827 Forest Code remained in place, and Francs-Comtois continued to leave the countryside in droves in search of urban opportunity. Even so, the bold defiance of Rouhe's residents and other Comtois communities compelled the state to acknowledge their concerns and gained them concessions they might not have gotten otherwise. After the hostage episode, Rouhe's previously denied customary claims on state and private lands were reexamined and converted to communal property. This maneuver, known as *cantonnement*, significantly reduced rights holders' overall forest access but accorded them full title to a fraction. Economically deleterious at the time because of the inadequacy of the reduced portion, Comtois communities' *cantonnements* earn them substantial revenues today through sales of surplus firewood and timber.[17]

15 *Cahier de doléances*, Rouhe, 18 March 1789, in Vion-Delphin and Lassus, *Le bailliage de Quingey*, 108.

16 AN BB^{30} 359, Rapport du procureur général près la Cour d'appel de Besançon, April 1848.

17 This revenue has earned them the reputation of "*communes riches*," free of municipal taxes and awash in perks like new appliances for residents. Philippe Klein, "La Franche-

In the process of exploring the forest struggles of the Revolution and their legacy, three main arguments emerge. First, as the case of Rouhe suggests, peasant opposition to state forest policies produced greater gains than previously appreciated, even when it was crushed at the time. The outbreak of revolution in 1789, 1830, and 1848 catalyzed and capacitated resistance by giving woodland citizens new reasons, opportunities, and political and rhetorical tools to press their case. Although they remained for the most part at a legal and governmental disadvantage and were inevitably outgunned when military force was brought to bear, their tenacity gained them tangible results, especially over the long term.

Second, environmental conservation as it emerged out of the political, social, and economic transformations of early modern and modern France had less to do with farsighted ecological protection – an assertion that is often made in hindsight about both the 1669 Ordinance and the 1827 Forest Code – than with extending state power, suppressing sedition, and substituting commercial exploitation for communal utility. Although conservation has a generally positive connotation for most present-day readers, in practice it had frequently negative consequences for human communities and their natural surroundings.[18] Promoted in the name of state stewardship but premised on suspicion of rural inhabitants, the policies crafted by French governments from Louis XIV forward failed to preserve forests as they were intended.[19] Instead, they intensified antagonisms and exacerbated woodland abuses by rupturing local management regimes, criminalizing customary usage, and forcing inhabitants

Comté touche toujours du bois," *Pays comtois*, March–April 1997, 56. See also discussion in Robert Layton, *Anthropology and History in Franche-Comté: A Critique of Social Theory* (Oxford: Oxford University Press, 2000), 83–84. According to Layton, some villages adopted the practice of selling *affouage* timber and apportioning the revenues directly to inhabitants from 1969 forward, but by 1995 they were no longer doing so.

18 This analysis is well established among social scientists of Asia and Africa. Most recently, see Oliver Wapulumuka, *Conservation Song: A History of Peasant-State Relations and the Environment in Malawi*, 1860–2000 (Cambridge: White Horse Press, 2011); Thaddeus Sunseri, *Wielding the Ax: State Forestry and Social Conflict in Tanzania, 1820–2000* (Athens: Ohio University Press, 2009); and Tim Forsyth and Andrew Walker, *Forest Guardians, Forest Destroyers: The Politics of Environmental Knowledge in Northern Thailand* (Seattle: University of Washington Press, 2008).

19 On views of the peasantry as inherently destructive, see Peter McPhee, "'The Misguided Greed of Peasants'? Popular Attitudes to the Environment in the Revolution of 1789," *French Historical Studies* 24, no. 2 (2001): 247–69; Patrick Matagne, "The Politics of Conservation in France in the 19th Century," *Environment and History* 4 (1998): 359–67; and Frédéric Chauvaud, "Les répresentations morbides de la forêt au XIX siècle," in *La forêt: Perceptions et représentations*, ed. Andrée Corvol et al. (Paris: L'Harmattan, 1997), 367–74.

into furtive and rapacious behaviors.[20] At the same time, espousing prevailing assumptions about the advantages of unencumbered, exclusive private property, the crown and subsequent legislative authorities began granting landed elites widening freedom to manage their woodlands as they saw fit. By handing over supervision of the greater part of the nation's forests to individuals whose primary focus was short-term profit rather than the perpetuation of resources for future generations, this practice accelerated timber sales and clearing and marginalized peasants who relied on customary rights for their survival. Codified in the new woodland laws of 1827, this form of woodland conservation went on to influence national forest policies around the world, most notably in the United States.[21]

Third, if my investigation of Franche-Comté directs a critical eye toward state-centered conservation, it is to point out that local, community-based arrangements for the management and use of natural resources, particularly communal property and customary rights, have been unfairly maligned, both in the historiography of the Revolution and in contemporary policymaking. All too often rural communities have been unproblematically blamed for France's woodland deterioration in the later eighteenth and early nineteenth centuries.[22] This view, framed in terms of promoting conscientious management and even social egalitarianism, was first put forward by agronomists, landowners, and administrators to justify divesting private lands of customary rights and partitioning uncultivated communal holdings.[23] In the twentieth century, the population ecologist Garrett Hardin tidily and influentially summarized the argument as the "tragedy of the commons," the belief that property held in common will inevitably be depleted by free-for-all

20 On this point see Paul Warde, "Fear of Wood Shortage and the Reality of the Woodland in Europe, c.1450–1850," *History Workshop Journal* 62, no. 1 (2006): 28–57; and Paul Warde, *Ecology, Economy, and State Formation in Early Modern Germany* (Cambridge: Cambridge University Press, 2006), 199–206.

21 Gifford Pinchot, the first head of the U.S. Forest Service and founder of the Yale School of Forestry, obtained his only professional forestry training in France. I address this issue at greater length in the Epilogue. See J. E. de Steiguer, "The French National Forestry School One Hundred Years after Pinchot," *Journal of Forestry* 92, no. 2 (1994): 18–20.

22 See, for example, Caroline Ford, "Nature, Culture and Conservation in France and Her Colonies 1840–1940," *Past and Present* 183, no. 1 (2004): 177.

23 The movement was analogous to nineteenth-century enclosure in England, but the form that it took in Revolutionary law (if not in actual practice) was more insistently democratic. On England, see Robert C. Allen, *Enclosure and the Yeoman: Agricultural Development of the South Midlands, 1450–1850* (Oxford: Clarendon Press, 1992).

competition among its users.[24] Yet as this book seeks to show, and as anthropologists, economists, political ecologists, and others have demonstrated in other parts of the world with increasing force over the past five decades, local institutions such as those that governed the woodlands of Franche-Comté were long effective at allocating natural resources in a sustainable fashion, and they did so in a way that allowed for greater biodiversity through mixed forms of use. The difficulty arose when new, extractive pressures on the forest that were oriented primarily toward external markets, including shipbuilding, salt making, and ironworks, were introduced to the region and rapidly expanded with little input or control on the part of the communities most affected.

Community-based arrangements were by no means perfect. As with so many Old Regime structures, local woodland customs frequently perpetuated socioeconomic inequity between landed and landless, men and women, married and widowed, legitimate and illegitimate offspring, and more. Likewise, the tremendous accomplishments of the forest administration in ameliorating deteriorated spaces – driving back the encroachment of sand dunes in the Landes and Sologne, and protecting upland watersheds in the Alps and Pyrenees – likely could not have been achieved without centralized coordination.[25] Nonetheless, by seeking to rationalize rural access and impose uniform restrictions upon woodland resources, often in cooperation with private or industrial interests, the state frequently aggravated the very problems it sought to solve.[26] In this way,

[24] Garrett Hardin, "The Tragedy of the Commons," *Science* 162 (1968): 1244–45. Hardin's brief but forceful essay spawned generations of rebuttals, including Elinor Ostrom et al., "Revisiting the Commons: Local Lessons, Global Challenges," *Science* 284, no. 5412 (1999); Louise Fortmann and John W. Bruce, "Why Tenure Matters," in *Whose Trees? Proprietary Dimensions of Forestry*, ed. Louise Fortmann and John W. Bruce (Boulder, CO: Westview Press, 1988), 1–9; and David Feeny et al., "The Tragedy of the Commons: Twenty-Two Years Later," *Human Ecology* 18, no. 1 (1990): 1–19.

[25] On reforestation, see Frédéric Fesquet, "L'arbre au secours des hommes: Les bienfaits de la forêt dans le discours forestier au XIX et XXe siècles," in *La forêt: Perceptions et représentations*, ed. Andrée Corvol et al. (Paris: L'Harmattan, 1997), 163–72; and Tamara Whited, *Forests and Peasant Politics in Modern France* (New Haven, CT: Yale University Press, 2000). The Landes and Sologne projects were largely undertaken by private landowners, but the state offered incentives and took an active role. *Les Eaux et Forêts du 12e au 20e siècle*, Histoire de l'administration française (Paris: Éditions du Centre national de la recherche scientifique, 1987), 579–81.

[26] As one group of scholars has observed, "When rules are imposed by outsiders without consulting those who are most affected, local users are more likely to become robbers, rather than cops, toward the resources they might otherwise have managed sustainably and to try to evade apprehension by the external authorities' cops." Clark C. Gibson, Elinor Ostrom, and Margaret A. McKean, "Forests, People, and Governance: Some

conservation and conflict were mutually constituted, and woodland ecosystems suffered the consequences.

Beginning with the physical geography and socioeconomy of Franche-Comté and its forests, the book follows a roughly chronological format. Chapter 1, "The Lay of the Land," introduces the forest's many stakeholders – rural communities, seigneurs, manufactures, and the state – and investigates the roots of Comtois antipathy toward French woodland intervention, focusing on the imposition of the 1669 Ordinance in the province.

Chapter 2, "'Agromania' and Silvicultural Science," examines the intellectual underpinnings of conservation, focusing on how Enlightenment thinkers in the second half of the eighteenth century addressed or dismissed anxieties about France's mounting wood and timber shortages. Expressed in a wide range of approaches, from detailed treatises on tree care to agricultural policies that exacerbated deforestation, these ideas influenced the bitter debates and forest upheavals to which the Revolution gave rise.

The next two chapters form the core of the book. Chapter 3, "'A necessity as vital as bread': Forest Crisis on the Eve of the Revolution," explores the forest-related concerns expressed in Comtois *cahiers de doléances* and contemplates their influence on local outlooks in the Revolution's early stages. Chapter 4, "'Seduced by the word liberty': Woodland Crisis and the Failure of Revolutionary Reform," looks at the Revolution's national context, tracing revolutionary legislators' repeated, ultimately futile efforts to create a comprehensive new forest code. As political factions emerged and conflict erupted over reform initiatives, the goal of producing a more efficient model of woodland management receded, with troubling consequences.

Initial Theoretical Lessons," in *People and Forests: Communities, Institutions, and Governance*, ed. Clark C. Gibson, Margaret A. McKean, and Elinor Ostrom (Cambridge, MA: MIT Press, 2000), 233. See parallels in Ramachandra Guha, *The Unquiet Woods: Ecological Change and Peasant Resistance in the Himalaya* (Berkeley: University of California Press, 1990); and K. Sivaramakrishnan, *Modern Forests: Statemaking and Environmental Change in Colonial Eastern India* (Stanford, CA: Stanford University Press, 1999), 34–71 and 121–44. For a look at the complexities involved, see Roderick P. Neumann, "Land, Justice, and the Politics of Conservation in Tanzania," in *People, Plants, and Justice: The Politics of Nature Conservation*, ed. Charles Zerner (New York: Columbia University Press, 2000), 117–33.

Chapter 5, "'Nothing is more respected . . . than the right of property': The Creation of the 1827 Forest Code," traces the development of new woodland initiatives, beginning with Napoleon's reinstatement of centralized forest oversight – mostly a paper achievement, but a sign of things to come – and culminating in the promulgation of a sweeping new Forest Code in 1827. The Code took an especially hard line on customary usage and communal property yet left private landowners largely free to pursue their own objectives, a stance that would lead to accelerated clearing and forest fragmentation. Chapter 6, "'Not even a branch of wood has been granted to us,'" discusses the Forest Code's enforcement and the incendiary violence and opposition that the law engendered from the 1830s onward.

The last chapter, "Homo is but Arbor Inversa," takes the form of an epilogue, exploring the decline of Franche-Comté's rural communities in the face of an increasingly powerful, technically proficient forest administration under Napoleon III. Despite the forest administration's achievements from the middle of the nineteenth century forward in restoring the upland watersheds of the Alps and Pyrenees, the state's indulgence of private ownership and commercial claims undercut its aims of woodland productivity and protection. In closing, the book turns to the present, which provides an instructive coda to the history of forest use in France. For two centuries, state legislators and foresters labored mightily, with mixed success, to stifle customary rights and exclusively control common pool resources. Today, however, amid increased ecological awareness, social and political nostalgia for rural life, and growing concern about the threat of climate change, the state's woodland policies have come full circle. Arguing that traditional forms of management are crucial for supporting biodiversity, suppressing wildfires, and increasing property values, the forest administration is now promoting the very activities it once sought to extinguish. Among other practices, silvopastoralism – once decried for causing erosion – is being hailed as the "ecological" answer to France's forests health and, by extension, the health of a planet in peril.[27] In this new concept of conservation, the woodland rights Franche-Comté communities fought so hard to preserve have been revived and relegitimated. How they will play out in the present century is yet to be seen.

[27] See, for example, the training event held in 2007 by the French animal husbandry institute: Institut de l'élevage, *Maîtriser par le pâturage les milieux embroussaillés: Intérêt et outils pour la gestion pastorale*, 5–7 June 2007, announced at www.inst-elevage.asso.fr/.

1

The Lay of the Land

> One can go everywhere under the cover of forest.
>
> Loys Gollut, *Les mémoires historiques de la République séquanoise*, 1592[1]

In late February 1765, two royal forest officers journeyed to the crown woodland of Chaux, in the French frontier province of Franche-Comté, intending to investigate a spate of wood thefts. When they entered the dark shadows of the forest's canopy, they were confronted by "two hundred men," "masked and armed with guns, pistols, and hatchets," who "gave chase" as they drove the terrified officials out. The rebels occupied the Forêt de Chaux for weeks, felling wood and funneling it out to supporters for sale and use, but in the end, they were forced from their posts when the crown sent its army to attack their homes and intimidate their families.[2]

This affair, to which I return later in this chapter, was but one of many woodland uprisings in Franche-Comté over the course of the eighteenth century. Though most of the clashes were smaller in scale, they shared a common catalyst: the dispossession of Comtois communities from their forests by the encroaching forces of French monarchical authority and seigneurial entrepreneurship. Far removed from centers of power and isolated by rugged terrain, Franche-Comté was for centuries a minor

[1] Loys Gollut, *Les mémoires historiques de la République séquanoise et des princes de la Franche-Comté de Bourgougne*, new ed., 2 vols. (Roanne: Editions Horvath / Bibliothèque municipale de Besançon, 1978 [1592]), 1:125.

[2] On the uprising, see Archives départementales du Jura (hereafter ADJ) C128 (dossier 7) and AN Q1 416, Procès-verbaux de la forêt de Chaux; also Vernus, *La vie comtoise*, 1:78–79.

player in the long-running drama of dynastic struggles between France, Burgundy, and the Holy Roman Empire. The region's inhabitants – landed and landless alike – enjoyed considerable self-determination in their internal governance, and their territory's name, *Franche-Comté de Bourgogne*, "Free County" of Burgundy, reflected this historical independence.[3] Though Franche-Comté became a Habsburg possession in 1477, the conditions of its territorial subordination remained comparatively slight. Other than the obligation of supplying money and men to support the Holy Roman Empire's wars, Comtois communities were largely left to adminster themselves, including the allocation and management of their natural resources.[4]

From the early modern period forward, however, Franche-Comté's effective autonomy – and with it, rural inhabitants' cherished forest freedoms – came increasingly under siege. On the economic front, rising commercial demand for commodities that required massive quantities of fuelwood to produce – particularly salt and iron – led seigneurs to infringe on villages' woodland possessions and restrict tenants' formerly generous use rights. Politically, Franche-Comté's privileged circumstances also changed. Louis XIV's conquest of Franche-Comté in 1674 and formal annexation of it to France four years later introduced a far more intrusive administration than the Francs-Comtois were accustomed to. The Sun King's attraction to the region was driven not only by the desire to create a buffer against Habsburg incursion but also to secure the province's ample natural riches, including salt, furs, iron, and timber, for France's benefit. The alternately abrupt and protracted process of implementing French rule in Franche-Comté fundamentally transformed the province. Under these combined pressures – rising industrial appetites for wood and the introduction of forceful French oversight – Comtois communities would find themselves locked in an ever-growing struggle to retain the forest rights critical to their survival.

A BOUNDLESS FOREST

Located on France's eastern border, Franche-Comté measures slightly more than 16,000 square kilometers, or about 3 percent of the surface

[3] Franche-Comté's political status was established by the first count of Burgundy in the eleventh century. The "free county" designation was bestowed in 1366. Roland Fiétier, ed., *Histoire de la Franche-Comté* (Toulouse: Edouard Privat, 1977), 122–35.

[4] Darryl Dee, *Expansion and Crisis in Louis XIV's France: Franche-Comté and Absolute Monarchy, 1674–1715* (Rochester, NY: University of Rochester Press, 2009), 16–17.

area of present-day continental France. The former provinces of Alsace, Lorraine, Châlons (Champagne), and Burgundy cradle the region to the north and west, Switzerland lies to the east, and Rhône-Alpes is located to the south. Prior to 1790, Franche-Comté contained three primary and eleven secondary administrative districts, known as *bailliages*. At the time of the Revolution, however, the province was divided into three uniformly sized departments: the Haute-Saône, Doubs, and Jura, plus the tiny Territoire de Belfort in its northeastern corner.[5] These administrative divisions remain in place today.

Franche-Comté's terrain is marked by enormous variety, from peat bogs and ponds in the Haute-Saône to the steep escarpments and snowy peaks of the Doubs and the Jura. Rising steadily in elevation from east to west, the plains bordering Burgundy give way to plateaus that split the province on a long transversal axis and divide its warm western flank from its icy eastern uplands. While the Revermont region in south-central Franche-Comté enjoys sun and temperatures mild enough to sustain a centuries-old viticulture, cold weather lingers in the recessed valleys and shadowed slopes of the Haut-Doubs. Even colder is the alpine Jura, home to the village of Mouthe, which regularly records the lowest winter temperatures in France.[6]

Regardless of elevation or climate, virtually every corner of Franche-Comté is blanketed with trees of one sort or another. Forty-four percent of its surface area is forested at present – far more than the national average of 29 percent – and the figure continues to grow as agriculture declines.[7] Much of the forest's dominance lies in its diversity. In the

[5] This figure, 16,202 square kilometers, includes the Territoire de Belfort. Because of its geographic isolation and historical dissimilarities with the rest of Franche-Comté, the Territoire de Belfort, annexed to France by the Treaty of Westphalia (1648), is not included in this study. During the early modern period, Franche-Comté measured 15,345 square kilometers excluding the Comté de Montbéliard (an independent enclave until 1793) and the Territoire de Belfort. P. E. Herbin de Halle and J. Peuchet, eds., *Statistique générale et particulière de la France et ses Colonies: avec une Nouvelle Description topographique, physique, agricole, politique, industrielle et commerciale de cet état*, 7 vols. (Paris: F. Buisson, 1803 [year XII]), 5:6, 5:57, 5:300.

[6] Remy Fière, "Pour aller à Mouthe, descendre sous le zero" (*Libération*, 3 February 2001), www.liberation.fr/guide/0101362860-pour-aller-a-mouthe-descendre-sous-le-zero.

[7] As of 2010, Franche-Comté followed closely behind Corsica and the Provence–Alpes–Côte d'Azur region in percentage of forest cover; however, reforestation (intentional and passive) is expanding rapidly in areas of France – particularly Languedoc-Roussillon, Corsica, and the southern Alps – where silvopastoralism has declined. Institut national de l'information géographique et forestière, Inventaire forestier national, *La forêt française: Les résultats issus des campagnes d'inventaire 2005 à 2009: Les résultats pour la*

Haute-Saône, fir, beech, and spruce mantle the weathered, dome-shaped *ballons* of the Vosges mountains; oak, ash, and maple dominate the Saône River basin; and pine, birch, willow, and alder jostle for space amid the department's abundant marshes. The departments of the Jura and Doubs – the latter named for the 430-kilometer river that snakes through it – also host distinct woodland ecosystems. On the floodplain of the Doubs and Loue rivers, groves of oak, beech, and hornbeam compose the sprawling Forêt de Chaux, the second largest woodland *massif* in France. Farther east, the coniferous plantations of Levier, Joux, Risoux, and Massacre cloak the uplands around Nozeroy and Champagnole, climbing ever higher until they reach the Jura's highest *haute joux* – wooded summits – beyond which lie Switzerland.

Describing Franche-Comté in the sixteenth century, the chronicler Loys Gollut observed, "From whatever spot one wishes, one can go everywhere under the cover of forest."[8] Looking out over Franche-Comté's wooded landscape today, that observation still appears apt. Yet beneath the apparent serenity of the forest's unbroken canopy is a much more complicated picture. Continuously exploited from the time of its earliest settlement, sometimes to the point of devastation, Franche-Comté's woodlands have been shaped and reshaped by shifting and rival interests. The forest, in turn, played a central role in shaping the political economy and identity of its inhabitants.

For thousands of years, humans transformed Franche-Comté's forests in an ever-shifting mosaic of settlement and clearings, heaths and hedges, and woods and waterways. The record of these transformations is inscribed on the land itself. According to landscape archeologists, the area's earliest human inhabitants, Neolithic pioneers who crossed into Franche-Comté from the east some six thousand years ago, would have encountered a lush array of deciduous and coniferous tree species. Over a span of several thousand years, these early settlers altered the forest's primordial composition by promoting certain varieties and winnowing out others. In place of inedible pine and poisonous yew they favored sessile oak and beech, whose nuts provided rich fodder for their semiwild herds of pigs. They also cleared away *bois blanc* – softwoods like alder, willow, and

region Franche-Comté, http://inventaire-forestier.ign.fr/spip/IMG/pdf/IFN_Publi_2010_Franche-Comte.pdf, 5; Institut national de l'information géographique et forestière, Inventaire forestier, *Le mémento: La forêt en chiffres et en carte*, http://inventaire-forestier.ign.fr/spip/spip.php?article583, 5, 27.

[8] Gollut, *Les mémoires historiques*, 1:125.

poplar – which had limited strength, produced no fruit, and gave little heat when burned. The imprint of these first modes of management is still visible today in the *massifs forestiers* of Chaux and La Joux, two forests that, for reasons of soil quality, climate, location, and utility, have persisted continually since the Neolithic period.[9]

Further changes came about with the Roman conquest in 58 BCE. Known to the Romans as Sequania, the region was at that time thinly inhabited by Gallic tribes, who put up a fearsome resistance to outside domination. Julius Caesar himself described the Sequani as bold fighters, but for his Roman soldiers, the region's "boundless forest" was as intimidating as its inhabitants.[10] Gradually, however, both the land and the Sequani succumbed to Roman colonization. Under Caesar Augustus, soldiers and settlers felled trees, planted crops, and constructed roads, increasing the occupied and arable portion of the terrain. From this point forward, clearing and cultivation of Franche-Comté would continue apace, punctuated by several notable war and plague-related reversals, until the mid-nineteenth century.

During the Carolingian period, from 750 CE forward, subsistence-oriented estates expanded beyond the old Roman towns, which had been abandoned amid the fifth- and sixth-century Frankish invasions. Among the newcomers were monastic brotherhoods, who established cloisters across the region, often in remote locations like Baume-les-Messieurs, tucked away in a cirque-shaped valley near Lons-le-Saunier, the present-day capital of the Jura.

To attract tenants who would clear their territories and make them profitable, lay and ecclesiastic seigneurs offered favorable conditions of *abergement*, the right to colonize new areas in exchange for rent and other obligations. Tenants also obtained liberal access to collective resources like forests and streams. These arrangements were confirmed in capitularies and charters as early as the eleventh century. While some of the provisos supported inhabitants' needs, including rights to gather firewood and graze flocks in the woods, others assured persistent inequity, like the obligatory labor-service known as the *corvée*, and mortmain, which guaranteed seigneurs' inalienable

[9] Jean Rousseau and Michel Bouvier, *La grande forêt de Chaux: Ses riverains, ses usagers, ses gardiens* (Dole: Les Presses Jurassiennes, 1980); René Schaeffer, "La forêt résineuse comtoise et la politique forestière au XVIIIème siècle," *Bulletin de la Société Forestière de Franche-Comté et des Provinces de l'est* 34, no. 6 (1969): 161–67; 242–48.

[10] Julius Caesar, *De Bello Gallico*, I, 39, cited in Vion-Delphin et al., *Les hommes et la forêt en Franche-Comté*, 53.

landownership.[11] The case of the abbey at Montbenoît, in the Saugeais valley near Pontarlier, is instructive. Founded by a single monk in 1050 CE, the monastery grew by the thirteenth century into a twenty-five-square-mile estate, thanks to bequests by the neighboring lords of Joux. The abbey granted tenants a variety of concessions, including access to pasturage in the understory of the nearby Bois de Ban, the right to gather deadwood "in exchange for an annual fee of one chicken per person at Pentecost," and the right to fish at certain spots on the Morteau River. Monastic authority and community stability were ensured through *mainmorte adoucie*, or limited mortmain, which permitted settlers to leave if they wished, but ensured that their lands would revert to the abbey upon their departure. Between 1200 and 1251, this combination of concessions and servitudes contributed to the establishment of twelve villages in the Saugeais.[12]

In this way, Comtois settlers gained rights to seigneurially controlled resources from the medieval period forward. Frequently, though not always, residents were able to secure their privileges via a communal covenant known as a *charte de franchise*, a written document confirming their village's obligations and easements. The villagers of Saugeais, for example, successfully negotiated a 112-article covenant in March 1459, which included forest usages among its many stipulations. As the population and demands on the forest grew in tandem, however, these covenants would come to hamper seigneurial and state efforts to extract profits and exercise sovereignty over the land.

EARLY MODERN MANAGEMENT, ORGANIZATION, AND EXPLOITATION

Like Franche-Comté's Neolithic predecessors, medieval and early modern communities also transformed the forest through their modes of use. To meet their demand for specific species and sizes of trees, Comtois

[11] On charters and capitularies, see Richard Keyser, "The Transformation of Traditional Woodland Management: Commercial Sylviculture in Medieval Champagne," *French Historical Studies* 32, no. 3 (2009): 353–84; Ellen Arnold, *Negotiating the Landscape: Environment and Monastic Identity in the Medieval Ardennes* (Philadelphia: University of Pennsylvania Press, 2012), 15–20.

[12] "Le coutumier du Val du Saugeois," 475–77, art. 93–94, in *Mémoires et documents inédits pour servir à l'histoire de la Franche-Comté* (Besançon: Académie de Besançon, 1900), 9:411–19; René Locatelli, "Les origines de Montbenoît et du Sauget (XIIe–XIVe siècles)," *Mémoires de la Société d'émulation du Doubs* 18 (1976).

inhabitants developed a regionally distinct method of selective felling, known as *jardinage comtois*, that allowed them to obtain mature timber for lumber while conserving a canopy of nut-bearing *fruitiers* like oak, beech, and walnut sufficient for satisfying porcine appetites. The prevalence of woodland pasturage kept the understory open and the canopy high, but inhabitants made sure to retain a range of fast-growing *morts-bois*, including hornbeam, maple, ash, and linden, to supply the small wood needed for making farm equipment and crafts. In accordance with their customary usages, communities could also gather *chablis*, trees and branches blown down by wind and storms, and *bois mort*, dead wood that had fallen or surreptitiously been shaken from senescent snags.[13]

With the aid of the forest's bounty, Franche-Comté's populace – approximately 400,000 in the mid-fifteenth century – eked out a fragile existence growing wheat and rye, dairy farming, raising cattle and pigs, and purveying timber, furs, salt and iron to French, Swiss, and Italian traders.[14] Farming predominated in the southwestern Bressane region, around the floodplains of the Doubs, Loue, Saône, and Ognon rivers. Farther south, in the Revermont, winegrowers produced *vin jaune*, a distinctive chardonnay renowned throughout Europe. Animal husbandry, especially dairying, was the primary activity of the *montagne*, where the short growing season, poor soil, and steep slopes hindered the cultivation of all but oats and barley, the hardiest grains, in long rotations with naturally regenerated pasturage. As one Comtois writer remarked, the high ridges and windy plateaus of the Haut-Doubs and the Haut-Jura were places more suited to "the happiness and well-being of cows" than humans.[15] Mountain residents also raised horses, sheep, goats, pigs, geese, and bees, or, as in the case of the pack mule herdsmen of Arinthod and carters and colporteurs of Grandvaux, exported their men half the year for seasonal work.[16]

[13] Vion-Delphin et al., *Les hommes et la forêt en Franche-Comté*, 78–79, 90.

[14] For population estimates, see Fiétier, *Histoire de la Franche-Comté*, 218, 226, 231; and Lucien Febvre, *Histoire de la Franche-Comté*, nouvelle ed. (Paris 1976 [originally published 1912]), 200–202.

[15] Auguste Bailly, "La Carcasse et le Tord-Cou: Roman inédite," in *Les Oeuvres libres* (Paris: Fayard, 1923), 26:191.

[16] Colette Merlin, *Ceux des villages ... La société rurale dans la "Petite Montagne" jurassienne à la veille de la Révolution*, Cahiers d'études comtoises 52 (Besançon: Annales littéraires de l'université de Besançon, 1994), 52, 205; Marie-France Briselance, *Voyage en Franche-Comté littéraire* (Baume-les-Dames: Éditions Cêtre, 1991), 39; and Alphonse Rousset, *Dictionnaire géographique, historique et statistique des communes de la Franche-Comté, Département du Jura* (Besançon: Bintot, 1853–57), 3:274.

Franche-Comté's diverse topography fostered distinct social, cultural, and organizational patterns. Upland communities tended to enjoy greater privileges, including freedom from mortmain, than lowland villages, perhaps in recognition of the increased challenges of surviving at higher elevation. Spatially, mountain homes and outbuildings tended to be bunched atop outcroppings as a wedge against avalanche, or clustered at the base of ravines to shelter from winter winds. Villages in the plains and valleys, by contrast, stretched along streams and thoroughfares and featured larger interiors.[17] In terms of trade and social relations, the deeply carved and folded terrain of the Haut-Doubs and the Haut-Jura inhibited communication even from one town to the next but encouraged entrepreneurial self-reliance and pragmatic communalism. In particular, it gave rise to a distinctive local institution, the *fruitière*, in which villagers grazed their animals collectively and pooled their milk to produce *vachelin*, a hard, Gruyère-like cheese in large wheel form that remained edible the whole winter. Villagers divided up the wheels according to the shares of milk each family had contributed, and they exported any surplus to southern ports, where it was loaded on ships for long voyages.[18]

Domestic construction, heating and cooking, and woodland pasturage long constituted the primary pressure on Franche-Comté forests, but from the fourteenth century forward, competition from rural industry posed a mounting challenge. It wasn't simply that manufactures – saltworks and ironworks especially – increasingly catered to external markets, but rather that these industries required a very different form of forest management to meet their ravenous appetites. This disjuncture would eventually lead to bitter conflicts between *usagère* (right-holding) communities and their seigneurs.

Early industry took three main forms in Franche-Comté: artisanal crafts, salt making, and metallurgy. Each benefited from the region's abundant resources: lime, clay, sand, and ores; natural brines and mineral springs; and navigable rivers and fast-flowing waters for turning millstones and hydraulic wheels. Most of all, Franche-Comté industries relied on the forest. Trees were employed from top to bottom: tanners used oak bark to soften hides, coopers fashioned barrels and buckets from

[17] Boichard, *Le Jura*.

[18] On the *fruitière*, see Merlin, *Ceux des villages*, 52; and Jean Boichard, "L'espace agricole comtois à travers le temps," in *Provinces et états dans la France de l'Est: Le rattachement de la Franche-Comté à la France*, ed. Maurice Gresset, Cahier d'études comtoises (Besançon: Annales litteraires de l'université de Besançon, 1977).

timber staves, weavers wove baskets from pliant willow osiers, and *sabotiers* carved pine blocks into clogs.

Wood-burning industries, by contrast, consumed trees almost exclusively as combustibles. Tileworks and potteries burned raw wood, while glassworks utilized wood charcoal. Working in close-knit family groups, itinerant charcoal burners (*charbonniers*) felled successive sections of forest and reduced it to charcoal in cone-shaped ovens, then returned and repeated the cycle once the trees had sufficiently grown back. (See Figures 1 and 2.) The Vieille-Loye glassworks, strategically located in the Forêt de Chaux between the alluvial outwash of the Doubs and Loue rivers, consumed charcoal so prodigiously that the area around it still remains bare to this day, a century after glassmaking ceased.[19]

Salt making required even greater quantities of fuel, leading to both ecological damage and socioeconomic disruption. One of the earliest trade commodities in Europe, salt was known as "white gold" because of its unparalleled importance as a preservative and flavoring and the arduousness of its production. Whereas coastal inhabitants produced salt comparatively easily by evaporating seawater in flat beds, landlocked populations either had to mine for salt underground or painstakingly boil brine from naturally occurring springs. Franche-Comté lacked salt mines but was geologically blessed with abundant sources of brine, or *muire*, bubbling from the earth. The plethora of salt-related place-names in the province – among them Lons-le-Saunier, Salins-les-Bains, and Soulce-Cernay – reflect this abundance.[20]

Salt manufacture initially occurred in each of these spots, but from the thirteenth century forward, under the influence of Jean "the Antique," powerful patriarch of the noble Chalons family, it became concentrated in just one site: Salins-les-Bains in the Jura. Salins-les-Bains had a number of natural advantages: its brine was exceptionally salty, extensive forests grew nearby, and its location at the base of a narrow pass rendered it easer to defend.[21] Enriched by their salt-making monopoly, Chalons

[19] Maurice Rey, "La forêt de Chaux au moyen âge: Étude de contacts" (paper presented at the Colloque sur la forêt, Besançon, 21–22 October 1966). On Franche-Comté glassworks and the distinctive vessel known as the *clavelin*, see Pierre Izibert, "La forêt de Chaux," *Pays comtois: Le magazine du patrimoine, de l'histoire et de l'art de vivre*, March–April 1998, 39.

[20] *Saunier* means salt maker or salt merchant, *salins* means salty, and *soulce* is a variant on *sel* (salt). René Locatelli et al., eds., *Les salines de Salins au XIIIème siècle: Cartulaires et livre des rentiers*, Annales littéraires de l'université de Besançon 448; Cahiers d'études comtoises 47 (Paris: Les Belles Lettres, 1991), 28–29.

[21] Locatelli et al., *Les salines de Salins*, 31–32; also *Les Eaux et Forêts*, 10.

FIGURE 1. Woodcutters working in a coppice plantation (*Travail des bûcherons d'une exploitation en taillis*)
Source: Henri-Louis Duhamel du Monceau, *De l'exploitation des bois* (Paris: H. L. Guerin & L. F. Delatour, 1764), vol. 1, p. 250, plate III. Courtesy of the Beinecke Rare Book and Manuscript Library, Yale University, 1976 1890 1.

descendants went on to wield enormous influence in Franche-Comté as well as in the wider Burgundy–Swiss trade region. As Salins-les-Bains' financial fortunes grew, however, the saltworks exacted a growing environmental and socioeconomic toll. By the sixteenth century,

FIGURE 2. Charcoal burners at work (*Travail des charbonniers*)
Source: Henri-Louis Duhamel du Monceau, *De l'exploitation des bois* (Paris: H. L. Guerin & L. F. Delatour, 1764), vol. 1, p. 250, plate V. Courtesy of the Beinecke Rare Book and Manuscript Library, Yale University, 1976 1890 1.

the hillsides around Salins were denuded of trees, and its river, La Furieuse, was notoriously polluted. As sixteenth-century humanist and Comtois chronicler Gilbert Cousin noted, the river was nicknamed "*la Foirie* or *Foirole*" (diarrhea-place) "because it is the sewer of

Salins."[22] Successive decrees by the Chalons from 1560 through 1601 extended Salins-les-Bains' claims on firewood to forests within a three-, four-, and even five-league (twenty-nine-kilometer) radius, thereby depleting an ever-widening area of Comtois woods and driving up fuel prices. Highly unpopular with local inhabitants, these demands made salt-making a focus of discontent that would only grow more intense after the French conquest.

By contrast with the Chalons family's concentrated monopoly on salt production, metallurgy in Franche-Comté was scattered across the province in a multitude of small- and large-scale operations. Nonetheless, forges and furnaces also came to have a significant impact – both large and localized – on the socioeconomy and ecology of Comtois forests.

Prior to the end of fifteenth century, metal production was relatively modest in Franche-Comté. Thereafter, however, two phenomena rapidly catalyzed iron production and its related wood consumption.[23] The Treaty of Senlis, signed in 1493, marked a cessation of France's claims on Franche-Comté and recognized it as a Habsburg satellite. The ensuing peace, paired with declining outbreaks of the plague, made way for a century-long golden age of relative peace and prosperity in the region, during which its tilled surface area increased, the number of rural industries swelled, and population nearly doubled.[24]

At the same time, new technology in the form of the blast furnace enabled a rise in the quality and quantity of iron production.[25] Cast in an array of forms, from firearms and breastplates to chimney plates and cooking pots, the increased output of iron met a growing military and domestic demand. However, it also consumed much more wood.

[22] Gilbert Cousin, *La Franche-Comté au milieu du XVIe siècle ; ou, Description de la Haute-Bourgogne connue sous le nom de Comté, par Gilbert Cousin, de Nozeroy (1552 et 1562)*, Émile Monot, ed., Publications de la Société d'émulation du Jura (Lons-le-Saunier: Lucien Declume, 1907), 183.

[23] François Lassus, "La sidérurgie comtoise d'Ancien Régime (XVIe–XVIIIe siècles)," in *De la mine à la forge en Franche-Comté: Des origines au XIXe siècle*, ed. Jean-Paul Jacob and Michel Mangin (Paris: Annales littéraires de l'université de Besançon [Belle Lettres], 1990).

[24] On the dynastic struggles between France and the Habsburgs for Franche-Comté and neighboring Burgundy, see Pierre Gresser, *Le crépuscule du Moyen Âge en Franche-Comté* (Besançon: Éditions Cêtre, 1992), 17. On the "Golden Age," see Fiétier, *Histoire de la Franche-Comté*, 220–24.

[25] On the shift from the "*bas fourneau*" (direct metallurgy) to the "*haut fourneau*" (blast furnace), so called for its tall, funnel-like shape, see Lassus, "La sidérurgie comtoise," 139–42.

Like glassworks, blast furnaces relied on charcoal, itself the product of slowly and laboriously oxidized cordwood, which lost as much as three-fifths of its weight when converted to charcoal.[26] One ironworks alone could devour 120 hectares of forest per year in charcoal; maintaining this supply required a minimum of 2,000 hectares of forest.[27] Thus, in the course of the two-stage process between charcoal production and iron manufacture, staggering quantities of natural and temporal capital were reduced to a relatively small amount of metal. Access to running water for cooling, waste disposal, and transport was also vital. Of the twenty-three forges and furnaces in operation in Franche-Comté in 1562, all but a handful were located along the Saône, Doubs, and Ognon rivers.[28]

Though not quite as powerful as the Chalons family, forge owners in Franche-Comté were almost exclusively affluent seigneurs, both clerical and lay, with extensive holdings of forests, waterways, and mineral deposits.[29] Even so, these resources began to fall short from the mid-sixteenth century forward. Seigneurs and their designated forge masters responded by intensifying fuel production on existing properties and expanding their claims on communal holdings.

Among the most palpable changes was the conversion of woodlands from multidimensional, biotically diverse spaces maintained as *futaie* (high forest) or *taillis sous futaie* (coppice with standards) to monocropped, even-aged coppices (*taillis*) – essentially fuel fields. (See Figure 3.)

[26] Estimate is from Duhamel du Monceau, *L'art du charbonnier*, 1761, cited in Jean-François Belhoste, "Une sylviculture pour les forges, XVIe–XIXe siècles," in *Forges et forêts: Recherches sur la consommation proto-industrielle de bois*, ed. Denis Woronoff (Paris: Éditions EHESS, 1990), 244.

[27] Belhoste, "Forges et forêts en révolution," in *Révolution et espaces forestiers: Colloque des 3 et 4 juin 1987*, ed. Denis Woronoff, Collection alternative[s] rurales (Paris: L'Harmattan, 1988), 127; C.A.R.A.N., ed., *Histoire de forêts: La forêt française du XIIIe au XXe siècle* (Paris: Centre historique des Archives nationales, 1997), 43; Serge Benoit, "La consommation de combustible végétal et l'évolution des systèmes techniques," in *Forges et forets*, ed. Denis Woronoff (Paris: Éditions EHESS, 1990), 103.

[28] François Vion-Delphin, "L'homme et la forêt dans les Vosges saônoises au XVIII siècle: L'exemple de la gruerie de Faucogney," in *Connaissance de la Franche-Comté: Régards sur les Vosges comtoises: À la découverte des terres comtoises du Parc naturel régional des Ballons des Vosges* (Besançon: Centre Universitaire d'Études Régionales [CUER], Université de Franche-Comté, 1991), 219–42.

[29] Lassus, "La sidérurgie comtoise," 140; Maurice Gresset, "Les débuts du regime français en Franche-Comté (1674–1675)," in *Provinces et états dans la France de l'Est: Le rattachement de la Franche-Comté à la France*, ed. Maurice Gresset, Cahier d'études comtoises (Besançon: Annales litteraires de l'université de Besançon, 1977), 19–38. See also Paul Walden Bamford, *Privilege and Profit: A Business Family in Eighteenth-Century France* (Philadelphia: University of Pennsylvania Press, 1989).

FIGURE 3. Modes of forest management
Source: Graphic prepared by Kevin Cross ©. Printed with permission.

Uniformity of size, age, and species in a coppice ensured that the charcoal produced from it would generate an even heat.[30] Once felled, the stumps resprouted, and the cycle of growth, felling, and oxidation recommenced.

In theory, coppicing for fuel production was perpetually renewable. In practice, however, the fifteen to twenty years needed for stems to reach optimum maturity took too long. As Franche-Comté's golden-age economy boomed, metal production increased and demand for charcoal grew. Charcoal burners began felling coppices on shorter and shorter rotations, resulting in poor-quality fuel and lasting damage to the forest. Without sufficient time to grow, shoots became stunted, stumps ceased to sprout, and barren patches became permanent.

Coppicing also impeded other woodland uses. At nearly every stage of the rotation, gathering and grazing were obstructed. Long-standing custom and newly issued decrees barred herders and their flocks from entering coppices until the tender young shoots resprouting from stumps, or "stools" were no longer vulnerable to animals' nibbling – a period of eight to twelve years. By that point, the woodlands were almost impenetrably brushy. Strict coppice regimes – oriented purely toward woodfuel rather than containing timber or seed-bearing standards – also eliminated large old trees, thereby reducing ecosystem diversity, impoverishing the soil, and limiting the availability of timber and wind-blown branches (*chablis*).

Beyond the shifts in woodland management, ironworks engendered other problems. Charcoal ovens triggered wildfires, blast furnaces created

30 Julian Evans, "Coppice Forestry: An Overview," in *Ecology and Management of Coppice Woodlands*, G. P. Buckley, ed. (London: Chapman and Hall, 1992), 18–27; Oliver Rackham, *Trees and Woodland in the British Landscape* (London: J. M. Dent, 1989), 63–66.

contaminated ore pits, and toxic slag choked and overflowed rivers. Because forges and furnaces employed relatively few local residents – most forge workers and charcoal burners came from elsewhere – communities had little incentive to tolerate their drawbacks.[31] Little wonder that Franche-Comté communities reacted with hostility when seigneurs sought to encroach on communal woods and take back long-standing use rights.

FOREST RIGHTS UNDER SIEGE

In his description of Franche-Comté in 1552, Gilbert Cousin remarked that "the inhabitants of this country, having endured the terrible rigors of the past, now enjoy the most perfect contentment. ... Where forests shaded the soil [one finds] sun-rich vineyards and fertile plains."[32] While across the border in France brutal warfare between Huguenots and Catholics bloodied the latter half of the sixteenth century, the Francs-Comtois – the vast majority of whom were devout Catholics – enjoyed a period of comparative peace and stability.[33] Inhabitants expanded their population and lands under cultivation. Comtois seigneurs in turn began seeking greater profits from their forests. But first they would have to grapple with communities' centuries-old charters and other traditional arrangements that governed their woodlands' use.

Seigneurs across the region had already begun monitoring their land more carefully in the fourteenth and fifteenth centuries. They established *grueries*, offices devoted to policing against poaching, unauthorized grazing, and *mésus de bois* (hacking off branches or felling a tree without permission). Guards and officers levied fines for infractions, collected fees for woodland pasturage, and branded and inventoried inhabitants' pigs prior to the annual *glandée* (nut mast grazing).[34]

31 Denis Woronoff, "La politique des autorisations d'usines et la question du bois," in *Forges et forêts*, ed. Denis Woronoff (Paris: Éditions EHESS, 1990), 62–65.

32 Cousin, *La Franche-Comté au milieu du XVIe siècle*; also Fiétier, *Histoire de la Franche-Comté*, 216.

33 Boichard, *Le Jura*, 95.

34 M. Guyot, ed., *Répertoire universel et raisonné de jurisprudence civile, criminelle, canonique et bénéficiale; ouvrage de plusieurs jurisconsultes* (Paris: Visse, 1784), 8:161–66; J. Gadant, "Le chêne aux XVIIe siècle: Bois de marine et arbre à fruits," *Revue forestière française* 29, no. 1 (1977): 57–59. On *grueries*, see François Vion-Delphin, "L'homme et la forêt dans le Haut-Doubs au XVIIIe siècle: L'exemple de la gruerie de Pontarlier," in *La Forêt, 113e congrès national des Sociétés savantes, Strasbourg, 1988*, ed. Andrée Corvol (Paris: Éditions du C.T.H.S., 1991), 260–71.

As commercial demands for wood and timber rose in the sixteenth century, seigneurs deployed new tactics to restrict customary usage. They issued detailed decrees that reinterpreted communal rights and prohibitions, and they devised legal and regulatory justifications for limiting rights holders' access. Among the first of these decrees was the *Ordonnance forestière*, issued in 1595 in Montbéliard, an independent enclave in the northeast corner of Franche-Comté. Montbéliard's ruler, Frédéric I, sought to gain greater control over his forests and conserve them for the future. "One quarter of the said Forests will be placed in reserve on the best grounds and [in the] most convenient locations," the 1595 ordinance stipulated. Its instructions indicated a preference for certain species: "In accordance with the designation of the Forest-master in concert with the Mayors and Counselors of each Community; this réserve ... will [consist], as much as possible, of Oak or Beech, to be made available in the case of Communities' needs."[35]

Other seigneurs tried punitive deterrence. Typical of the *ordonnances forestières de 1619* in Doubs was a proviso that threatened a fifty-*livre* fine against "peeling & barking any standing oaks, apple trees, pear trees, *faugs* and other fruit-bearing wood either in the Forêt de Chaux or in other [woods] belonging to the Sovereign."[36] As emphatic as these edicts were in writing, however, on-the-ground enforcement fell short. Despite the establishment of *grueries*, most seigneurial woodlands were too large for the small number of poorly paid, untrained guards to monitor effectively, and supervision remained minimal.[37]

More effective, at least from seigneurs' perspective, were systems of land division known as *cantonnement*, *triage*, and *tiercement*, which conferred control of a portion of a woods upon *usagères* while liberating

[35] *Ordonnances de tres-haut, tres-illustre, puissant prince et seigneur, Frideric ... Duc de Vvirtemberg & Teck, Comte de Montbéliard, Touchant les bois et Forests* [sic] (Montbéliard: Jacques Foillet, 1595). Montbéliard was joined to Franche-Comté in 1793. Rey, "La forêt de Chaux au moyen âge"; and Vion-Delphin et al., *Les hommes et la forêt en Franche-Comté*, 88–89.

[36] *Ordonnances forestières de 1619*, in *Recueil des arrêts, ordonnances, et lois de la Chambre des comptes du Parlement de Dole*, book 7, title XLV, art. MDV, p. 303. *Faug* likely referred to *fagus*, or beech.

[37] Guards were usually paid from the fines they collected, but they had little incentive to levy them if they could cut better deals with inhabitants. By the late seventeenth century, most *grueries* were closed due to ineffectiveness or endured only as pretexts for personnel to collect bribes and conduct illicit timber sales. François Vion-Delphin, "Les forêts du Nord de la Franche-Comté à la veille de la Révolution d'après les cahiers de doléances des bailliages d'Amont et de Baume," in *La Franche-Comté à la veille de la Révolution*, ed. Maurice Gresset (Paris: Les Belles Lettres, 1988), 41–42.

the remainder from customary obligations. Complex and often contentious, these arrangements greatly increased in incidence and significance over the eighteenth and nineteenth centuries.[38]

For communal rights holders, *cantonnement* was the most favorable form of proprietary concession, because it granted them title – though they still had to pay rent – to roughly two-thirds of the portion of seigneurial owned woods to which they had access. The seigneur in turn gained exclusive control over the remainder. Practiced from the twelfth century forward, *cantonnement* was widespread in Franche-Comté, particularly in the uplands where physical barriers blocked seigneurs from undertaking more lucrative ventures like timbering and manufacturing.[39]

More controversial was *triage*, a method of woodland partition developed in the mid-sixteenth century. Predicated on the notion that all lands originated with a king or noble lord, *triage* permitted seigneurs to acquire one-third of a communal woods, even if they or their ancestors had formally conceded the woods centuries before to the community. *Triage* signaled seigneurs' increasing assertiveness amid swelling golden-age demands for forest resources.[40]

A third means of partition, known as *tiercement*, went further still in extending seigneurs' proprietary claims. For *usagères*, it was a devastating alternative, for it restricted all or part of their use rights to one small segment of the forest – generally one-third while seigneurs took the remaining two-thirds – while also denying them any sort of title to their share. Seigneurs were unable to reject entirely the web of traditional rights that linked communities to the land, but *tiercement* gave them the freedom to cut, burn, fell, or sell their holding as well as maintain control over the whole.[41]

As these measures suggest, the days when seigneurs eagerly solicited peasant settlement through generous use rights were over. Though even the most unfavorable mode of reapportionment still permitted inhabitants

[38] Michel Devèze, "Superficie et propriété des forêts du Nord et de l'Est de la France vers la fin du règne de François 1er (1540–1547)," *Annales E.S.C.* 3 (1960): 485–92; also Whited, *Forests and Peasant Politics*, 35–37. On *cantonnement*, *triage*, and *tiercement*, see *Les Eaux et Forêts*, 222–23. On their legal origins and implications, see G. Antonetti, "Le partage des forêts usagères ou communales entre les seigneurs et les communautés d'habitants," *Revue historique de droit français et étranger* 41 (4th ser.), no. 2 (1963): 238–442.

[39] Devèze, "Superficie et propriété des forêts," 485–92; also Whited, *Forests and Peasant Politics*, 35–37.

[40] Antonetti, "Le partage des forêts usagères," 261–64. *Les Eaux et Forêts*, 222–23.

[41] *Les Eaux et Forêts*, 223.

some forest access, a substantially different outlook on woodland possession was clearly emerging. Landlords increasingly shifted away from paternalistic relations and the shared forms of land use that had shaped the medieval socioeconomy, and instead embraced an ideology of resource commodification based on individualism and exclusion – concepts that would become more clearly articulated during the Enlightenment. This trend and the conflicts it engendered intensified following France's annexation of Franche-Comté in 1678, which put into place a still more authoritarian approach to woodland management.

1678 AND ITS AFTERMATH: CONQUEST, REFORM, AND THE 1669 ORDINANCE

Franche-Comté was long subject to external control. From the first century CE, Romans, Franks, and Burgundians in turn acquired and ruled the region. Toward the end of the fifteenth century, it came under the sphere of the Habsburgs. Nonetheless, the province retained its historical liberties through each change of rule. Under the Habsburgs, Franche-Comté was administered by a governor chosen from among the local nobility and had its own *parlement* at Dole, which served both judicial and political advisory functions. Taxation was approved by Estates representing the three orders.[42] These liberties continued through the rule and abdication of the Holy Roman Emperor Charles V, grandson of Ferdinand and Isabella and heir to both the Spanish crown and the Habsburg territories.[43]

The ascension of Charles V's son, Philip II of Spain, in 1556 marked the start of Franche-Comté's tumultuous "Spanish period," which unleashed the Inquisition upon the region's largely Catholic faithful. Caught up in the frenzy, the humanist Gilbert Cousin was killed and innocents were accused of sorcery.[44] Philip II required Comtois communities to garrison Spanish and Italian troops en route to religious warfare,

[42] Boichard, *Le Jura*, 88; also Guy J. Michel, *La Franche-Comté sous les Habsbourg (1493–1678)* (Wettolsheim: Mars et Mercure, 1978).

[43] Charles V (d. 1558) abdicated in 1555 and decamped to a monastery. He left part of his territories to his son, Philip II (d. 1598), and the remainder to his brother, Ferdinand I. Fiétier, *Histoire de la Franche-Comté*, 223–25; also J.-F. Solnon, *Quand la Franche-Comté était espagnole* (Paris: Arthème, Fayard, 1983).

[44] Lucien Febvre, "Un sécretaire d'Erasme: Gilbert Cousin et la réforme en Franche-Comté," *Bulletin de la Société de l'histoire du protestantisme français* 56 (1907): 97–148; Aristide Dey and Jules Finot, *Histoire de la sorcellerie au comté de Bourgogne: Procès de sorcellerie au bailliage de Vesoul de 1606 à 1636*, ed. Michel Collée (Marseille: Lafitte Reprints, 1983; reprint of 1861 [Dey] and 1875 [Finot]).

but offered Franche-Comté no protection when its northern Saônoise plains were raided by foreign Protestants.[45] In addition to these woes, the Francs-Comtois suffered recurring plagues and famines from the mid-sixteenth century forward, conditions brought on, or at least exacerbated by, the cooling of the global climate.[46]

The Thirty Years' War (1618–48) was a particularly low point. Though not a central theater of battle, Franche-Comté was subject to repeated invasion by French, German, and Swedish armies who crisscrossed the territory, besieged towns, appropriated supplies, and leveled villages. By the war's end, an estimated 270,000 people – more than half the Comtois population – had fled or died, and the region's thriving economy had collapsed.[47] Nonetheless, the calamities had a silver lining for the forest: if wood consumption had continued at its earlier rate, Franche-Comté would have faced a fuel and timber crisis as great as the one emerging in neighboring France. Instead, amid the chaos and abandonment, trees regained ground lost to cultivation in the sixteenth century.[48]

Compared with the Habsburgs, whose interest in Franche-Comté was perennially distracted by war elsewhere, France's ambitions for the province were sharp and clear. The crown – initially Henry II in 1595, and subsequently Louis XIV in 1668 and 1674 – sought to create a geographic buffer around France while also expanding its natural resources. Though Louis XIV's conquest and annexation of Franche-Comté were not formalized until 1678 via the Treaty of Nijmegen, inhabitants began seeing changes as soon as Franche-Comté's first *intendant*, Camus de Beaulieu, was installed in 1674. To punish the city of Dole for having fought the conquest, the crown moved Franche-Comté's judiciary, *parlement*, and university away from the provincial seat and installed them in Besançon instead, thereby disrupting the elites' carefully tended power relations.[49]

[45] Boichard, *Le Jura*, 93–95; Lucien Febvre, *Philippe II et la Franche-Comté, la crise de 1567, ses origines et ses conséquences: Étude d'histoire politique, religieuse et sociale*, abridged ed. (Paris: Flammarion, 1970).

[46] Andrew B. Appleby, "Epidemics and Famine in the Little Ice Age," in *Climate and History: Studies in Interdisciplinary History*, ed. Robert I. Rotberg and Theodore K. Rabb (Princeton, NJ: Princeton University Press, 1981), 63–84; Brian M. Fagan, *The Little Ice Age: How Climate Made History 1300–1850* (New York: Basic Books, 2000).

[47] On the demographic impact of the Thirty Years' War on Franche-Comté, see Dupâquier, *Histoire de la population française*, 2:165, 2: 200–201.

[48] Boichard, *Le Jura*, 96–97; Febvre, *Histoire de la Franche-Comté*, 201–2; and P. Gresser, ed., *Franche Comté* (Le Puy-en-Velay: C. Bonneton, 1983), 39–40.

[49] Boichard, *Le Jura*, 98–99. On integration and resistance to French rule after annexation, see Dee, *Expansion and Crisis in Louis XIV's France*, 38–84.

Over the next quarter century French regulations and jurisprudence would be implemented in the region step by step, including ordinances on criminal law (1679), civil law (1684), and commercial law (1700). The crown offset and profited from its administrative expenses in the new territory by making all Comtois offices venal.[50]

In 1694, the crown turned its eye toward Franche-Comté's forests, replacing the province's traditions of communal possession, customary rights, and seigneurial oversight with the uniform stipulations of its 1669 *Ordonnance des eaux et forêts* and with agents who reported to the crown.[51] This extension of the state into communities' most critical resource sowed the seeds of upheaval. Outraged at the exactions of an administration they still regarded as "foreign" a century after annexation, residents around the Forêt de Chaux and throughout Franche-Comté vigorously resisted subjugation, using lawsuits, subterfuge, and occasionally open rebellion to preserve their cherished woodland liberties.

From the 1669 Ordinance's inception, the law and its administration were charged with irreconcilable objectives. On one hand, the crown wished to protect the realm's timber resources by preventing abuses and regulating felling – goals laid out in the very first forest ordinance in 1219.[52] On the other hand, French monarchs tended to regard forests as inexhaustible sources of revenue that they could dip into whenever they needed. These two outlooks were in constant tension. Time and again the crown's long-term agenda of conserving timber, particularly for the navy, was quite literally undercut by short-term imperatives, especially military spending.[53] Also problematic was the crown's belief that the forests' management should be self-supporting, leading to a remuneration system in which

50 Maurice Gresset, *L'introduction de la vénalité des offices en Franche-Comté: 1692–1704*, Annales littéraires de l'université de Besançon 394 (Paris: Les Belles Lettres, 1989).

51 Hereafter referred to as the 1669 Ordinance. *Édit portant règlement général pour les Eaux et Forêts, Saint-Germain-en-Laye, 31 août 1669*. Edict no. 571; reprinted in François-André Isambert, Athanase-Jean-Léger Jourdan (Decrusy), and Alphonse-Honoré Taillandier, eds., *Recueil général des anciennes lois françaises, depuis l'an 420 jusqu'à la Révolution de 1789*, (Paris: Belin-Leprieur, 1821–33), 18:219–311.

52 The first law concerning all French woodlands, royal and otherwise, was a 1219 ordinance issued by Philippe-Auguste (1180–1223). The term "Eaux et Forêts" derives from this law. In its medieval sense, "*eaux*" referred to rivers and streams, while "*forêt*" concerned only areas where the king reserved his right to hunt, including deer parks and rabbit warrens (*garennes*). *Les Eaux et Forêts*, 12–13, 48.

53 As early as 1346, Philip VI undertook a broad reorganization of the forest administration in order to rebuild the French fleet. *Les Eaux et Forêts*, 55–56, 62; Michel Devèze, *La vie de la forêt française au XIVe siècle* (Paris: S.E.V.P.E.N., 1961), 1:64–72.

most woodland personnel were paid from the income they generated. The approach tended to encourage rather than rein in timber cutting.

Under Henry III (1574–89), venal office holding became the norm. Beset by ongoing civil war between French Huguenots and Catholics, Henry III raised money by multiplying and selling state offices, including those of the Eaux et Forêts. In place of one *grand maître* (director), he created six heritable positions, each with a steep price tag. The move was not entirely fiscally motivated; new forest regulations enacted by Henry's predecessor, Charles IX, far exceeded the capacities of one "grand master" alone. But as with other French offices, venality in the Eaux et Forêts gave rise to superfluity and corruption. By 1595, the number of *grands maîtres* had almost tripled, and personal profit-taking was on the rise.[54]

The cost of a *grand maître* office was high – about 350,000 livres from Louis XIV forward – but the officers and their subordinates received an annual salary plus a share of the fees generated by marking, measuring, and delivering timber from the royal domain.[55] This, plus the fact that they hailed from the wealthiest noble families, was supposed to divert them from misconduct. The same held true of the *grands maîtres*' subordinate agents, the *maîtres particuliers*: though they came from a lower economic stratum, their independent means were supposed to divert them from misconduct. However, with little supervision or accountability, negligence, fraud, and incompetence ran unchecked.[56]

France's woodlands fared poorly in the wake of Henry III's expansion of venal offices. Sabotage by warring religious factions and industrial and naval expansion caused added pressure. Concerned that the forests would be irreparably degraded, the next on the throne, Henri IV, decreed numerous reforms, but they remained unrealized for want of funding. The troubles only grew worse under his successor, Louis XIII, who scarcely concerned himself with the forest during the whole of his thirty-three-year reign.[57]

Finally, in 1661, the Eaux et Forêts came under the command of Jean-Baptiste Colbert (1619–83), Louis XIV's energetic and visionary minister

[54] *Les Eaux et Forêts*, 91–94, 102–3, 120; also G. Geneau, "La législation forestière sous l'Ancien Régime," *Revue des Eaux et Forêts* 174 (1940): 32–35, 171–74.

[55] *Les Eaux et Forêts*, 188–90, 200–203.

[56] On the rampant corruption, see Henri Sée, "Les forêts et la question du déboisement en Bretagne à la fin de l'Ancien Régime," *Annales de Bretagne: Revue publiée par la faculté des lettres de Rennes* 36 (1924–25): 15–16. On the *grands maîtres*, see Jean-Claude Waquet, *Les grands maîtres des Eaux et Forêts de France de 1689 à la Révolution, suivi d'un dictionnaire des grands maîtres* (Geneva: Droz, 1978).

[57] *Les Eaux et Forêts*, 116, 217.

of finances. Having earlier overseen the reorganization of forests in the Nivernais province southeast of Paris, Colbert was well aware of the problems plaguing France's woodlands even before he took up his new post. So vital did he consider forests to the nation's well being that he was barely in office a month before developing a far-reaching plan to restore them. With the king's approval, Colbert moved quickly to assess woodland conditions and problems throughout the country. In keeping with his administrative approach in other areas, he chose to commission his own fact-finding force rather than use the *grands maîtres*, who had a vested interest in maintaining the status quo. The commissioners were empowered not only to carry out investigations but also to enact reforms once they had been determined.[58]

For Colbert, the forest's significance lay neither in its importance as the nation's foremost energy source nor in its utility as a place of pasturage, nor certainly in its ecological value as home to a wide array of flora and fauna. Rather, he viewed the forest as an essential component of his plan to expand France's military and economic might. In Colbert's estimation, France could only succeed as a late-entry competitor for Asian and Atlantic trade if the crown marshaled its timber resources and expanded and reinforced its naval and commercial fleets.[59] The Dutch and the English, whose successful trading monopolies Colbert hoped to unseat, relied on timber imports from Russia, the Baltics, and, increasingly, North America, to construct their "floating fortresses." France alone among the western European powers had the potential to meet its own maritime needs. To do so, however, would involve restructuring the material, legal, and technical foundations of the nation's woodland administration.[60]

[58] On Colbert's information system, see Jacob Soll, *The Information Master: Jean-Baptiste Colbert's Secret State Intelligence System* (Ann Arbor: University of Michigan Press, 2009). On the 1669 Ordinance and the production of state knowledge, see Hamish Graham, "Fleurs-de-lis in the Forest: 'Absolute' Monarchy and Attempts at Resource Management in Eighteenth-Century France," *French History* 23, no. 3 (2009): 311–35.

[59] Jonathan R. Dull, *The Age of the Ship of the Line: The British and French Navies, 1650–1815*, Studies in War, Society, and the Military (Lincoln: University of Nebraska Press, 2009), 12–16; also Andrée Corvol, "Un nouveau Colbert," in *Actes du colloque pour le tricentenaire de la mort de Colbert*, ed. Roland Mousnier (Paris: C.D.U. / S.E.-D.E.S., 1983), 230–31; and J.P. Husson, *Les forêts françaises* (Nancy: Presses universitaires, 1995), 124–25.

[60] Paul Walden Bamford, *Forests and French Sea Power, 1660–1789* (Toronto: University of Toronto Press, 1956); A. R. M. Lower, *Great Britain's Woodyard: British America and the Timber Trade* (Montreal: McGill-Queens University Press, 1973); and Martine Acerra et al., eds., *Les marines de guerres européenes, XVIIe–XVIIIe siècles* (Paris: Presses de l'Université de Paris-Sorbonne, 1985). See also Jan Glete, *Navies and Nations:*

With this goal in mind, Colbert and his commissioners drafted a sweeping new body of woodland law: the *Ordonnance des eaux et forêts*, promulgated in August 1669. In true mercantilist fashion, the Ordinance demanded more efficient regulation and strict protection of forest resources, and it made clear that the crown's needs took precedence over all other domestic or industrial applications. In particular, it endowed the navy with near-monopolistic authority over the nation's timber, granting naval purveyors the right to inspect, mark, and reserve trees they considered seaworthy, be they on crown, communal, or seigneurial land.[61] The Ordinance and its accompanying reforms also vetted and reorganized the office of the *grand maître*, transforming it into an appointed, rather than venal, position. Though most of the edict was devoted to detailed instructions regarding restrictions of various activities, penalties for misconduct, and the sale and felling of timber, its few technical directives reflected Colbert's concern with maritime reserves.[62] Surveyors preparing timber sales, for example, were instructed to reserve "ten trees in each [lot] of lofty growth, great vigor, and fine proportions, of oak, and if possible of good wood and appropriate size."[63] With such guarantees in place, Colbert believed that France's present and future forest needs would be assured and the nation would at last achieve supremacy on the seas.

The reforms and safeguards enacted by the Ordinance enjoyed an initial flurry of success in the years after 1669. For the first time in decades, wood sales from the royal forests earned the crown substantial revenue, while the establishment of timber reserves across the nation seemed to promise an end to perennial "wood famines." By the time of Colbert's death in 1683, however, Louis XIV had already begun chipping away at his minister's scheme in order to finance his ongoing military entanglements. By 1715, when the crown passed to Louis XIV's great-grandson, venality, corruption, and negligence had been fully reestablished in the Eaux et Forêts. Yet even in the face of these setbacks, the driving philosophy behind the Ordinance persisted and even gained ground. Over the next century and a half, the 1669 edict evolved into the single most influential

Warships, Navies and State Building in Europe and America, 1500–1860 (Stockholm: Almqvist & Wiksell International, 1993).

61 1669 Ordinance in Isambert, vol. 18, title 21, art. 1–3; and title 26, art. 3.

62 On limits on various activities, especially customary usage, see 1669 Ordinance in Isambert, vol. 18, title 27, art. 1–46; title 28, art. 1–4; title 19, art. 1–15. On sales and marking of timber, windfall, and other wood, see 1669 Ordinance in Isambert, vol. 18, titles 15–17 (complete); on penalties, see vol. 18, title 32, art. 1–28.

63 1669 Ordinance in Isambert, vol. 18, title 15, art. 11.

forest law – indeed, one of the most influential laws, on any subject – in the history of France. By attempting to assure the nation's economic and political future through the long-term protection of naval timber, Colbert introduced a novel, nation-building, narrowly focused form of conservationism to French statecraft. This outlook ranked the production of high-quality timber above virtually all other woodland uses except industrial fuel. Henceforth, the overseers of the forest, from the lowliest guard to the king himself, would be unable to ignore the long-term significance of the forest's welfare to the state's well-being. While this approach helped contain wanton felling and indiscriminate clearing, in the long run its emphasis on national strategy over local and domestic needs threatened the viability of rural communities already struggling under the pressures of industrial and demographic growth. The promises and pitfalls of the law were especially clear in Franche-Comté, where the 1669 Ordinance, introduced in fits and starts over a half a century, supplanted traditional forms of forest management that were based on the life of the community rather than the imperatives of commerce and war.

FOREST TRANSFORMATION IN FRANCHE-COMTÉ

To Louis XIV, the thickly forested, sparsely populated Franche-Comté must have seemed like an answer to his country's woodland woes. Cash-strapped and timber-poor France and its prominence as a European power were threatened without an infusion of foreign resources like those that had benefited the Spanish, Dutch, and English. Accordingly, while Colbert sought in the 1660s to expand France's reach on the seas, Louis XIV, like Henry IV before him, looked eastward and saw where he might broaden his continental grip. When he attacked Franche-Comté in 1668 and again in 1673, natural resources – both ligneous and mineral – were a significant part of the attraction. As soon as the region's conquest was complete, Louis XIV sent surveyors to take stock of the forests. Per the dictates of the 1669 Ordinance, the *grands maîtres* in charge of the project, Bardon de Moranges and Jean Sanguinière, mapped and marked the crown's new possessions, scrutinized the validity of communities' customary rights, and assessed the forest's potential for profit.[64] A second survey in 1680 specifically targeted naval timber. By 1686, naval timber

[64] Bardon de Moranges began the project in 1679 and was succeeded by Jean Sanguinière from 1686 to 1689. AN E 564A, fol. 58, *Arrêt du Conseil du 1er octobre 1686, rendu à Versailles*, cited in Waquet, *Les grands maîtres des Eaux et Forêts*, 253, 226.

purveyors were sending thousands of cubic feet of wood per year from Franche-Comté to the royal shipyard at Toulon.[65] Toulon's shipbuilders were eager to have it, for the naval expansion that Colbert had set in motion was exceeding prior supply zones. As the director of the Toulon arsenal, de Vauvré, observed in 1688, "the oak woods in forests of the Comté ... are virtually the only ones on which we can rely in the future, Burgundy being practically depleted."[66]

The officials who investigated Franche-Comté during the initial years of French rule were struck by its forests' magnitude as well as their mediocre condition. The 1698 comments of Jean-Baptiste Desmarets de Vaubourg, one of the province's earliest *intendants*, were typical: "Although a great quantity of wood has been cleared since the province came under the domination of the king and [although] the forges and furnaces consume a great deal daily, there are few lands as covered [in forests] as this." Nonetheless, Desmarets de Vaubourg was worried about the woods' long-term viability. Too little was being done to protect them, he hinted, and the navy's hunger for timber was taking its toll. He observed:

> It is to be feared that the woods may soon become rare, not only [because] of the damage committed in them, but also by the carelessness and disorder ... seen in the fellings; already it is difficult to find proper construction wood near the towns, most is beech; there is also a lot of oak and pine, and a rather large amount of all sorts is taken by the navy.[67]

France's timber appetite was just one of any new pressures on Comtois forests at the turn of the eighteenth century. In this newly annexed region, the venality of the Eaux et Forêts spawned dishonesty and administrative negligence just as it did elsewhere in the realm. In one especially egregious

65 AN Marine D3 (1664–1704) and AN Marine B3 329 (1716), fol. 243–44 / 253–54–1; cited in François Vion-Delphin, "La fourniture des bois de marine en Franche-Comté de la conquête à la Révolution" (paper presented at the 103rd Congrès national des Sociétés savantes, Nancy-Metz, 1978), 442. On the impact of timber harvesting for the navy, see *Les Eaux et Forêts*, 176, 214.

66 AN Marine D3 13 (no. 17), 5 April 1688.

67 *Mémoire de l'Intendant de Franche-Comté, publié d'après un manuscrit de la Bibliothèque de Besançon*, ed. Maurice Pigallet (Paris: Librarie ancienne Honoré Champion, 1914), 18. There is some disagreement about the author of this text. J. Gauthier, the original editor of the piece in the *Annuaire du Doubs* of 1901, believed it was written by d'Harouys, *intendant* of Franche-Comté from 1700 to 1702. But because the manuscript copies are consistently dated 1698, I have concluded that the author was Jean-Baptiste Desmarets de Vaubourg, *intendant* from 1698 to January 1700. The text was addressed to the dauphin for his instruction in state affairs.

case, the *grand maître* of Burgundy and Franche-Comté from 1690 to 1703, Guillaume Perrault, regularly and openly defrauded the state, to the great detriment of the forests. Ousted on several occasions for malign disregard of his duties, Perrault always managed to return because he owned the title to his office. Finally, in January 1714, the crown's council issued an *arrêt du Conseil*, banning him for life from being a *grand maître*.[68] While this move put a stop to Perrault's troublemaking, it did nothing to change the conditions that made it possible.

The tremendous challenges of imposing the 1669 Ordinance on Franche-Comté meant that even devoted officers of the Eaux et Forêts struggled with the task. Commencing in 1717 with the Forêt de Chaux, Franche-Comté's largest crown woodland, and continuing through 1726 for smaller forests, the reform of the region's royal holdings took several decades to complete. The process took even longer in Franche-Comté's communal woods, not only because of their complexity and larger extent but perhaps also because municipal authorities and inhabitants were reluctant to supply information that might lead to their exclusion.[69] In Vesoul, for example, the community's forest reform began in 1700 and dragged on to 1765.[70]

The duration of these projects is perhaps not surprising given the enormity of the task. For each forest, whether royal or communal, reform commissioners first had to make sense of the diverse local context, if only to be able to dismantle it more effectively.[71] Understanding the particularities of each parcel was doubly difficult given that forest possession in many cases was in flux. For centuries, Franche-Comté inhabitants had understood woodland possession more as a bundle of rights than as exclusive territorial dominion. True, the increasingly common modes of partition like *triage* and *cantonnement* had the effect of limiting *usagères*

[68] See table and notations on the *grands maîtres* of Franche-Comté in Waquet, *Les grands maîtres des Eaux et Forêts*, 253–54.

[69] Christelle Bertrand, "Les hommes et la forêt dans les communautés villageoises de la maîtrise particulière des Eaux et Forêts de Vesoul, au XVIIIe siècle," *Haute-Saône S.A.-L.S.A.* 26 (1994): 154–55.

[70] The initial efforts of the crown's forest surveyors were interrupted by the War of the League of Augsburg (1688–97) and War of Spanish Succession (1701–14). Reforms began again only after Louis XV came to the throne. Vion-Delphin, "Les Forêts du Nord de la Franche-Comté," 42–44.

[71] See for comparison the reform in the Midi-Pyrenees, which featured crude cartography but efficient management assessments. Chandra Mukerji, "The Great Forestry Survey of 1669–1671: The Use of Archives for Political Reform," *Social Studies of Science* 37, no. 2 (2007): 227–53.

to a specific section of woods, but these arrangements did not fully abstract the terrain from the human activities upon it, as the ideas of demarcation and commodification espoused by the Ordinance did. In a 1738 lawsuit over a contested *triage*, for example, the inhabitants of Rougnon in the Doubs defined their woods not by spatial indicators, but by what the seigneur could not do there: "in the parts reserved for the inhabitants ... the seigneur will not have any right nor usage, no firewood nor pasturage, neither for himself nor for his bakehouse, [nor for] his tenants, domestic beasts and horses, nor the power to take poles for his vines and fasteners for his tenants' sheaves."[72]

The crown had two main objectives in Franche-Comté's forests. One, it wished to assess the condition and scope of individual woodlands, and two, it sought to reorient forest exploitation away from agricultural and domestic use toward more lucrative pursuits, namely naval timber and industrial fuelwood. To that end, *réformateurs* were extremely successful. Between 1722 and 1764, Franche-Comté's woodland revenues rose from 35,000 livres to 200,000 livres.[73] Much of this income came from timber sales, but a significant portion also derived from royal licenses that allowed forge masters to harvest fuelwood in the crown woodlands. For Francs-Comtois already struggling under *triage*s and *tiercements*, however, the Eaux et Forêts' intervention only intensified their troubles.

Among the most vexing aspects of the reform was the *quart en réserve*, which required all communities, whether religious or lay, to set aside one-quarter of their most promising timber trees to grow into high-quality, big-timber *haute futaie*.[74] While this provision was intended to guarantee that France would always have wood appropriate for shipbuilding and construction, it significantly reduced the resources available to inhabitants. If villagers wanted to obtain timber from their *quart en réserve*, they faced time-consuming requirements, including persuasive testimony explaining their need for the wood. Acceptable reasons included having to repair the parish church or replace barn roofs after a storm. Even in urgent cases like these, however, actually procuring the wood could take years while the *intendant*, officers of the *maîtrise*, and municipal

[72] AN Q1 182. [73] Vion-Delphin et al., *Les hommes et la forêt en Franche-Comté*, 96.

[74] In private woodlands, the *quart en réserve* was only advised, not required. Michel Devèze, "La crise forestière en France dans la première moitié du XVIIIe siècle et les suggestions de Vauban, Réaumur, Buffon." in *La forêt et les communautés rurales-XVIe-XVIIIe siècles: Recueil d'articles parus en revue* (Paris: Publications de la Sorbonne, 1982), 75. See also discussion in Andrée Corvol, *L'homme et l'arbre sous l'Ancien Régime* (Paris: Economica, 1984), 163–71.

authorities negotiated. In the interim, frustrated inhabitants watched the reserves grow old and decay.[75]

Landowners, too, found the reforms aggravating. Though their holdings were not subject to the same level of scrutiny as royal and communal forests, seigneurial, ecclesiastical, and bourgeois proprietors nonetheless operated under comparable restrictions concerning felling and selling, and they were required to permit surveyors from the Eaux et Forêts and purveyors of naval timber to inspect their woods for desirable specimens.[76] If there was sufficient cause, a *grand maître* could also deny landowners the right to fell a certain area of forest or even a specific tree.[77]

Nonetheless, the commissioners' efforts to protect timber were complicated by the crown's competing interests in ironworks in the region. Following the French conquest, forges and furnaces proliferated at an even faster pace, especially in the Saône basin around Gray, where the crown granted forge masters generous concessions of fuelwood in return for iron. This subsidy of Comtois industry contrasted sharply with royal restrictions elsewhere in France. In 1723, the crown had issued an edict insisting that coppices be at least ten years old before felling and requiring all new furnaces, forges, and glassworks to be approved by *lettre patente*.[78]

[75] See, for example, the 1783 petition of Mont-de-Laval in the Doubs, discussed in AN Q1 183.

[76] 1669 Ordinance in Isambert, vol. 18, titles 24 and 26. See also Hamish Graham, "Greedy or Needy? Forest Administration and Landowners' Attitudes in South-western France during the Eighteenth Century," *Rural History* 16, no. 1 (2005): 1–20. On the relevant laws, see Bernard Lutun, "Les approvisionnements de la Marine (1661–1761): De la règlementation. Règles générales (1674)," *La revue administrative* 55, no. 327 (2002): 292–311.

[77] *Ordonnance de Louis XIV pour les armées navales et arcenaux de marine*, 15 April 1689, book XXIII, title 1, art. 1 (Paris: Chez Estienne Michallet, 1689), 447. A November 1683 decree required that landowners obtain royal authorization to fell more than twenty-five arpents (thirteen hectares) of *futaie* within fifteen leagues of the coast or six leagues of navigable rivers. Subsequent decrees in 1685 and April 1689 barred selling or exploiting woods within these limits. Owners of timber woods who wished to fell more than twenty-five arpents at once had to advise the *grand maître* and controller general one year in advance. Paul Walden Bamford, "French Forest Legislation and Administration, 1660–1789," *Agricultural History* 29 (1955): 100; Devèze, "La crise forestière," 75, n. 5.

[78] *Arrêt du conseil qui défend de couper aucun bois taillis qui n'ait au moins l'âge de dix ans*, 19 July 1723; and *Arrêt du conseil qui [fait deffenses à toutes sortes de personnes, communautez regulières seculières, d'établir à l'avenir] d'aucuns fourneaux, martinets, forges et verreries, sans lettres patentes*, 9 August 1723, in Isambert et al., *Recueil général des anciennes lois françaises*, 21:257, edicts 289 and 290. See related decrees in Jacques-Joseph Baudrillart, *Traité général des eaux et forêts, chasses et pêches*, part 1, tome 1, *Recueil chronologique des réglemens forestiers: contenant les ordonnances, édits et déclarations des rois de France; les arrêts du conseil et des cours souveraines; les lois,*

Yet the requirement was only loosely applied in Franche-Comté and Lorraine, the sole areas of France where new forges and furnaces were established after 1730.[79]

While the comparative abundance of wood, iron ore, and water played a primary role in the promotion of ironworks in Franche-Comté, the crown's willingness to relax its industrial restrictions and inconvenience the local populace also signified the unequal treatment the province received as a recently defeated domain. With Louis XIV's conquest, the Francs-Comtois saw their *parlement* enfeebled, their system of justice replaced, and their natural resources carted off. In addition, because of its status as a "foreign" province within the realm, trade between Franche-Comté and the rest of France was hampered by internal tolls and other customs barriers.[80] Together, these irritations contributed to the persistence of Comtois antipathy toward French rule in the eighteenth century.

Chafing against their new ruler's constraints, Comtois inhabitants used a variety of tactics to resist. The municipal authorities of Epenoy in the Doubs, for instance, took nine years to respond when asked by an officer of the reform for information about their communal woods. When they finally did, they underreported the size and condition of Epenoy's holdings. Their deception was discovered in a follow-up investigation that revealed that instead of 199 *arpents* of scrub and bramble, the village had nearly 405 *arpents* of healthy forest.[81]

Even more common were "weapons of the weak," through which inhabitants maintained, albeit illicitly, their former woodland practices.[82] These included snapping off branches to create gatherable "dead" wood, bribing guards to allow their flocks into restricted areas, and clearing trees

arrêtés du gouvernement, décrets, ordonnances du roi, arrêts de la Cour de Cassation, décisions ministérielles, circulaires et instructions administratives (Paris: Imprimerie de Madame Huzard, 1821; hereafter cited as *Recueil chronologique*, tome 1), 156, 157, 226, 229. For discussion, see Woronoff, "La politique des autorisations d'usines," 58–59.

79 Fiétier, *Histoire de la Franche-Comté*, 272.

80 On taxation in Franche-Comté and its status as a "province réputée étrangère," see Boichard, *Le Jura*, 99–100, and Jean-Luc Mayaud, *La Franche-Comté au XIX siècle*, L'histoire de la Franche-Comté 7 (Wettolsheim: Mars et Mercure, 1979), 29.

81 See discussion in Richard Moreau and René-André Schaeffer, *La forêt comtoise* (Besançon: Centenaire de la Société Forestière de Franche-Comté et des Provinces de l'Est, 1990), 50–51.

82 James C. Scott, *Domination and the Arts of Resistance: Hidden Transcripts* (New Haven, CT: Yale University Press, 1990); also James C. Scott, *Weapons of the Weak: Everyday Forms of Peasant Resistance* (New Haven, CT: Yale University Press, 1985).

along forest edges to enlarge their fields. While this behavior had always been a part of the push and pull of rural life, the introduction of the 1669 Ordinance and its bureaucratic apparatus unintentionally fostered rapacity and exacerbated tensions by removing the woods from the realm of communal authority and rendering everyday activities illegal. Having been dispossessed of their traditional rights, inhabitants were left with the choice of acceptance and submission or disobedience and delinquency.[83] Nor was resistance limited to the humble or impoverished; as the infraction registers of Eaux et Forêts guards attest, seigneurs were just as likely as peasantry to flout the Ordinance's rules. Indeed, landowners themselves were behind some of the most audacious offenses, like hauling off high timber from a community's reserve, and clearing whole woods for planting.[84]

Though ephemeral in their efficacity, these insubordinate acts nonetheless put forest administrators on notice that compliance would not come easily. This was all the more true of the revolt in the Forêt de Chaux with which I began this chapter. Perhaps the longest single instance of coordinated opposition in Franche-Comté, this uprising was ruthlessly quelled after two months. In the long term, however, the historical memory of the affair lingered and worked to *usagères*' advantage, as they recalled how resistance had kept the forest administration at bay and won them recognition of their communal rights.

The Forêt de Chaux was and remains one of the oldest extant forests in Franche-Comté, but even in the eighteenth century its "old growth" days were well over.[85] Located on a floodplain between the Loue and the Doubs rivers, this forest, because of its poor drainage and acidic soil, was spared from being cleared for agriculture as so many other lowland forests were during the early period of Comtois settlement, but it was nonetheless heavily cut over by charcoal burners and regularly grazed by hundreds of communal flocks. Until the French takeover, the forest figured prominently in the holdings of the dukes of Burgundy, but after 1678 Chaux became part of the royal domain. By the start of the eighteenth century, it extended over some 40,000 arpents (20,680 hectares) and was a major source of firewood, timber, and charcoal in central

[83] For a present-day comparison, see Gibson, Ostrom, and McKean, "Forests, People, and Governance," 233.

[84] Seigneurs regularly paid as much in fines as the communities. Merlin, *Ceux des villages*, 52, 116–17. See also Hamish Graham, "Policing the Forests of Pre-industrial France: Round Up the Usual Suspects," *European History Quarterly* 33, no. 2 (2003): 157–82.

[85] *Chaux* refers to the lime in the soil.

Franche-Comté. Nearly forty villages enjoyed privileges there, from pasturage and *glandée* to the gathering of mushrooms, berries, bark, and dry leaves. As François Joseph Legrand de Marizy, the *grand maître* in charge of bringing the forest into conformity with the 1669 Ordinance, noted in 1738, neighboring communities "carried out [their usages] without constraint [and] because the rules to which they were subjected were scarcely troublesome, everyone overlooked them. … As long as they did not abuse their rights through expropriations … the inhabitants … lived in the greatest abundance."[86]

Legrand de Marizy hoped to alter this situation. His initial reform plan established Chaux as a timber reserve with fellings scheduled at one-hundred-year cycles, intending for the forest to become a major supplier of high-quality *haute futaie* to the navy. The prospect of century-long rotations dismayed forge masters and salt makers, however, who protested that the arrangement would decimate their fuelwood supplies. In response, Legrand de Marizy organized the forest into coppices on thirty-year rotations, augmented by a timber reserve on a much longer timetable. Though the communities surrounding Chaux also objected to Legrand de Marizy's reforms, they were less successful than the industrialists at wresting concessions. Instead, they waged lawsuits to maintain their rights of entry and, while the crown was distracted by war from 1741 forward, openly disobeyed the reforms and went about practicing their use rights as they had in the past.[87]

Other phenomena conspired to reduce access to Chaux's resources, however. In 1739, a prolonged freeze killed off many trees and shrubs, driving up firewood prices and imperiling the region's shivering inhabitants. Firewood costs rose even higher after 1753, when the crown granted

[86] AN Q1 416, Procès-verbaux de la forêt de Chaux, 6 September 1738. Marizy was commissioned to carry out special reforms of the Forêt de Chaux and other sites to supply fuelwood to Salins' saltworks. He was later *grand maître* of Burgundy from 1754 to 1789.

[87] Several of the lawsuits concerning the Forêt de Chaux lasted into the 1830s. See *Jugement du Tribune de première instance de Dole (Jura) pour les communes riveraines de la forêt de Chaux [les maintenant dans l'exercice de leurs droits de panage, pacage et affouage dont elles jouissaient avant et lors de la promulgation du Code forestier de 1827] contre l'Etat (du 28 août 1839)* (Dole: Imprimerie Pillot, 1842), 64; *Arrêt de la Cour royale de Besançon, pour les communes riveraines de la forêt de Chaux, contre l'Etat: 23 décembre 1841* (Besançon: Imprimerie de Bintot, 1842), 24. Throughout the 1740s and 1750s, France was embroiled in the War of Austrian Succession (1740–48) and the Seven Years' War (1756–63). On the impact of these conflicts on woodland reformations throughout France, see *Les Eaux et Forêts*, 199.

a fuelwood concession in Chaux to the saltworks at Salins. While some villages had *affouge* rights (the collective right, common in eastern France, to take an annual allocation of firewood from communal woods), most had to buy their wood on the open market. The insertion of Salins' charcoal burners into this tense economy further exacerbated the competition for fuelwood among inhabitants and the ironworks, glassworks, potteries, and tanneries that ringed the forest's perimeter.[88] Residents of the nearby city of Dole were particularly hard hit by the price increases. Unable to provision themselves illicitly as their rural neighbors did, the Dolois launched a protracted legal battle that would last into the 1830s to regain their "ancient firewood rights" in Chaux.[89]

By 1760, the dual demands of industrial and domestic usage had upended the tenuous balance of communal consumption and resource viability in Chaux. Together with environmental factors like climatic instability, they threatened to exhaust the forest permanently. Chaux's deteriorating condition was apparent in the steady spread of *vides*: barren, degraded patches caused by excessive cutting and pasturing. In 1763, administrators from the Eaux et Forêts began to enforce Legrand de Marizy's woodland reforms in earnest. The shift came about as part of a new, nationwide effort to rejuvenate the navy in the wake of the Seven Years' War, in which France had fared miserably. Spearheaded by the duc de Choiseul, Louis XV's highest minister, the project echoed Colbert's goals a century earlier: to rid the *maîtrises* of corruption, boost timber production, implement stricter oversight, and improve policing. This new effort, however, raised tensions between forest officers and local villagers, who found themselves being fined and imprisoned for long-standing practices like gathering wood. On 19 February 1765, the hostilities boiled over into revolt.

[88] Forge masters routinely petitioned their local *maîtrises* for permission to cut fuelwood and produce charcoal in royal and communal woodlands. These grants were in addition to woods that the forge master or owner possessed individually. See, for example, Archives départementales de la Haute-Saône, Vesoul (hereafter ADHS) B 3233 12 May 1790, Autorisation donnée par François Joseph LeGrand, de Marizy, grand maître des eaux et forêts de France, à Jacques-Antoine Prailleur, propriétaire des forges du Magny Vernois, de construire des fours de charbon dans les coupes destinée aux usances des communautés de Saulnot, Crevans, Secenans, Grange, Lachapelle, Malval.

[89] ADJ C 128, dossier 15, *Mémoire sur les difficultés qu'éprouve la ville de Dole de s'approvisionner de bois de chauffage, et sur les moyens d'y remédier*, 1768; see also Vuillier, *Mémoire concernant les droits d'usage indéfini de la ville de Dole, dans la forêt impériale de Chaux* (Dole: Imprimerie de Joly, 1812).

The incident was set in motion by an investigation into woodland infractions. It was no spontaneous protest, however. Two hundred men were involved. In addition to arming themselves, they donned a makeshift disguise – vaguely feminine in appearance – that eventually earned them the title of "demoiselles." Like the later demoiselles of the Pyrenees who chased officers out of the woods in the 1820s, the moniker was perhaps ironic.[90] Either way, the demoiselles – mostly potters, artisans, and peasants, all inhabitants of four nearby villages – took coorduinated action to claim what they regarded as theirs.[91] Holding off officers who came to investigage, they occupied the northern part of the forest, harvested wood, and funneled it to their families for sale to manufacturers and residents of Dole.

For nearly two months, the forest administration sought to regain control, but they were aided little by local guards, many of whom chose to relinquish their posts rather than confront their countrymen. As the forest authorities at Besançon noted on 11 March 1765, "the officers of the *maîtrise* have ordered the arrests of many known individuals; but the terror is so great, no bailiff wants to carry [them] out." Anyone caught buying wood from the rebels, the authorities warned, would be pursued and punished as a delinquent. The demoiselles themselves were threatened with "pain of death." Concluding that "an excess as unrestrained as this can only bring about the total ruin of this vast forest, and accustom the inhabitants of ... neighboring communities to violence and crime," the Eaux et Forêts called on the aid of the king's army to put down the revolt. A regiment arrived in late March. Unwilling to engage

90 On the uprising in Chaux, see ADJ C 128, dossier 7, 1765; Vernus, *La vie comtoise*, 1:78–79. The name *demoiselles* is widespread in the secondary literature, but I have not found clear evidence showing that it was used at the time. Scholars may be simply invoking the image of the nineteenth-century demoiselles of the Ariège. See, for example, Georges Plaisance, "La révolte des Demoiselles," *Barbizier almanach populaire comtois* 5 (1951): 483–86; Vion-Delphin et al., *Les hommes et la forêt en Franche-Comté*; and Rousseau and Bouvier, *La Grande forêt de Chaux*. On the Ariège uprising, see AN BB^{18} 1308; also John M. Merriman, "The Demoiselles of the Ariège, 1829–1831," in *1830 in France*, ed. John M. Merriman (New York: New Viewpoints, 1975), 87–118; Whited, *Forests and Peasant Politics*; and Peter Sahlins, *Forest Rites: The War of the Demoiselles in Nineteenth-Century France* (Cambridge, MA: Harvard University Press, 1994).

91 The rebels hailed from the towns of Salans, Rans, Étrepigney, and La Bretenière, all on the northeastern edge of the forest. See comments of the 1 March 1765 letter from the *subdelegué* of Dole to the *intendant* of Franche-Comté: "the inhabitants ... have as [their] goal, not only to devastate this forest, but still more to destroy it, and the guards and officers of the *maîtrise*." ADJ C 128. On the *réformation* of the Forêt de Chaux, see AN Q1 416, Procès-verbaux de la forêt de Chaux.

the demoiselles in the woods, however, the soldiers viciously attacked their families and pillaged their villages until the men finally emerged and surrendered.[92]

Though pitilessly suppressed, the insurrection was not without success. To mollify inhabitants and avert further troubles, the Conseil d'état ordered a revision of the annual firewood allotments for the villages surrounding Chaux. Completed the following year, the hard-won revisions constituted an enduring material as well as moral achievement.[93] As the forest's *usagère* communities continued to struggle over the next century against unfavorable legislation and industrial usurpation, the precedent set by the demoiselles would embolden others to fight in defense of their woodland way of life.[94]

A century after the French conquest, the cultural memory of Franche-Comté's former autonomy persisted. In rural communities, this memory took the form of hostility to the controls imposed by the forest administration. Though inhabitants did not all agree about the allocation of their woodland resources – indeed, as external, commercial pressures on the forest mounted, community consensus deteriorated – together they put forward a markedly different vision from the crown and its Ordinance concerning how natural resources should be used and who should benefit.[95] Over centuries of exploitation, Franche-Comté's settlers had incontrovertibly changed the forest, but they had also developed methods of conservation and management aimed at sustaining their vital needs. This resourcefulness and self-determination was stifled by the intervention of the Eaux et Forêts and its contrasting emphasis on timber and fuelwood. Paradoxically, the loss of local oversight and the introduction of

[92] *Arrêt du Parlement rendu en la Chambre souveraine des Eaux et Forêts au sujet des attroupemens [sic] & violences des riverains de la forêt de Chaux, du 11 mars 1765*. The court was based in Besançon. On the outcome, see Vion-Delphin et al., *Les hommes et la forêt en Franche-Comté*, 96–98.

[93] Archives départementales du Doubs (hereafter AD Doubs) P 579, Le contingent annuel des communes riveraines de la forêt de chaux après le nouvel aménagement du 19 août 1766.

[94] Among the many who invoked the memory of the uprising, see Cl. P. Jos. Balland, *Observations sur l'administration des forêts* (Paris: Imprimerie du Cercle Social, 1791), 20.

[95] On increasing social differentiation in the eighteenth century, see Hamish Graham, "Profits and Privileges: Forest and Commercial Interests in Ancien Regime France," *French History* 16, no. 4 (2002): 388; Andrée Corvol, "L'affouage au XVIII siècle: Intégration et exclusion dans les communautés d'Ancien Régime," *Annales E.S.C.* 36, no. 3 (1981): 390–407; and Olwen Hufton, *The Poor of Eighteenth-Century France, 1750–1789* (Oxford: Clarendon Press, 1974).

the French rule helped trigger the decline of forest cover in Franche-Comté. The trend was driven in part by demographic growth after 1700 and by the imposition of royal taxes requiring payment in grain, which led to an increase in clearing for agriculture.[96] France's woodland policies also played a role by introducing techniques that were out of step with the environment and by privileging naval timber and industrial expansion over the needs of the populace.

These shifts were just the beginning of a much longer struggle for the forest in Franche-Comté, a struggle that suggests how the loss of control by local stakeholders can radically affect the human and natural ecology of a region. Over the course of the eighteenth century, as the tactics of Franche-Comté's new forest masters evolved in concert with the interests of the crown, a new conservationist ideal would emerge to reinforce the Ordinance's mission, that of marshaling the country's resources for the service of the state.

[96] On taxation and its effects, see Colette Merlin, "Impositions, charges et résistance paysanne dans la Petite Montagne jurassienne à la veille de la Révolution," in *La Franche-Comté à la veille de la Révolution*, ed. M. Gresset (Besançon: Annales littéraires de l'université de Besançon, 1988), 24.

2

"Agromania" and Silvicultural Science

Conservation's Intellectual Underpinnings

> Indeed is it not obvious that a country denuded of Wood would be uninhabitable, and that if one did not have the resource of Coal and Peat, it would not be possible to prepare food and protect oneself from the rigors of winter? ... Already in some districts of the Realm, they are reduced to burning dry grasses and animal dung to support their most pressing needs.
>
> Henri-Louis Duhamel du Monceau, *Du transport, de la conservation et de la force des bois*, 1767[1]

The century spanning the enactment of the 1669 Ordinance and the eve of the Revolution marks a formative period in the development of French environmental discourse and conservationist policy. During those critical years, demands on France's forests mounted as never before. As elsewhere on the continent, the country's woodlands steadily retreated in the face of demographic and agricultural expansion, chronic warfare, urban growth, and industrial production.[2] At the same time, spurred by wood's increasing commodity value, many forest owners, including the crown, began to seek ways of reducing or eliminating the complex web of customary rights that burdened their properties, thereby aggravating local shortages.

[1] Henri-Louis Duhamel du Monceau, *Du transport, de la conservation et de la force des bois; ou l'on trouvera des moyens d'attendrir les bois, de leur donner diverses courbures, sur-tout pour la construction des vaisseaux; faisant la conclusion du Traité complet des bois & des forests* (Paris: L. F. Delatour, 1767), viii.

[2] Warde, "Fear of Wood Shortage," 38–46. David Bruce Young, "A Wood Famine? The Question of Deforestation in Old Regime France," *Forestry* 49, no. 1 (1976): 45–56; Edward A. Allen, "Deforestation and Fuel Crisis in Pre-Revolutionary Languedoc, 1720–1789," *French Historical Studies* 13 (1984): 455–73.

Because of wood's essential role in agriculture, industry, and maritime trade and defense, the increasing incidence of severe, albeit sporadic fuelwood and timber shortages from the mid-seventeenth century forward provoked widespread anxiety. Many observers warned that a wood famine was imminent and that it could cripple the nation's economy. Together with concerns about France's agricultural productivity, the shortages prompted an outpouring of pragmatic and philosophical writings concerning humans' relationship with the environment. Part of the broader intellectual efflorescence of the Enlightenment, these essays, edicts, and commentaries on topics like agricultural improvement, forestry, landscape gardening, and the despoliation of nature reflected both the scientific optimism of the age and a growing unease about the effects of environmental exploitation in France and its colonies. At the same time, rural communities and a small number of forest administrators called for adaptations and improvements based on their experience of woodland challenges on the ground. Their insights received less attention.

Focusing on the writings of state administrators, naturalists, agronomists, and nascent romanticists from the twilight of Louis XIV to the start of the Revolution, this chapter explores the emergence of a discourse of environmental conservation in tension with forest practice. The gulf between them would grow wider at the time of the Revolution, with cataclysmic results.

SEVENTEENTH-CENTURY ORIGINS

Two works produced within seven years of each other profoundly influenced later ideas about forest conservation in Europe: Colbert's 1669 Ordinance, and *Sylva*, a guide for private forest owners published by the English writer John Evelyn in 1662. Though one was a work of policy and the other a disquisition with literary ambitions, both were prompted by concern about timber shortages and their nations' naval and military defense. By insisting on the importance of afforestation and long-term woodland planning and protection, they enduringly shaped not only principles of forest management but also concepts of sustainability still in force today.[3]

[3] On Evelyn and Colbert's significance, see S. Ravi Rajan, *Modernizing Nature: Forestry and Imperial Eco-development 1800–1950* (Oxford: Clarendon Press, 2006), 23.

When Jean-Baptiste Colbert took office as Louis XIV's minister of finances in 1661, he faced a double challenge: industrial growth was decimating the nation's forests, and there were fewer and fewer *futaie*, the tall forest trees demanded for their size and strength by shipbuilders. These problems had a long history. Since the late 1400s, the proliferation of blast furnaces had enabled the smelting of greater quantities of higher-quality iron, but only by consuming prodigious amounts of fuelwood. Over time, this intense exploitation drained forests of quality timber, leaving only the scrubby masses of coppice that charcoal burners preferred.[4] For Colbert, the decline of big timber constituted a national emergency, particularly in light of French naval preparedness and maritime competition with the English and Dutch. In the brief period of peacetime that followed the Thirty Years' War (1610–48), Colbert launched an aggressive campaign to improve France's woodland fortunes. Starting in 1661, he appointed forest reformers to map, measure, and redistrict woods throughout the realm, with instructions to pay special care to the condition of standing timber. He also replaced venal officeholders in the Eaux et Forêts with more capable, less corrupt candidates.[5] By 1668, these efforts were already paying off: despite ongoing felling, timber stocks in the royal forests multiplied. Woodland revenues, meanwhile, grew from 169,000 livres in 1661 to 1,029,000 in 1683, and the profits reaped from fines on forest offenses more than paid for the project's implementation.[6]

As with his other grand reform projects, among them standardizing civil procedure and compiling the notoriously harsh *Code noir* on the status of slaves, Colbert ensured his accomplishments' survival by inscribing them in law.[7] A blend of technical, fiscal, and criminal regulations,

4 Blast furnaces consumed twelve to seventeen cords per *millier* (500 kilograms). To operate at full capacity indefinitely, an iron producer needed as much as 2,000 hectares of coppice cut on continuous rotation. The wood was then converted into charcoal. Belhoste, "Forges et forêts en révolution," 127.

5 Charles Woolsey Cole, *Colbert and a Century of French Mercantilism* (repr., Hamden, CT: Archon Books, 1964), 2:311, 2:348, 2:453–56; C.A.R.A.N., *Histoire de forêts*, 96–97.

6 Thanks to forest revenues, Colbert balanced France's revenue and expenditures in 1672 for the first time since his predecessor had done so. *Table des ventes des bois du roi, 1660–1689*, reproduced in Gustave Huffel, *Économie forestière*, 2nd ed., vol. 3 (Paris: Librairie Agricole de la Maison Rustique, 1926), 3:264.

7 H. A. de Colyar, "Jean-Baptiste Colbert and the Codifying Ordinances of Louis XIV," *Journal of the Society of Comparative Legislation* 13, no. 1 (1912): 56–86. The 1669 Ordinance appears in François-André Isambert, Athanase-Jean-Léger Jourdan (Decrusy), and Alphonse-Honoré Taillandier, eds., *Recueil général des anciennes lois*

the *Ordonnance des eaux et forêts*, signed by Louis XIV on 31 August 1669, governed all the woodlands of the realm, including individual holdings. This extent in itself was an unprecedented – and to seigneurs, unwelcome – innovation. Earlier forest decrees were largely limited to crown and communal possessions. It was in its silvicultural prescriptions, however, that the 1669 Ordinance was most pathbreaking. Shifting from the crown's traditional interest in forests as royal hunting ground, the Ordinance instituted regulations aimed at increasing timber and fuel-wood production. The *quart en réserve*, a designation that set aside one-quarter of every crown, communal, and private forest to mature into large trees, promised an ongoing supply of hulls, bows, sprits, and spars to feed the French navy's ravenous consumption of timber. An estimated 2,800 century-old oaks and 6,338 pines went into building one seventy-four-gun ship of the line alone.[8] At the same time, the system known as *taillis sous futaie*, coppices mixed with mature trees for reseeding the forest, was intended to reconcile the need for timber with mounting demands for industrial fuel and domestic heating. (For *taillis sous futaie*, see Figure 3.)

Despite the Ordinance's initial success, however, its implementation and enforcement soon faltered, a casualty of Colbert's fall from political favor and Louis XIV's resumption of woodland office sales to pay for his latest military conflict, the War of the League of Augsburg (1689–97). The decision reflected an ongoing pattern in the Old Regime, in which fiscal urgency repeatedly undercut the crown's own efforts at rational reform.

The Ordinance faced other challenges as well. In addition to being unpopular with seigneurs, who resented the state intruding on their territorial authority, the law's emphasis on naval timber and firewood consolidated a trend toward privileging military, urban, and industrial needs over those of rural communities.

Like the 1669 Ordinance, John Evelyn's *Sylva* also placed a high emphasis on timber production. Evelyn's recommendations on woodland

françaises, depuis l'an 420 jusqu'à la Révolution de 1789, 29 vols. (Paris: Belin-Leprieur, 1821–33), 18:219–311 (cited hereafter as the 1669 Ordinance in Isambert, vol. 18).

8 Martine Acerra, "Marine militaire et bois de construction: Essai d'évaluation (1779–1789)," in *Révolution et espaces forestiers: Colloque des 3 et 4 juin 1987*, ed. Denis Woronoff (Paris: L'Harmattan, 1988), 114; Michel Devèze, "Les forêts françaises à la veille de la Révolution de 1789," in *La forêt et les communautés rurales-XVIe-XVIIIe siècles: Recueil d'articles parus en revue* (Paris: Publications de la Sorbonne, 1982), 194–95.

management moreover remained influential more than a century after its initial publication. As a reviewer of the 1776 edition observed, "no English book on the subject hath been more sought after, or consulted to better effect."[9] Evelyn himself boasted that his tome had "been the occasion of propagating many millions of useful timber-trees throughout this nation."[10]

Sylva's impact extended well beyond England. Hans Carl von Carlowitz, credited with the earliest articulation in print of the concept of sustainability, drew on both *Sylva* and the 1669 Ordinance in his *Sylvicultura Oeconomica* (1713), Germany's first book on forest management and a cornerstone of scientific forestry.[11] *Sylva* likewise found an enthusiastic audience in France among the scholars of the Jardin du Roi (Royal Garden), who plumbed its pages for the insights it offered on woodland improvement and profitability.[12] English treatises on botany, agriculture, and jurisprudence were in general popular among Enlightenment *philosophes*. As one reader enviously noted, England was "a rival Nation, that prides itself on always beating us in matters of utility and reason."[13]

As a practical guide to planting, pruning, felling, and exploiting different tree species, *Sylva* provided guidance that the Ordinance did not. Colbert's legislation left aside any discussion of the needs of specific species, much less differences among ecosystems.[14] Fixated on mercantilist and maritime self-sufficiency, the Ordinance instead prescribed a monodimensional formula for managing the nation's forests, one that

9 *The Weekly Magazine, or Edinburgh Amusements*, Wednesday, 14 January 1778, 94.

10 John Evelyn to Anne Spencer, Countess of Sunderland, letter of 4 August 1690, quoted in John Evelyn, *Sylva, or A Discourse Of Forest-Trees, and The Propagation of Timber In His Majesties Dominions: To which is annexed POMONA; Or an Appendix concerning Fruit-Trees in relation to Cider, the Making and several ways of Ordering it* (London: Royal Society of London for Improving of Natural Knowledge, 1664), in *The Writings of John Evelyn*, edited by Guy de la Bédoyère (Woodbridge, UK: Boydell Press, 1995).

11 Hans Carl von Carlowitz, *Sylvicultura Oeconomica* (Leipzig: Braun, 1713).

12 Richard H. Grove, *Green Imperialism: Colonial Expansion, Tropical Island Edens, and the Origins of Environmentalism, 1600–1860* (New York: Cambridge University Press, 1995), 160.

13 Ethis de Novéau, "Essai sur cette question: S'il seroit plus utile en Franche-Comté de donner à chacun la liberté de clorre ses héritages pour les cultiver à son gré, que de les laisser ouverts, pour le vain [*sic*] pâturage, après la récolte des premiers fruits?" in *Mémoires de l'Académie de Besançon* (Besançon: Imprimerie de Cl. Jos. Daclin, 1767), 2.

14 The Ordinance identifies specific tree species by name only once. See the prohibition against uprooting oaks or hornbeam without permission, "under threat of exemplary punishment and five hundred livres fine," in 1669 Ordinance in Isambert, vol. 18, title 27, art. 11.

severely restricted other vital woodland activities like grazing and gathering and interpreted the forest primarily as a live storehouse for timber. This failure to recognize the forest's broader social, cultural, and economic values would ultimately undercut the law's effectiveness.

Despite their differences in approach, the 1669 Ordinance and *Sylva* reflected the certainty that rational, top-down regulation and scholarly erudition were the answers to forests' deterioration. Forest reformers on the ground and rural inhabitants had a different view.

ADMINISTRATORS, LOCAL INTERESTS, AND NATURAL PHILOSOPHERS

Following Colbert's death in 1683 and the resumption of war in 1688, France's forests began a long decline that would continue, with only short reprieves, through the Revolution and into the first decades of the nineteenth century. Between 1689 and 1714, the Eaux et Forêts conducted no new woodland *réformations* and made no amendments to the original Ordinance.[15] With the start of Louis XV's regency in 1715, the crown attempted halfheartedly to bring the forest administration back into line, but the Ordinance itself remained unmodified, its mandates increasingly out of date as the years passed.

A number of factors toward the end of Louis XIV's rule conspired to set back the gains Colbert had made. Ecologically, a spate of calamities, including rains and ruined harvests in 1693–94 that led to an estimated ten thousand deaths, triggered an increase in the clearing and conversion of woods to agriculture. During the horrifying winter of 1709 as many as one million people lost their lives. Trees perished, too, freezing where they stood, their barren and brittle limbs breaking off easily at the hands of desperate foragers.[16] Industrial pressure on forests also mounted, especially in French-controlled Lorraine and the recently annexed Franche-Comté. The greatest challenge, however, lay in the crown's ongoing auction of woodland offices, which caused the Eaux et Forêts to balloon out of proportion. The sale of sixteen newly created *grand maître* positions, each with their own district, netted almost two million livres in

[15] *Les Eaux et Forêts*, 168.

[16] On increased clearing, see *Les Eaux et Forêts*, 171. On freezes and famines, see Pierre Goubert, *The French Peasantry in the Seventeenth Century* (Cambridge: Cambridge University Press, 1986), 93; Pierre Goubert, *The Ancien Régime: French Society, 1600–1750*, trans. Steve Cox (New York: Harper & Row, 1973), 41.

three years. Military expenses again multiplied during the War of the Spanish Succession (1701–14), leading the crown to become more inventive still: alongside "supplementary" *maîtrises*, *gardes générals*, and surveyors, it created a lucrative roster of "alternates."[17] Louis XIV used the money from the sale of these superfluous positions to fend off his Austrian Hapburg, English, and Dutch enemies, but the forests – poorly managed and often abused by officers more interested in cultivating a return on their investment than taking care of trees – paid the price.

Effects of the damage were soon seen. In 1701, the crown's newly created Conseil du Commerce launched a countrywide investigation to measure forest decline throughout the realm. Provincial administrators were asked to comment on availability of timber in their districts, as well as identify *terres vagues* – fallow "wastes" – that could be planted with trees. According to the twenty-odd reports that have survived, deficits of fuel and timber were common in many parts of the country. Woodlands in the central basin of Orléans, Berry, and Touraine had suffered considerable decline, as had the plateaus and slopes of Auvergne, Montauban, and Béarn. Reasons for the shortages varied. In Bordeaux, the *intendant* (governor) asserted that his district had "suffered more than any other department from wood shortages" because of ongoing clearing for vineyards.[18] The *intendant* of the Dauphiné explicitly blamed "manufacturers of steel, anchors, [and] sword blades," for the loss of big trees. "One is rightly alarmed," he wrote, "[for] there are no provinces where there had been a greater quantity of wood than the Dauphiné, nor where one could obtain serviceable [wood] for masts and construction. But the trouble that has been noticed in other provinces has now reached as far as this."[19] Despite Colbert's efforts thirty years earlier, timber for buildings, equipment, wagons, and, above all, naval and merchant ships was now scarcer than ever.[20]

Other areas fared better. According to the *intendants* of Burgundy, Franche-Comté, and Lorraine, the problem was not so much a dearth of

17 C.A.R.A.N., *Histoire de forêts*, 93; and Corvol, *L'homme et l'arbre*, 395, 414.

18 The *intendant* of Bordeaux is cited in Devèze, "La crise forestière," 64–65.

19 *Intendant* Bouchu to the Conseil du Commerce, in Devèze, "La crise forestière," 67, 64.

20 On the gamut of responses to the Conseil de Commerce's 1701 investigation, see A. M. Boislisle, ed., *Correspondance des contrôleurs généraux des Finances avec les intendants des provinces, publiée par ordre du Ministre des finances, d'après les documents conservés aux Archives nationales*, vol. 2, *1699 à 1708* (Paris: Imprimerie nationale, 1883), item 355 and note to item 355, 100–102.

timber, but how to convey it from the region's remote mountain slopes to the distant naval yards at Toulon, Rochefort, and Brest.[21] In this way, isolation and ecological conditions that favored forest growth helped preserve France's eastern woodlands long after its southern slopes were denuded.

Because of their far-reaching implications, France's woodland troubles attracted the concern of many other commentators, including the military engineer Sébastien le Prestre de Vauban (1633–1707).[22] Acclaimed for his fortifications and innovations in organizing artillery fire, Vauban needed reliable sources of timber to build his battlements. Responding to the Conseil du Commerce's investigation in 1701, Vauban noted that some areas of the country were "entirely lacking in wood," and that even the crown's forest officers were brazenly flouting restrictions on felling. "Many older forests, if not to say all," he lamented, "for want of attention and having been extremely neglected, are ruined and in a pitiful state."[23] Unlike most provincial governors and even many royal forest administrators, Vauban had personally visited woods throughout France to obtain construction materials, and his assessment reflected this knowledge. In his opinion, the solution to France's timber problem lay in the strict enforcement of the Ordinance and in voluntary tree planting by wealthy landowners, who presumably had acreage to set aside for woods. Echoing Evelyn's *Sylva*, Vauban declared,

> If His Majesty saw to it that the officers of the Eaux et Forêts performed their duties well regarding the observation of ordinances, that the seigneurs of his court and other affluent gentlemen of the realm did their best to replenish the vacant places on their lands with woods, and that the rich and well-off clerical communities did the same, there would be plenty of new forests, incomparably more beautiful and better than those that presently exist.[24]

This vision notwithstanding, Vauban recognized that few proprietors could afford to wait decades – lifetimes, even, depending on the tree's type and intended purpose – before realizing a return on their investment. Fewer still would embark on what might only amount to ancestral largesse compared with the annual returns from grain or even coppices. He resigned himself to hoping that landowners would "do their best," while conceding that

[21] Devèze, La crise forestière, 64. [22] Anne Blanchard, *Vauban* (Paris: Fayard, 1996).
[23] "Traité de la Culture des Forêts," 14 October 1701.
[24] Husson, *Les forêts françaises*, 129.

plantations were really "an activity of the King," for only the crown had the authority and incentive to cultivate timber over the long term.[25]

Like Vauban, René-Antoine Ferchault de Réaumur, a natural scientist who later gained renown for his pioneering work in entomology, also addressed France's forest problems with a critical eye. At Louis XIV's request, Réaumur had reported in 1710 on the country's industries and natural resources. He integrated those findings in a later essay of 1721, "Reflections on the Conditions of the Woods of the Realm," in which he pointed out the weaknesses of the Ordinance's management mandates.[26] He was particularly critical of the practice of *taillis sous futaie*. As a compromise between growing high-volume fuelwood (*taillis*) and tall, full-grown timber (*futaie*), Réaumur argued, the approach failed at both. Coppice stems sprouted thinly and feebly in the shady understory, while the tall trees left in the canopy became contorted and sheared by the wind, developed knots, and lost their value.[27] If the crown wanted quality timber, Réaumur asserted, it needed to use different methods: remove tall trees from coppices, let the most vigorous coppices grow, and burn the rest to prepare the ground for planting trees specially picked for local conditions. Knowledge of the local environment was essential to determining which species to plant and where, as well as when to fell mature specimens. "Everything depends on the species, and above all on the terrain," he emphasized. While seemingly sound and even obvious advice, this site-specific approach would not be officially taken up by the state until after 1825.[28]

In the meantime, forest officials on the ground, most of whom had no training in natural history, botany, or even the law, either ignored the silvicultural responsibilities of their office in favor of levying fines to recoup their purchase price, or looked to long-standing local practice to guide their work.

Exemplary among the latter was Louis Marie Maclot, the *grand maître* assigned to reform the forests of the Jura in Franche-Comté, who put his peers' venal preoccupations to shame by devising management guidelines

[25] Devèze, "La crise forestière," 73.

[26] "Réflexions sur l'état des bois du royaume et sur les précautions qu'on pourrait prendre pour en empêcher le dépérissement et les mettre en valeur," *Comptes rendus de l'Académie des sciences*, 24 December 1721.

[27] Réaumur, *Mémoire*, discussed in Belhoste, "Une sylviculture pour les forges," 242.

[28] E. Lachaussée, "L'évolution de la sylviculture en France depuis 100 ans, ou des principes de Lorentz et Parade à la sylviculture à but industriel et commercial," *Bulletin de la Société Forestière de Franche-Comté et des Provinces de l'Est* 33, no. 1 (1966): 5–10.

specific to the region's coniferous uplands. A former counsel to the *parlement* of Metz, Maclot had purchased his first Eaux et Forêts office, the position of *grand maître* of Champagne, in 1701. In 1724, he was appointed by the crown to go to Franche-Comté and take on the formidable task of renovating the woods that supplied Salins' saltworks with fuel and timber.[29] Franche-Comté's saltworks had become royal possessions after the French conquest – as spoils of war go, "white gold" was prized indeed. But the area was widely deforested and had been that way long before Louis XIV's annexation of the province. For centuries, the Chalons family, who owned the saltworks, had been exacting a punishing tribute in wood and timber from the communities around Salins. The saltworks' wood purveyors had to go farther and farther afield to find the fuel they needed, and rural inhabitants were not happy about the increasing demands.

Implementing the management prescriptions of the 1669 Ordinance, with its overwhelming emphasis on timber rather than firewood, would not only fail silviculturally in the woods around Salins, Maclot reasoned, but might also affect salt production. Echoing Réaumur, Maclot had little positive to say about *taillis sous futaie*, and he was nearly as negative about *tire et aire*, or shelterwood felling, through which timber was cut in successive strips, or parcels. These methods were well-suited to the broadleaved woodlands of the Paris basin, with its mild climate and modest elevation. In the coniferous forests of the Haut-Jura and Haut-Doubs, by contrast, frigid temperatures inhibited growth and steep slopes defied regimented, plot-based fellings on rotation.[30] Putting *taillis sous futaie* into practice there would needlessly harm healthy trees and impede the regeneration of new ones, Maclot reasoned. *Tire et aire* was even worse, for it provoked soil erosion on slopes and exposed remaining trees to sun and wind, causing them to branch out and thicken rather than continue to grow straight and tall, the optimal shape for construction wood.

Maclot's observations echoed the concepts and complaints of local inhabitants. In 1719, the monks of Montbenoît noted that the

[29] On Maclot's reforms of the forests of Levier and la Joux, which supplied the Salins saltworks, see *Arrêts du Conseil des 18 janvier et 2 mai 1724, rendus à Versailles*. AN E 2055, fol. 18 & 115.

[30] On *taillis sous futaie* and *tire et aire*, see 1669 Ordinance in Isambert, vol. 18, titles 15 and 17. In *tire et aire* (shelterwood system), forests are divided into long parallel plots that are successively felled over the course of a multiyear rotational cycle. Trees on the edge of newly opened clearings become exposed to sudden light and wind. See David Smith, *The Practice of Silviculture*, 8th ed. (New York: Wiley, 1986), chap. 14.

1669 Ordinance was inappropriate for local conditions. "Pines do not sprout from the stump and experience shows that young [saplings] do better in the shadow of old [trees] than outside of it because, growing among rocks, the great heat dries them out when they are still too vulnerable," they protested in a letter to the *intendant.* As for the trees left standing as seed-bearers in *taillis sous futaie*, they remarked, "It is clear that all the pine-standards that we are obliged to leave in the plots will be broken or uprooted by the violence of the winds. . . . these trees are fragile and do not have deep roots."[31]

Maclot had no specific silvicultural training, but he recognized the utility of the monks' observations. Modeling his reforms on the selection system known as *jardinage comtois* rather than the shelterwood and coppicing methods prescribed by the Ordinance, Maclot divided the woods around Salins into twelve lots, from which a handful of trees – generally the smallest, spindliest specimens – would be culled each autumn. He also formally authorized the long-standing Comtois tradition of *bois en ban*, informal reserves in which suitable specimens were set aside annually to grow into timber. Together, these native approaches enabled Maclot to assure Salins' wood supplies while also meeting the Ordinance's demand for timber.[32]

THE IMPACT OF "AGROMANIE" AND PHYSIOCRACY ON THE FORESTS

With France's entry into the War of the Austrian Succession in 1741, the crown's modest efforts against corruption in the Eaux et Forêts abruptly stopped. Though individual proprietors' interest in woodland profits continued, the renewed demands of military financing, in combination with an emerging agronomic movement, weakened the state's resolve to continue forest reforms. The forest reforms that had proved so contentious in Franche-Comté also ebbed, as the concerns of royal administrators and self-described natural philosophers increasingly shifted to agriculture in the interest of raising and feeding a larger population.[33] Between 1700 and 1790, France's population grew from

[31] The *abbé commandataire* of Montbenôit, to the *Intendant*, Courbouzon, 1719, cited in Moreau and Schaeffer, *La forêt comtoise*, 58.

[32] Lucien Turc, "L'aménagement des bois de sapin de Franche-Comté," *Revue forestière française* 9 (1950): 445–57. On Maclot, see also Jean-Claude Waquet, *Les grands maîtres des Eaux et Forêts*, 254, 387–88.

[33] *Les Eaux et Forêts*, 199.

roughly twenty-one million inhabitants to more than twenty-eight million. Though this increase is unremarkable if averaged over the whole period, much of it appears to have been concentrated in the third quarter of the century. Gains were especially great in the northern and eastern countryside and in the maritime ports of the west.[34] Franche-Comté's population more than doubled between the French conquest in 1678 and the eve of the Revolution. While the increase was partly a recovery to previous population levels prior to the Thirty Years' War, it nonetheless placed enormous pressure on the soil and contributed to a spirit of "agromania" among landowners, who sought to increase their holdings' productivity.[35]

Adding to the "agromania" was the concomitant rise in new landowners: merchants and plantation owners involved in France's burgeoning colonial economy who hastened to buy large estates, and the noble titles that went along with them, as soon as they had made their fortunes. Unlike many more esteemed, longer-standing noble families, this newly propertied merchant class had the money – and the initiative – necessary to undertake major improvements. They surveyed their new holdings with a critical eye and set about turning them to account, not only for their own profit, but also in the name of the public good.

The second half of the eighteenth century saw a torrent of treatises on techniques for agricultural improvement, like irrigation, enclosure, drainage, and, above all, clearing of land for cultivation. While some landowners balanced their interests between creating new fields and conserving their forests, others, notably the influential group of economic theorists known as *économistes* or, colloquially, "physiocrats," were little concerned with the problems of deforestation or timber shortage.

For nearly two decades beginning in the 1750s, these early economic thinkers persuaded the crown to push agricultural expansion at the expense of the forest, a move that ended up incurring floods and erosion.

According to François Quesnay, physiocracy's intellectual founder, France could only grow through agricultural expansion.[36] Quesnay's somewhat convoluted economic logic was inspired by his reading of Englishman William Harvey's *Essay on Circulation* (1628), which

[34] Dupâquier, *Histoire de la population française*, 2:52–78.

[35] Franche-Comté's population grew from roughly 330,000 in 1688 to more than 678,000 in 1783. Dupâquier, *Histoire de la population française*, 2:75–78.

[36] The term referred to a government in consonance with nature. Liana Vardi, *The Physiocrats and the World of the Enlightenment* (Cambridge: Cambridge University Press, 2012).

explored the workings of the human circulatory system.[37] A doctor himself – Quesnay was the personal physician of Madame de Pompadour and a court doctor for Louis XV – he believed a circulatory model could be applied to France's chronic provisioning problems. To make France healthy, the circulation of grain needed to be assured throughout the realm. To this end, Quesnay called on the crown to liberate the grain trade from its "unnatural" impediments – tolls, fees, and taxes – and unfetter landowners from the common rights and collective usage that rural communities relied on. Doing so, he argued, would foster competition, stimulate production, provide opportunities for laborers, generate surpluses, and trigger the growth of an industrious population.[38]

Believing, as Quesnay put it, that "land is the unique source of wealth, and agriculture is what increases it," other physiocratic theorists who had greater influence on the crown, including the future minister of finance, Anne Robert Jacques Turgot, pushed for royal policies that would maximize all available arable soil – even if it came at the expense of forests and fallows used for grazing.[39] Yes, wood was needed for shipbuilding and construction, heating and crafts, the physiocrats acknowledged, but if profits and population growth could be attained from putting more land under cultivation, then landowners should be encouraged to do so. As the Marquis de Mirabeau, an outspoken proponent of agricultural reform, explained in his 1756 *Ami des hommes*, "Where there are fields, there are men, ... where the fields produce the most, there are more men."[40] Besides, he and Quesnay observed, making trees more scarce might encourage the discovery of "peat and coal and other riches hidden for us, so useful to our neighbors."[41]

These views had significant consequences for the countryside. Under the leadership of Henri Bertin, a reform-minded minister closely linked to

[37] William Harvey, *De circulatione sanguinis. The circulation of the blood. Two anatomical essays by William Harvey, together with nine letters written by him*, trans. Kenneth J. Franklin (Oxford: Blackwell Scientific Publications, 1958).

[38] Quesnay, *Maximes générales du gouvernement économique d'un royaume agricole* (1758), no. XXV, in François Quesnay, *Oeuvres économiques et philosophiques*, ed. Auguste Oncken (New York: Burt Franklin, 1969), 336.

[39] Quesnay, *Maximes*, no. III, in Quesnay, *Oeuvres économiques*, 331.

[40] Victor de Riqueti, marquis de Mirabeau, *L'ami des hommes, ou Traité de la population* (Avignon, 1756), part I, chap. 3, 81–82.

[41] Victor de Riqueti, marquis de Mirabeau, and François Quesnay, *Philosophie rurale, ou, Économie générale et politique de l'agriculture réduite à l'ordre immuable des loix physiques & morales, qui assurent la prospérité des empires*, vol. 2 (Amsterdam: Chez les Libraires associés, 1763), 132–34.

the new economic thought, more than a dozen physiocratic initiatives were enacted between 1754 and 1787. Five liberalized domestic and foreign trade in grains, while another seven encouraged enclosure and offered tax incentives for clearing.[42] Physiocrats were confident these laws would, in the words of one agronomist, "see barrenness and ruin succeeded by fertility, abundance, and improvement."[43] In reality, however, the results were more ambiguous. Overall, an estimated 600,000 hectares were cleared in response to the initiatives, amounting to a 3 percent increase in France's cultivable soil.[44] More land was cleared without authorization. Nonetheless, the total was far short of the eleven million hectares Quesnay hoped for, and the clearings' distribution was markedly uneven. In Languedoc, small landholders cleared and planted wine grapes, olive trees, and mulberry bushes, all well suited to the region's arid, rocky slopes.[45] In Burgundy, administrators embraced clearing and enclosure, increasing arable acreage by 8 to 10 percent. In Franche-Comté, by contrast, the decrees' impact had comparatively slight impact, but enclosure of former commons, as the region's *cahiers de doléances* later attested, accelerated.[46]

Contemporary observers like Arthur Young were quick to point out physiocracy's failings. On his agricultural tour of France in 1788, the Englishman derided the theorists as quacks. "The nobility in France have no more idea of practicing agriculture ... than of any other object the most remote from their habits and pursuits," he remarked. "I do not

[42] See especially *Arrêt du Conseil qui accord des encouragements à ceux qui défricheront les terres*, 16 August 1761; and *Déclaration qui accorde des encouragements à ceux qui défricheront les lands et terres incultes*, 13 August 1766. The latter awarded exemption from the *dîme*, *taille*, and other taxes to clearers of *landes*.

[43] *Le journal d'agriculture*, cited in Georges Weulersse, *La physiocratie à l'aube de la Révolution, 1781–1792*, ed. Corinne Beutler, vol. 4 (Paris: Éditions de l'École des Hautes Études en Sciences Sociales, 1985), 92, n. 308.

[44] Discussed in Gille, *Les sources statistiques*, 72–73; also Peter McPhee, "'The Misguided Greed of Peasants'? Popular Attitudes to the Environment in the Revolution of 1789," *French Historical Studies* 24, no. 2 (2001): 262–63.

[45] Noelle L. Plack, "Agrarian Reform and Ecological Change During the Ancien Régime: Land Clearance, Peasants and Viticulture in the Province of Languedoc," *French History* 19, no. 2 (2005): 189–210.

[46] On clearing, see Weulersse, *La physiocratie*, 4, 93; Anatoli Ado, *Paysans en révolution: Terre, pouvoir et jacquerie, 1789–1794 (Traduction revue et complétée d'après la seconde édition russe, 1987)*, trans. Serge Aberdam (Paris: Société des Études Robespierristes, 1997), 37–39; and Pierre de Saint Jacob, *Les paysans de la Bourgogne du nord au dernier siècle de l'Ancien Régime*, Publications de l'université de Dijon 21 (Paris: Société les belles lettres, 1960), 337–43, 31–58, and 61–63. Complaints about enclosure appear frequently in the Comtois *cahiers de doléances*, but clearing is not mentioned at all.

so much blame them for this neglect, as I do that herd of visionary and absurd writers on agriculture who, from their chambers in cities, have, with an impertinence almost incredible, deluged France with nonsense and theory, enough to disgust and ruin the whole nobility of the kingdom."[47]

Others voiced their dismay at the policies' environmental impact. In an article denouncing the "fashion for clearing" in the *Encylopédie méthodique*, the abbé Tessier ventured that while the acts "without doubt contributed greatly to the conversion of *landes*, brush, and *vagues* into cultivated fields, ... all things have a limit." "That which is good, becomes bad, when one overdoes it," he added, observing that woods and pasture had been "sacrificed" out of greed, with little regard for the consequence.[48]

Physiocracy's mixed results aside, its underlying assumptions nonetheless had a lasting impact on the discourse of resource use in France. From this point forward, supporters of customary rights, who believed in apportioning the fruits of the land rather than the ownership of the soil, would have to defend themselves against the assertion that individuated ownership was not only more efficient, but essential to productivity.

As one writer contended in his 1775 "Patriotic Essay Proving the Worthlessness of the Commons," "it is natural to have much desire and diligence to make what we own flourish," but "one takes no care to conserve that which [is held] in common."[49] By insisting that national regeneration could come about only by agricultural growth, the physiocrats ensured that land reform would figure strongly in the broader political aims of the Revolution.

FOREST IMPROVERS AND SILVICULTURAL SCIENCE

While agriculture was the primary focus of the midcentury improvement efforts, most estates of any size also contained woods, bosks, and orchards from which their owners might profit. In stark contrast to the

[47] Arthur Young, *Travels in France during the years 1787, 1788, and 1789* (Garden City, NY: Doubleday, 1969), 100.

[48] L'abbé Alexandre-Henri Tessier, "Défrichement," in Tessier et al. eds., *Encyclopédie méthodique: Agriculture*, vol. 4 (Paris: Agasse, 1796), 32–38.

[49] M. le Baron Scott, *Essai patriotique, ou Mémoire pour servir à prouver l'inutilité des communaux, l'avantage qu'il y auroit de les défricher, ainsi que toutes les terres incultes; celui que l'état retireroit de la protection accordée à l'agriculture, & les causes qui en empêchent les progrès* (Geneva, 1775), 13, 17.

physiocrats' advocacy of grain cultivation and disregard for forests, Henri-Louis Duhamel du Monceau, a contemporary of Quesnay and shared enthusiast of agricultural improvement, believed that France's forests could and should flourish. Duhamel du Monceau is regarded by many as the founding father of modern silviculture thanks to his pioneering work on plant physiology and forest systems, which influenced early German arborists and, through them, gave rise to scientific forestry in the nineteenth century.[50] Noting an alarming lack of technical know-how among Eaux et Forêts officers and private landowners alike, Duhamel du Monceau produced the *Traité complet des bois et des forêts*, a massive, eight-volume guide to woodland management published. The work covered every conceivable detail one should know to make a forest prosper, from site preparation and seedling protection to hauling logs and sawing timber.[51] So painstaking was the project, remarked Duhamel du Monceau, he never would have started it if he had known how large it would be.[52] Nonetheless, the project's importance sustained him. "Is not obvious," Duhamel du Monceau declared, "that a country denuded of Wood would be uninhabitable, and that if one did not have the resource of coal or peat, it would not be possible to preserve oneself from the harshness of winter [nor] cook one's food?"[53]

The crowning achievement of a career devoted to scientifically grounded schemes of improvement, the *Traité complet des bois et forêts* epitomized Duhamel du Monceau's conviction that the starting point of national reform lay in the dissemination of knowledge. His pragmatism made for arid prose; unlike John Evelyn's *Sylva*, Duhamel du Monceau's encyclopedic study of trees focused completely on utility and avoided digressions on arboreal beauty. As he declared in the preface to *De l'exploitation des bois* (1764), "The work that I am presenting to the public is one of the most useful that one could want on the subject of the Forests."[54]

[50] Duhamel's writings were "epoch-making," according to Joachim Radkau, *Wood: A History*, trans. Patrick Camiller (Malden, MA: Polity Press, 2012), 194.

[51] Five titles in eight volumes: *Traité des arbres et arbustes* (1755), *La physique des arbres* (1758), *Des semis et plantations des arbres* (1760), *De l'exploitation des bois* (1764), and *Du transport, de la conservation et de la force des bois* (1767).

[52] Duhamel du Monceau, *Du transport, de la conservation et de la force des bois*, vi.

[53] Ibid., viii.

[54] Henri-Louis Duhamel du Monceau, *De l'exploitation des bois; ou, Moyens de tirer un parti avantageux des taillis, demi-futaies et hautes-futaies, et d'en faire une juste estimation: avec la description des arts qui se pratiquent dans les forêts: faisant partie du Traité complet des bois & des forests* (Paris: H. L. Guerin & L. F. Delatour, 1764), 1:xvii.

Duhamel du Monceau sought above all to turn his readers into productive proprietors. While not as hostile to communal usage as one might expect – his works discussed customary rights in more sympathetic tones than other *philosophes* – he nevertheless believed that such rights "prodigiously degraded the Forests."[55] He was especially critical of woodland grazing. "Nothing is more contrary to the growth of woods," he remonstrated, "than to introduce nibbling beasts into them." The best solution would be to "always bar animals' access" to the forest, but in light of the peasantry's reliance on woodland pasture and the crown's need to "facilitate the increase of livestock," he recognized this was unfeasible.[56] Instead, he offered tips to minimize grazing's impact. Duhamel du Monceau's openness to compromise reflected his long experience in the field. Most forest theorists who came after him would be less accommodating.

WOODLAND ROMANTICS AND THE NATURAL IDEAL

Even as physiocrats and foresters like Duhamel du Monceau were promoting landed improvement, a counter vision of the public good and of humankind's relationship to nature was also emerging. It originated with observers like the abbés Galiani and Mably – men of the cloth who were also *philosophes* – who expressed unease at the technocratic strategies being pursued by the crown's advisors. Arguing that humans' estrangement from their environment reflected a deeper moral degeneracy, a growing number of individuals, among them Jean-Jacques Rousseau, sought spiritual rejuvenation in the natural world. Taking John Evelyn's rhapsodic descriptions of trees a step further, these writers envisioned a wild nature of passion and imagination as opposed to a "scientific" nature of decipherable logic and attainable order. This shift in outlook signaled the beginnings of the Romantic movement and would enduringly alter the direction of conservationist thought.

The movement began with debates over *luxe*, a term that expressed a wide range of ideological and cultural concerns in the eighteenth century.[57]

[55] Henri-Louis Duhamel du Monceau, *Des semis et plantations des arbres et de leur culture, ou, Méthodes pour multiplier et élever les arbres, les planter en massifs & en avenues: former les forêts & les bois: les entretenir & rétablir ceux qui sont dégradés: faisant partie du Traité complet des bois & des forêts* (Paris: H. L. Guerin & L. F. Delatour, 1760), 339.

[56] Ibid., 330–31, 333.

[57] Jeremy Jennings, "The Debate about Luxury in Eighteenth- and Nineteenth-Century French Political Thought," *Journal of the History of Ideas* 68, no. 1 (2007): 79–105.

From the start of the century forward, commentators debated the morality of conspicuous displays of wealth. Observing Versailles' ostentation, the clergyman François Fénelon, for example, argued that the crown must repudiate its devotion to excess if the country is to remain healthy. Especially discomfiting for some was the embrace of opulence among lower ranks of the social order, whose behavior threatened carefully wrought distinctions between privileged and plebeian. *Luxe*'s impact, furthermore, extended beyond philosphical debate. As France's manufacture and trade in luxury goods grew, items of all sort became accessible. Silks, brocades, and cottons grew more common in bourgeois wardrobes, while aristocrats festooned themselves with the imported feathers of soon-to-be extinct birds. The period also saw an upsurge in the fabrication of other mostly decorative objects, like inlaid dressing tables, curio cabinets, and divans.[58]

The new acquisitions and affectations placed greater demands on natural resources. Already in 1721, Réaumur had warned, "we are building, ornamenting, and heating more living quarters than our fathers."[59] To Mirabeau's physiocratic sensibilities, the increased demand for fuel and timber further jeopardized agriculture. "Wood is becoming the most profitable merchandise," he complained. "Everyone is rushing to plant it, and thus strip the nourishment of men from a part of the patrimony."[60] In truth, however, what little planting there was could scarcely keep pace with the booming construction of chateaus, abbeys, and colleges, which took masses of wood to build and tons more to heat. Despairing that "luxury in buildings and in the decoration of rooms has grown to an excessive degree," the Chevalier Grignon, a Champagne forge master, condemned his countrymen's greed. "Society has created for itself unnatural needs," he noted, "which have forced open the canals and the highways to draw timber from the depths of the provinces: the forests have been depleted of nearly all the high trees[,] and the felling of coppices and saplings has had to be rushed to meet the enormous consumption of firewood by the people and townsmen."[61] Ironically, city dwellers often enjoyed more wood-fueled warmth than inhabitants of the countryside,

58 Daniel Roche, *France in the Enlightenment*, trans. Arthur Goldhammer (Cambridge, MA: Harvard University Press, 1999), 563–64.

59 Belhoste, "Une sylviculture pour les forges," 242.

60 Mirabeau, *L'ami des hommes*, 50.

61 AN F^{10} 403 Le Chevalier Grignon, *Observations sur la dégradation des forêts, particulièrement de celles de la Bourgogne, du Lyonnais, du Beaujolais et du Dauphiné et sur les précautions à prendre pour leur conservation*, 1778.

who could do little to stop seigneurs from shipping local wood supplies to lucrative urban markets.

Population growth played a role in this trend, but many attributed it more to *luxe*. Mirabeau mused, "Without entering into historical speculations, let us examine only whether men in early times consumed as much of the earth's produce as today. ... Did they burn up so much wood as in our day? ... In less than ten years Paris's consumption has grown by 200,000 loads, an increase of nearly a third. I do not think the number of inhabitants [has] risen that much."[62]

Longtime forest administrator Léon de Perthuis was more blunt. The problem was not population, but desire. Drawing a direct link between the new materialism and the forest's retreat, he remarked, "With more income, one had more needs: thousands of new hearths gave off heat, luxury became ascendant."[63] Delisle de Moncel, a prominent wolf hunter, similarly noted, "In the time of our fathers – brave but simple Frenchmen – one or two hearths at most sufficed ... : *luxe*, like a ruinous flood toppling all, has changed the state of things; today the citizen with the least bit of wealth lights four or five fires in his home."[64]

By the end of the eighteenth century, *luxe* came to be associated not only with economic excess, but also with unpatriotic self-indulgence that squandered the country's resources and threatened its future for the sake of short-term gratification. Citizens who were unable to withstand cold weather the way their forefathers had, Delisle de Moncel hinted, could justifiably be suspected of aristocratic arrogance.

[62] Mirabeau, *L'ami des hommes*, 49.

[63] Léon de Perthuis de Laillevault, *Traité de l'aménagement et de la restauration des bois et forêts de la France* (Paris: Mme Huzard, 1803 [year XI]).

[64] Delisle de Moncel, *Mémoire sur le repeuplement, l'augmentation & la conservation à venir des bois dans les départemens de la Meurthe, Moselle, Aisne, Meuse, Marne, &c* (Nancy: Chez H. Haener, Imprimeur du Roi, 1791), 1–2.

3

"A necessity as vital as bread"*

Forest Crisis on the Eve of the Revolution

> It is inconceivable how and how much the forests have declined in the last thirty years: one will no longer be able to find in woods that once furnished resources to the navy, a single timber ... or [stick] of firewood, that crucial item of daily need.... [This] makes the indigent unable to procure it, and by the indigent we mean three-quarters of the people of the countryside and perhaps of the cities.
>
> *Cahier de doléances* of Bourbévelle, Haute-Saône, 17 March 1789[1]

By the last quarter of the eighteenth century, forests were in decline in many parts of France. Assessments made at midcentury indicated that the country's woodlands had shrunk by as much as 10 percent since the century before, and the rate of deforestation continued to accelerate as would-be improvers felled trees to create fields; industrialists and timber merchants pushed into previously uncut areas; and the growing rural populace, squeezed by seigneurial avarice and royal reforms, pushed the boundary between possession and poaching.[2] On the eve of the Revolution, less than 13 percent of the country's surface area – a mere

* The quotation in the chapter title is from *Cahier de doléances*, Bourbévelle (Haute-Saône), 17 March 1789, in Godard and Abensour, *Cahiers de doléances du bailliage d'Amont*, 1:218.

[1] *Cahier de doléances*, Bourbévelle, 17 March 1789, art. 2, in ibid., 1:217–18. The village is on the banks of the Saône River, in the northern Franche-Comté.

[2] Wood sales greatly increased in this period, particularly from 1750 forward. See discussion in Devèze, "La crise forestière," 83; also François Vion-Delphin, "Forêt, frontière et commerce du bois: Noirmont et Val de Joux à la fin du XVIIIe siècle," *Société d'émulation du Jura: Travaux* 1990 (1989): 199–217.

6.5 million hectares – remained forested.[3] Of this, less than 600,000 hectares contained the tall timber trees sought by shipbuilders for their size and strength.[4] Coppicing for forges and furnaces transformed much of the remainder into dense thickets, while expanded clearing for agriculture eliminated many woods altogether. With forests dwindling and demand for wood and timber mounting, the price of woodlands soared after having remained low for decades. In Burgundy and Lorraine, the cost of a single arpent more than doubled between 1769 and 1789.[5]

To take stock of the crisis, the crown carried out several country-wide surveys in the 1780s. Prompted by fuel shortages in cities across France, the first of these surveys was initiated by the controller general in August 1783 and addressed all *intendants*.[6] As had been true of Orry's grand statistical projects in the 1740s, the value of the survey was tempered significantly by the inability and occasional disinclination of provincial officials to provide reliable figures. Asked to gauge the amount of fuelwood consumed in his district per domestic hearth as well as by iron, glass, and salt manufacturers, the beleaguered *subdélégué* of Lons-le-Saunier groused, "I will not hide it from you, sir, I had the hardest time imaginable obtaining the information. … It was nearly impossible for me … to give an exact account of the

[3] Estimates of France's Old Regime wooded acreage vary. Arthur Young calculated 8 million hectares based on Cassini's *carte générale de la France* (1740–82). But Michel Devèze argues that this figure is too high because it counts spaces that Cassini marked as "wastes" (*vides*, *friches*, *broussailles*). Devèze thinks the figure calculated by Telles d'Acosta, the *grand maître des eaux et forêts* of Champagne, is more reasonable: 6,428,000 hectares. See Jean-Pierre Husson, "L'histoire du paysage forestier français: Les trois ages de la forêt," *Gavroche* 62 (1992): 9; Devèze, "Les forêts françaises à la veille de la Révolution," 179, 181–82; D. A. Telles d'Acosta, *Projet d'une nouvelle administration pour les forêts de France et servir de supplément à notre "Instruction sur les bois"* (1791). As a comparison, the wooded surface of metropolitan (continental) France in 2013 covered 16.4 million hectares.

[4] On the desirability of *futaie* as naval timber, see *Les Eaux et Forêts*, 49, 53–55; Acerra, "Marine militaire et bois de construction"; and Bertrand, "Les hommes et la forêt dans les communautés villageoises."

[5] All conversions are based on the *arpent des eaux et forêts* (0.517 hectare or 1.278 acres). Land prices in general were up in the 1750s, but forest prices escalated comparatively higher. Fernand Braudel and E. Labrousse, *Histoire économique et sociale de la France*, vol. 2, *Dès dernier temps de l'âge seigneurial aux préludes de l'âge industriel, 1660–1789* (Paris: Presses Universitaires de France, 1970), 478–79.

[6] Responses for only nine of thirty-two *généralités* have been located, including the departmental archives of the Doubs. See discussion in Gille, "L'enquête sur les bois de 1783," 629–30.

quantity and quality of the forests."[7] Others felt their efforts would serve no purpose; the *subdélégué* of Vitry, in the Marne, complained, "I regard this entire operation as virtually useless."[8] Part of the problem lay with municipal officials, who were reluctant to impart any information that might lead to new taxation or, worse, the loss of communal holdings. Dozens of parishes whose woods had undergone reformation a generation before were still wrangling with the crown over their reduced access rights.

Despite the problems entailed in responding to the 1783 survey, the scope of its questions suggests that the crown was serious about confronting the problem of fuel shortages. In addition to inquiring about domestic and industrial wood use, the controller general asked about alternative fuels, like coal and peat. In regions like the Haut-Doubs, however, even these resources were running out. As the *subdélégué* of Pontarlier noted, "The dearth of heating is frightening, because the peat bogs are being exhausted and it takes a century to regenerate them."[9] The Eaux et Forêts was at least partly to blame for the region's fuel woes, he contended, because of a "lack of policing" in the woods.

Overall, the survey's responses demonstrated that deforestation was severe around urban areas and waterways but varied in the uplands, where inaccessibility prolonged forests' survival. Even there, however, woodland quality had declined since the 1660s and 1710s, when surveys were carried out in preparation for implementation of the 1669 Ordinance. The enormous consumption of timber for the duc de Choiseul's midcentury naval reconstruction efforts was partly to blame, as was the demand for new vessels and repairs during the American War of Independence.[10] In some areas, administrators reported that *haute*

[7] AD Doubs 1C 1365, *État général des bois et usines existants dans la province de Franche-Comté.* Observations of M. DeLeschaux, Lons-le-Saunier, 19 March 1784. DeLeschaux protested that the project ought to have been assigned to the officers of the *Eaux et Forêts.*

[8] AD Marne C 1133, cited in Gille, "L'enquête sur les bois de 1783," 631. Some respondents apparently feared painting too rosy a picture lest they be called on to deliver more wood to the capital.

[9] AD Doubs 1C 1365, *État général des bois et usines.*

[10] Étienne-François de Choiseul (1719–85), minister of foreign affairs from 1758 to 1770 and naval minister from 1762 to 1770. See discussion in *Les Eaux et Forêts*, 176; Acerra et al., *Les marines de guerres europées*; Acerra, *Rochefort et la construction navale française: 1661–1815* (Paris: Libraire de l'Inde, 1993); also Bamford, "French Forest Legislation," 97–107.

futaie were being cut after only twenty years – a far cry from Colbert's vision of centenarian oaks bearing the French navy to victory.[11]

Yet despite escalating military pressures, industrial exactions and domestic hearths were clearly the biggest factors in the forests' decline.[12] To meet the needs of a growing population and expanded manufacturing, woodcutters harvested coppices in cycles as short as eight years. Inevitably, the green and immature stems produced only a weak, smoldering charcoal, which promoted further overcutting because poor-quality fuel required manufacturers to use more.[13] In Burgundy, where wood shortages were even more acute, forge masters had success using coal, but similar efforts in Franche-Comté were disappointing.[14] "The castings that issued from [these experiments] produced harsh and brittle iron of the worst quality," noted the *subdélégué* of Gray, adding that "new experiments could have some success, but the forge masters think the cost of carrying them out will be too great."[15]

Facing deepening subsistence crises and industrial recession, the crown launched yet another forest survey five years later, in 1788. Initiated by the Bureau du Commerce, the *enquête sur les bouches à feu* focused exclusively on industry – both major fuel consumers like iron and saltworks and minor manufacturers like tile makers, tanneries, dye works, distilleries, clockmakers, and soap factories.[16] According to its findings, France had more than a thousand metallurgical establishments

[11] Gille, "L'enquête sur les bois de 1783," 634–35. Twenty years was the optimal growth for *taillis* according to Henri-Louis Duhamel du Monceau, in "Art du charbonnier ou manière de faire le charbon de bois," in *Descriptions des Arts et Métiers faites ou approuvées par Messieurs de l'Académie royale des Sciences*, 1761.

[12] France produced an estimated 30 million *stères* (cubic meters) of wood per year, at an average of three to four cubic meters per hectare. One-third (10 million cubic meters) was used by the iron industry, 5 million cubic meters went to urban heating, and 1.5 million cubic meters was consumed by glassworks. Jean-Francois Belhoste, "La sylviculture du XVIe au XIXe siècle," *Sciences et techniques en perspective* 36 (1996): 53.

[13] On industrial consumption in Franche-Comté, see Général Jacques Paul de Vergnes, *Mémoire sur la statistique du département de la Haute-Saône*, (Paris: Imprimerie des Sourds-Muets, 1801 [year IX]), 11–12. On domestic usage, see J.-C. Demard, "L'utilisation du bois dans la tradition paysanne et artisinale comtoise," in "Société et forêts," special issue, *Revue forestière française* (1980): 281–300; also Vion-Delphin, "Les Forêts du Nord de la Franche-Comté," 45–46.

[14] They were experimenting with bituminous coal. Given the choice, manufacturers and artisans preferred iron made with charcoal. Duhamel du Monceau, "Art du charbonnier."

[15] AD Doubs 1C 1365, *État général des bois et usines*, *subdélégation* of Gray.

[16] *Enquête sur les bouches à feu*, 1788. For responses, see AN F^{12} 680 and H. and G. Bourgin, *L'industrie sidérurgique en France au début de la Révolution* (Paris, 1920).

in 1788, which burned nearly the same amount of wood every year as the nation's four million domestic hearths.[17]

This intense competition for resources, together with the uncompromising and often unsuitable regulations of Colbert's Ordinance, vexed forest users to no end. Invited by the crown to submit their grievances to the forthcoming meeting of the Estates General, which had last gathered in 1614, rural inhabitants seized the once-in-a-lifetime opportunity to detail complaints of every dimension, from the material to the philosophical. Among their most emphatic concerns was the forest, especially for communities in Franche-Comté, one of the last provinces to be annexed by France and thus among the last to be governed by the Ordinance. Their distress reflected the forest's vital importance in daily life as well as structural inequities that gave rise to inequitable allocations of forest resources.

In this chapter, I investigate France's woodland troubles on the eve of the Revolution and analyze the bitter struggles that ensued, particularly in Franche-Comté, over who would control the forests and how. Beginning with an overview of conditions and concerns in the province in the decades before 1789, I examine the intense rivalries that arose among rural communities, seigneurs, and the state, over the region's declining woodlands. The chapter ends with a look at the royal saltworks at Arc-et-Senans, a folly-ridden project that, in both metaphorical and tangible senses, crystallized inhabitants' anger toward the crown regarding access to and control of the Comtois forests.

THE LANDSCAPE OF DISCONTENT

Though annexed by France a century earlier, Franche-Comté remained in many ways a region apart at the end of the eighteenth century.[18] Physically, the province was isolated by arduous travel conditions. Road construction spearheaded by *intendants* Barthélémy de Vanolles (1734–44) and Charles André de Lacoré (1761–84) improved internal

Some consider the Bourgin collection unreliable. See Gille, *Les sources statistiques*, 66–67.

[17] One blast furnace in 1780 consumed 4,000 cords per year, or the equivalent of 133 arpents (68.8 hectares, or 170 acres) of fifteen-year-old coppice. Schaeffer, "La forêt résineuse comtoise," 167.

[18] Franche-Comté was one of approximately thirty-nine provinces in the latter half of the eighteenth century. William Doyle, *The Oxford History of the French Revolution* (Oxford: Oxford University Press, 1989), 2.

communications, but voyaging from Besançon to Paris still took an average of six days at the end of the 1780s.[19] Commercially, trade remained stifled by customs barriers segregating the province from the rest of France, as well as from its longtime markets in Switzerland, Germany, and Lorraine.[20] Nonetheless, while traffic in traditional goods like cheese, wine, leather, and salt grew relatively slowly over the course of the century, metallurgical manufacturing surged, spurred in part by the increased demand for armaments.[21]

According to a 1784 study, metalworks, particularly iron forges, furnaces, and foundries, accounted for 40 percent of all the wood utilized in the province. Much of this usage was attributable to the production of charcoal, the fuel favored for blast furnaces because of its slow, steady heat. By contrast, other major wood-burning industries, including saltworks, glassworks, tanneries, and saltpeter production, together consumed less than 8 percent of fuelwood.[22] Comtois seigneurs were the primary force behind the new industrial expansion, for they had both the financial and landed resources necessary to undertake such projects.[23] The crown also encouraged metallurgy by granting forge masters fuelwood concessions in the royal forests.[24]

Agriculture in Franche-Comté also shifted during the eighteenth century. While villages in the fertile Bressane plains had long cultivated grain, the residents of Franche-Comté's plateau and mountain districts traditionally favored animal husbandry and dairying, in large part because the heavy, clayey soils of the uplands impeded agriculture. With the introduction and augmentation of royal taxes, however, rural

[19] Boichard, *Le Jura*, 104; also Colin Jones, *The Cambridge Illustrated History of France* (Cambridge: Cambridge University Press, 1994).

[20] Brossault notes, "En effet, la province est 'réputée étrangère', ce qui entraîne le maintien les barrières douanières." Colette Brossault, *Les intendants de Franche-Comté 1674–1790* (Paris: Boutique de l'Histoire, 1999). See also Boichard, *Le Jura*, 100–102.

[21] By the late eighteenth century there were forges in all the river valleys of Franche-Comté: the Saône, Doubs, Loue, Ognon, and Ain. Fiétier, *Histoire de la Franche-Comté*, 272–73. On the importance of Franche-Comté, along with western Normandy and the Paris basin, for the production of cast iron and bar iron, see Hugh D. Clout, "Industrial Development in the Eighteenth and Nineteenth Centuries," in *Themes in the Historical Geography of France*, Hugh D. Clout, ed. (London: Academic Press, 1977), 453–54.

[22] Thérèse Sclafert, *Cultures en Haute-Provence: Déboisements et pâturages au Moyen Age*, Les hommes et la terre 4 (Paris: S.E.V.P.E.N., 1959), 200.

[23] Lassus, "La sidérurgie comtoise," 148–49.

[24] AD Doubs 1C 1365; also Vion-Delphin et al., *Les hommes et la forêt en Franche-Comté*, 92–93. On saltpeter, see Colette Merlin, *Le pain de melée: Campagnes jurassiennes dans l'ancienne France* (Bourg-en-Bresse: Taillanderie, 1994), 84.

communities increasingly took up the plow. Like the tithe, most royal obligations had to be paid in wheat. The resulting woodland clearing put new pressure on higher-elevation ecosystems. The switch to a grain-based economy notwithstanding, however, the Francs-Comtois were slow to enact improvements like enclosure and drainage that had proven successful in France's northern provinces.

The persistence of mortmain, a form of serfdom in which inhabitants were barred from selling their land, contributed to the lag in improvements.[25] If an individual lacked appropriate heirs, possession would revert after his death to the "first occupant" – the seigneur. Especially common in the Haut-Jura cantons of Mouthe and Saint-Claude, mortmain deterred landed investment, encouraged emigration, and drove away potential newcomers.[26] The practice endured despite the crown's efforts to suppress it. In 1779, for example, the Besançon *parlement* stubbornly refused to register an edict ending mortmain in the royal domain. Influenced by physiocratic theory, the crown argued that eliminating this onerous custom would promote agricultural improvement and lead to higher rents and greater returns. Condemning mortmain as a "vestige of a rigorous feudality" that "caused industriousness to languish," it offered tax incentives to seigneurs to follow suit on their estates.[27] But the *parlement* – many members of which hailed from recently ennobled families – vigorously resisted these reforms, asserting that mortmain was needed to shield peasants from being bought out by greedy bourgeois. That there was some truth to this did not alter the fact that under mortmain seigneurs also benefited from access to *vaine pâture* (communal grazing on harvested or fallow private land) and other lucrative common rights that would be hindered by a shift toward individuation and fortified freeholder rights.[28] By refusing to abolish this and

[25] On mortmain, see D. G. Sutherland, *France, 1789–1815: Revolution and Counterrevolution* (New York: Oxford University Press, 1986), 71.

[26] More than a third of peasants in the Jura, especially in the mountain villages around Saint-Claude, were subject to mortmain. Boichard, *Le Jura*, 91–92. Of the 212 *cahiers de doléances* of the *bailliage* of Baume-les-Dames, 204 *cahiers* condemn the practice. Jouvenot, Robert, ed., *Le bailliage de Baume-les-Dames en 1789: Les cahiers de doléances* (Paris: Les Belles Lettres, Annales littéraires de l'université de Besançon, 1985), 9.

[27] *Édit ... portant suppression du droit de main-morte et de servitude dans les domaines du Roi et dans tous ceux tenus par engagement, et abolition générale du droit de suite sur les serfs et main-mortables ... Registré en Parlement le 10 [août 1779]* (Paris: Imprimerie royale, 1779).

[28] On the Besançon *parlement*'s resistance to Turgot's reforms, see Fiétier, *Histoire de la Franche-Comté*, 258–60.

other established privileges, the seigneurs of Franche-Comté, along with those of Burgundy, Lorraine, and Berry, upheld their province's reputation as a last bastion of feudalism.

Conservatism in spirituality and religious practice was another characteristic for which Franche-Comté was known. Historically fervent Catholics – their provincial piety would later be skewered by Stendhal in *Le rouge et le noir* – the Francs-Comtois for the most part remained impervious to the development of anticlericalism and secularism during the Enlightenment. This was true for both notables and peasants. Jansenism's austere, cerebral religiosity likewise attracted few converts.[29] As in the western region of the Vendée, Catholic conservatism and geographic isolation would contribute to the development of antigovernment sentiment in Franche-Comté during the Revolution. Nearly three-quarters of the local clergy would refuse the Civil Oath in 1791, and all three departments would become hotspots of government opposition in 1793.[30]

Nonetheless, the province was not entirely immune to the intellectual trends of the age. Besançon boasted its own Masonic order, as well as an Académie des Sciences, Belle-lettres et Arts. Both were encouraged by the *intendant*, Lacoré, whose enthusiasm for landed improvement was reflected in the topics the Académie chose for its essay competitions, like the value of enclosing fields and the best form of forest management for Franche-Comté. As a rule, the essayists – forest officers, forge masters, and others – held customary rights in low regard, accusing them of being "contrary to the common and individual good" and of having "no other merit than their antiquity."[31]

For their part, the Comtois peasantry focused on meeting the considerable challenges of their daily existence through tradition and adaptation. The nature and range of these challenges are suggested by the *cahiers de doléances* of the Third Estate. Though numerous *cahiers* simply parroted the templates that had been distributed in their district by municipal officials and lawyers, a substantial number of parishes also added a panoply of local concerns. To their relatively uniform demands for more equitable taxation and equal representation of the Third Estate in the

[29] See discussion in Fiétier, *Histoire de la Franche-Comté*, 309–18; and Timothy Tackett, *Religion, Revolution, and Regional Culture in Eighteenth-Century France: The Ecclesiastical Oath of 1791* (Princeton, NJ: Princeton University Press, 1986), 129.

[30] Tackett, *Religion, Revolution, and Regional Culture*, 53, 123–24; and Jones, *Cambridge Illustrated History of France*, 189.

[31] Quote from Ethis de Novéau, "Essai sur cette question," 1.

Estates General, inhabitants appended a catalog of woes, ranging from hailstorms and floods to accusations of corruption, detailed in passionate if plainspoken prose.[32] The more common complaints included the costly and cumbersome obligation of billeting the king's troops (a frequent occurrence in frontier provinces like Franche-Comté); the unwelcome *tirage au milice*, or mandatory militia service (which, lamented the *cahier* of Sénargent, "removes from the bosom of their family young men who should be farmers"); and an infinite variety of feudal dues, paid in cash or in kind.[33] The latter weighed especially heavily in estates where mortmain was still in force. In addition to costs associated with seigneurial *banalités* – monopolies on essential activities like milling, bread baking, and wine pressing – inhabitants had to pay a multitude of fees for "onerous and ridiculous [seigneurial] rights" like the *angal*, a wine tax; the *marque de cuisse*, a marriage tax based on the seigneur's time-honored prerogative to bed new brides; and the *éminage*, a tax on units of grain sold.[34] They were also called on for such duties as housing the seigneur's hounds, guarding his chateau, and even beating the ponds at night to prevent the noisy croaks of amorous frogs from disturbing the seigneur's slumber.[35]

Still, these obligations paled in comparison with the staggering fiscal charges parishes had to pay. As a group, the Comtois peasantry was far from uniform. They ranged from the landless poor, who scraped by as day laborers, pieceworkers, and seasonal harvesters, to the relatively well-off *cultivateurs* and *laboureurs*, who possessed enough land to turn a profit. Nonetheless, they suffered collectively under a tax load from which the nobility and bourgeoisie were largely exempt. The *cahier* of Verne, a fifty-household village on the plateau north of the Doubs River, detailed just how expensive these charges could be:

> [The Vernois] pay annually to the prince of Montbéliard and to the monastery at Baume a tithe in the value of 2,500 livres in grain, in addition to revenues in the range of 2,000 livres that the *curé* of Verne draws from his parish lands,

[32] On the *cahiers* of the *bailliage* of Amont, see Jean Girardot, *Le département de la Haute-Saône pendant la Révolution*, 3 vols. (Vesoul: Société d'agriculture, lettres, sciences et arts de la Haute-Saône, 1973–74), 1:86–90.

[33] *Cahier de doléances*, Sénargent, 22 March 1789, in Jouvenot, *Le bailliage de Baume-les-Dames*, 534.

[34] *Cahier de doléances*, Beutal, in Jouvenot, *Le bailliage de Baume-les-Dames*, 158.

[35] On frog-beating, see *cahier de doléances*, Moimay, in Godard and Abensour, *Cahiers de doléances du bailliage d'Amont*, 2:199. On the many obscure seigneurial rights in Franche-Comté, see Jouvenot, *Le bailliage de Baume-les-Dames*, 41–42.

> [an amount] which exceeds by half the royal taxes that the people of the countryside pay annually, and in addition to the local charges that amount each year to more than 6,000 livres.[36]

Distress over the tithe was especially acute in areas where parishes felt they received little in return for their contributions. In the Haut-Jura village of Charency, for example, inhabitants expressed dismay at the amount they owed annually to the priests of their district, though their own church was in such disrepair that no *curé* would hold services in it. Parishioners had to walk more than a league "on very bad roads" to attend mass in the next town, noted the Charency *cahier*, a distance that exhausted the village's elderly, women, and children, especially in winter.[37]

As the Vernois *cahier* suggests, royal taxation in Franche-Comté often amounted to less than the total owed to ecclesiastical and lay seigneurs. Even so, the crown's exactions deeply irritated inhabitants, who had benefited from lower charges and a less efficient collection system under the Habsburgs. Nobles, too, were aggravated, because unlike their compatriots in the interior of France, they enjoyed no exemption from the main direct tax, known as the *taille*. Thus, while the rate of royal taxation rose everywhere in France in the second half of the eighteenth century, its effects were all the more noticeable in Franche-Comté.[38]

As much as the Francs-Comtois deplored the burden of fines, fees, and feudal dues, they reserved some of their bitterest criticisms for the woodland decline that had been going on in the region ever since the French conquest. The majority of the woods in Franche-Comté – 70 percent, according to a 1784 survey – were owned by lay communities; nonetheless, over the course of the century many villages had seen their access to these woods diminish.[39] Pressured on all sides by population growth, royal and seigneurial exactions, and changing forms of exploitation, the forest had become a tinderbox of conflict. The *cahiers de doléances* gave voice to these tensions. In the northern and eastern *bailliages* of Vesoul and Baume-les-Dames, for example, 93 percent of *cahiers* enumerated

36 *Cahier de doléances*, Verne, 18 March 1789, in Jouvenot, *Le bailliage de Baume-les-Dames*, 554.

37 ADJ Lp1616, *Cahier de doléances*, Charency, 19 March 1789.

38 On royal tax increases in Franche-Comté, see Fiétier, *Histoire de la Franche-Comté*, 245. On the *taille*, see William Doyle, *Origins of the French Revolution*, 3rd ed. (Oxford: Oxford University Press, 1999), 112.

39 Vion-Delphin, "Les Forêts du Nord de la Franche-Comté," 45.

woodland grievances.[40] Many reported pervasive shortages, like the *cahier* of Corre, on the Saône River, which noted that inhabitants had "no wood at all" and "could hardly get it except from Champagne or Lorraine." Putting the blame on charcoal-burning industries, the Corréens warned that even those two provinces would "soon be lacking woods because like everywhere forges and glassworks are too widespread."[41] Numerous other *cahiers* condemned seigneurs for curbing customary rights and usurping communal woods. Most frequently of all, the *cahiers* called for the overhaul or even elimination of the Eaux et Forêts, which peasants, religious communities, and seigneurs alike deplored for its convoluted rules and endemic corruption.[42] To obtain a clearer picture of the woodland discord that would erupt during the Revolution, the remainder of the chapter investigates the forest struggles highlighted by the *cahiers*.

SEIGNEURIAL USURPATIONS

According to *cahiers de doléances* across France, seigneurs had for years been abusing the very essence of the feudal bond, namely, their promise to protect vassals' welfare in exchange for material contributions and labor. The grievance petitions teem with indignation against seigneurs for having replaced the reciprocal principles of feudal life with unfettered greed.[43] The trend was apparent in all aspects of nobles' behavior, but it was especially pronounced with regard to peasants' use of seigneurial forests and fields, which became the focus of increasing repression in the eighteenth century. As Jeremy Hayhoe has noted for northern Burgundy, nobles began using lawsuits as a kind of cudgel to rein in illegal pasturing and wood theft and aggressively punish offenders. Prosecutions for forest

[40] Ibid., 48. In the *bailliage* of Baume, 182 of 212 published *cahiers* mention the forest. For the *bailliage* of Vesoul, the figure is 357 of 370.

[41] *Cahier de doléances*, Corre, 20 March 1789, in Godard and Abensour, *Cahiers de doléances du bailliage d'Amont*, 1:398.

[42] François Vion-Delphin, "Forêts et cahiers de doléances: L'exemple de la Franche-Comté," in *Révolution et espaces forestiers: Colloque des 3 et 4 juin 1987*, ed. Denis Woronoff, Collection alternatives rurales (Paris: L'Harmattan, 1988), 13–14; and Jouvenot, *Le bailliage de Baume-les-Dames*, introduction, 59–60.

[43] These criticisms were levied against clerical and nonclerical seigneurs alike. See "Protestation d'un serf du Mont-Jura. Contre l'Assemblée des Notables, le Mémoire des princes du sang, le clergé, la noblesse & le Tiers-Etat. Au roi," (1789), 24 (French Revolutionary Research Collection, Fiche B2313 9.2/29 Fiche 2; also available on the Bibliothèque nationale website at http://gallica.bnf.fr/ark:/12148/bpt6k67044).

and field offenses in that region quintupled between the 1750s and the 1780s.[44] In Franche-Comté, the pattern was similar. Anxious to increase revenues and guard their properties against the encroachment of a growing populace, Comtois nobles cracked down on forest offenses and repudiated long-standing use rights, even as they defended and tried to expand their own claims of traditional usage.

Over the course of the eighteenth century, it became common for seigneurs to commandeer part of their parishes' annual wood harvest, or even whole sections of the forest itself, without compensation.[45] Such was the case in a dozen villages clustered in the broad agricultural plain northeast of Besançon. In their *cahier de doléances*, the inhabitants of Laissey noted that their seigneur, the baron Iselin de Lanans, had not only seized part of their communal woods; he had also confiscated the complaint they had filed against him and used his guards to make "threats and reports." "And as the inhabitants have no money available," the Laissey *cahier* observed, "[they] were forced to abandon the present proceedings."[46] The *cahier* of Le Puy, a fifteen-hearth hamlet, likewise condemned the baron for appropriating much of the community's woods via triage. Although this move had considerably reduced Le Puy's forest benefits, the baron had continued to collect the annual fee (*cens*) for woodland access. This, the petitioners emphasized, was "a considerable injustice toward the community."[47]

Subjected to similar infringements by their seigneur, the villagers of Froidefontaine, in the Jura uplands, were more irate:

> To constrain the communities who have use rights in the forests of the Seigneur to pay the *cens* every year and yet give them no wood, what an injustice? To overwhelm them with fines for having taken a few cartloads of rotten wood to warm themselves and bake their bread, [wood that is an] item of absolute necessity above all in the Mountains for making cheese, what an aggravation?[48]

To reclaim their rights, peasants could turn to the law, but the courts governing such matters generally answered to the seigneur himself. As the

[44] Jeremy Hayhoe, *Enlightened Feudalism: Seigneurial Justice and Village Society in Eighteenth-Century Northern Burgundy* (Rochester, NY: University of Rochester Press, 2008), 187–94.

[45] See discussion in Jones, *Peasantry in the French Revolution*, 51.

[46] *Cahier de doléances*, Laissey, art. 27, 18 March 1789, in Jouvenot, *Le bailliage de Baume-les-Dames*, 583.

[47] *Cahier de doléances*, Le Puy, art. 9, 20 March 1789, in Jouvenot, *Le bailliage de Baume-les-Dames*, 576–77.

[48] ADJ Lp1618, *Cahier de doléances*, Froidefontaine, March 1789.

inhabitants of Vregille observed, "in eating the soup of the seigneurs, the judicial officers of the district and lawyers associated with the seigneurs ... promised to make [us] do what they want."[49]

The lure of industrial profit especially impelled seigneurial usurpations in Franche-Comté.[50] Encroachment on communal forests was widespread in the industrialized *bailliages* of Vesoul and Gray, where since the start of the eighteenth century ironworks had proliferated on the Durgeon and Saône rivers, consuming crown, communal, and seigneurial woods for fuel. Seeing the profits that sales of fuelwood could bring, many Comtois communities initially took advantage of the situation by adapting their forests to the needs of manufacturers. Villages in the *bailliage* of Gray, for example, reorganized their woods from 1700 forward into *coupes réglées*, orderly plots of coppice that yielded a more or less reliable source of burnable energy when felled in rotation.[51] Communities in the nearby *bailliages* of Vesoul and Baume-les-Dames followed suit. With the money earned, municipal leaders paid down tax debts and funded civil improvement projects like public fountains, bridges, and churches. Nonetheless, many villages came to regret their market embrace. In restructuring their woods for industrial and urban fuel production, they limited woodland pasturage and eliminated the tall, older *futaie* that gave them timber and *chablis* (fallen branches for firewood).

Though the substitution of coppice for timber trees directly undermined the crown's naval concerns, and the suppression of inhabitants' use rights contravened the claims that royal *réformateurs* had affirmed in their scrutiny of Comtois *chartes de franchises,* the local administration of the Eaux et Forêts appeared unable or unwilling to stop it. That, at least, is what a note scribbled in the margin of the *cahier de doléances* of Navenne suggests: "Throughout the present year [the seigneur] has committed infractions that the inhabitants have been unable to stop with reports to the *maîtrise*, because the seigneur has no fear of what the officers might do and obeys no law in these matters."[52]

One of the laws defied by seigneurs, who constituted the majority of Comtois industrialists, was a royal edict of 1723 requiring new

49 *Cahier de doléances*, Vrégille [*sic*], 20 March 1789, in Godard and Abensour, *Cahiers de doléances du bailliage d'Amont*, 2:520.

50 On seigneurial efforts to profit from rising wood prices, see Sutherland, *France, 1789–1815*, 70, 72–73.

51 Belhoste, "Une sylviculture pour les forges," 247.

52 *Cahier de doléances*, Navenne, 18 March 1789, art. 12 (margin note), in Godard and Abensour, *Cahiers de doléances du bailliage d'Amont*, 2:242.

establishments to provide all or most of their own fuelwood.[53] The edict was intended to limit the amount manufactures could purchase, thereby reducing pressure on the market for wood. As conversion of Franche-Comté's once-impressive high forest into *coupes réglées* proved insufficient; however, industrializing seigneurs turned increasingly aggressive. Through *triages* and *cantonnements*, they liberated the major portion of their woods from use rights, clearing the way for them to be converted into fuel.[54] Under license of their authority, they strong-armed appropriations from communal woods. "Without any rights," complained the villagers of Molay, north of Gray, "the seigneur took part of [our] communal woods, which he kept for himself; independent of that, [he] still demands as primary resident a part of [our] shared woodlot."[55]

The most unscrupulous seigneurs deployed their guards to harass inhabitants. As the villagers of Melincourt, at the edge of the Vosges mountains, complained,

> the seigneur ... has three guards who [oversee] the whole extent of the territory, although they should not because in compliance with the laws of our sovereigns the community each year names three woodsmen and two watchmen for the protection of the communal woods and for the benefit of the territory. The [seigneur's] guards make a great many citations, for which the inhabitants cannot pay the fines, because the great majority [of them] are barely able to get by.

The seigneurs' agents also infringed on Melincourt's communal woods, taking "nearly half of it" each year and forcing residents to buy wood, an outlay of "at least 1500 livres."[56] In this way, seigneurial appropriations of the forest stimulated anti noble resentment well before 1789. Having earned ill will with his bullying ways, the seigneur of Melincourt would find no shelter in 1794, when he was guillotined for helping nobles slip out of the country.[57]

[53] Baudrillart, *Recueil chronologique*, tome 1, 235.

[54] *Triages* and *cantonnements* multiplied in Franche-Comté from the 1730s forward. Vion-Delphin, "Forêts et cahiers de doléances," 16–17.

[55] *Cahier de doléances*, Molay, 17 March 1789, in Godard and Abensour, *Cahiers de doléances du bailliage d'Amont*, 2:169, 2:202.

[56] *Cahier de doléances*, Melincourt, 15 March 1789, in Godard and Abensour, *Cahiers de doléances du bailliage d'Amont*, 2:169, 2:71.

[57] Dominique, prince de Broglie, *Les Broglie, leur histoire* (Paris: Éditions du Palais Royal, 1972). On seigneurial aggression and antiseigneurialism, see McPhee, "'The Misguided Greed of Peasants'?," 257; and Hayhoe, *Enlightened Feudalism*, chaps. 6 and 7.

INDUSTRIAL HARM

Environmental damage, which went hand in hand with the region's industrializing economy, was another prominent source of complaint in the *cahiers*. Throughout the province as a whole, iron and other metallurgical manufactures devoured more than 37 percent of the total fuelwood produced annually.[58] The *bailliage* of Vesoul alone contained nineteen blast furnaces, twenty-four forges, and five tilt hammers, which together burned more than 180,000 cords of wood each year.[59] This industrial appetite spurred wanton overcutting and severe local shortages, as in the parish of Les Magny on the Ognon River, where inhabitants complained that their seigneur "cut the forest so often that he left inhabitants no portion living or dead."[60] By monopolizing the market, ironworks drove up fuelwood prices and "shut the mouth" of all other consumers, lamented the petitioners of Bourbévelle.[61] The Saône River village of Betaucourt, just downstream from Bourbévelle, similarly suffered. Betaucourt was "surrounded by ten forges and furnaces," and as a result, its *cahier* explained, fuel costs had risen "to an exorbitant price" because the seigneur had only enough woods of his own to sustain the ironworks for six weeks out of the year.[62]

Beyond occasioning overcutting, ironworks also wreaked havoc by releasing a surging, toxic effluent that poisoned rivers and flooded fields. The inhabitants of Aroz groaned, "the community is burdened by three [blast] furnaces ... as well as the forges that result from them. ... The masters of these manufactures [dig] mines and rip up our soil, which causes the ruin of our lands, and [they] even discharge [the tailings] in the waters of the village and entirely drown our pastures as well as the cultivable land[,] causing the death of the animals who drink the water."[63]

58 A total of 477,728 cords, mostly in the form of charcoal. AD Doubs 1C 1365, *État général des Bois et Usines*, 1784; also Vion-Delphin, "Les Forêts du Nord de la Franche-Comté," 45–46.

59 180,000 cords equals 700,000 cubic meters of wood. See discussion in Vion-Delphin, "Les Forêts du Nord de la Franche-Comté," 45, and AN F12 680, *Enquête sur les bouches à feu*, 1788.

60 *Cahier de doléances*, Les Magny, 17 March 1789, in Godard and Abensour, *Cahiers de doléances du bailliage d'Amont*, 2:133.

61 *Cahier de doléances*, Bourbévelle, 17 March 1789, in Godard and Abensour, *Cahiers de doléances du bailliage d'Amont*, 1:220.

62 *Cahier de doléances*, Betaucourt, 20 March 1789, in Godard and Abensour, *Cahiers de doléances du bailliage d'Amont*, 1:177.

63 *Cahier de doléances*, Aroz, 20 March 1789, in Godard and Abensour, *Cahiers de doléances du bailliage d'Amont*, 1:20.

In an effort to end their industrial woes and ward off future ills, communities across Franche-Comté used their *cahiers* to propose a radical solution: shut down ironworks for good. Several variants of this idea circulated. Some stressed the harm manufactures caused, like the petition of Nans-sous-Sainte-Anne, northeast of Salins, which called for the elimination of forges "useless and pernicious to the continuation of woods necessary to the commerce and needs of the interior of the province."[64] Others specified that only manufactures with enough wood of their own to operate for six months a year should remain open.[65] Still others offered no compromise, demanding instead the closure of all ironworks built within the past thirty or even one hundred years.[66] Given that this was precisely the period in which metallurgy had expanded in the province, many Francs-Comtois seemed to want to turn back the clock, or at least halt it. The *cahier général* for the district of Salins summed up this outlook, asserting, "There [should] henceforth be no permissions nor *lettres patentes* granted for the establishment of furnaces, forges, and tilt hammers in the province, and all those that were established or reestablished in the last thirty years will be forbidden and suppressed."[67]

Even communities relatively unharmed by ironworks made a point of denouncing them. The *cahiers* of the *bailliage* of Baume-les-Dames, for example, make little mention of specific abuses: with only one forge, one blast furnace, and three tilt hammers, the district was far less industrialized than Vesoul or Gray, and the price of an arpent of woodland in the district had risen by only two livres between 1733 and 1763. By contrast, the equivalent price in the *bailliage* of Vesoul nearly doubled, from thirty-two to sixty-two livres, during the same period. Nonetheless, fourteen of Baume-les-Dames' villages sought to ward off any future problems, insisting, as the petitioners of Pierrefontaine-les-Varans did in a scrawl

[64] ADJ Lp1618, *Cahier de doléances*, Nans-sous-Sainte-Agne [Nans-sous-Sainte-Anne], 23 March 1789.

[65] Nearly a third of the *cahiers* of the *bailliage* of Vesoul included a clause to this effect. Godard and Abensour, *Cahiers de doléances du bailliage d'Amont*.

[66] The *cahier* of Vregille demanded the "destruction of all factories for which the owners do not own sufficient woods ... or at least the destruction of those built less than a century ago." *Cahier de doléances*, Vrégille [*sic*], 20 March 1789, in Godard and Abensour, *Cahiers de doléances du bailliage d'Amont*, 2: 526.

[67] ADJ Lp1618, *Cayer Genneral ... des Remontrances, plaints, et Doleances du Tiers Etat dudit bailliage [de Salins]*, 27 March 1789.

at the bottom of their *cahier*, on the "suppression of all newly established forges and furnaces."[68]

Strikingly vocal on the problems ironworks caused, the *cahiers* were virtually silent on the subject of ironworks' benefits. Few if any noted the significance of manufactures as a source of employment, nor did they acknowledge the income that fuelwood sales generated, thereby funding the construction of public fountains and other municipal improvements. Rather, most shared the outlook of the *cahier* of Cubry-les-Soing, a settlement on an oxbow of the Saône River: "Whatever benefit may be due to the manufacture of iron in the realm, the excessive increase of factories is clearly harmful to agriculture and is noticeably leading to a scarcity of wood, and a disastrous costliness of this staple good."[69]

To some extent, the *cahiers*' unmitigated criticism of industry can be attributed to their genre: the intent of a grievance petition is to grieve, and this the *cahiers de doléance* did in abundance. Yet inhabitants had further reasons for disliking industry, some of them difficult to articulate within the *cahier* form. These reasons included dislike of the soot, noise, stench, and heat generated by ironworks, as well as a general wariness of commercial metallurgy's social influence.[70] While tileworks, potteries, glassworks, and tanneries also consumed significant quanitities of wood, they employed local residents and fabricated goods primarily for the community. By contrast, the ironworks that had lately proliferated in Franche-Comté produced items mostly for export or for the crown, and they often employed outsiders, particularly single men whose presence villagers regarded as disruptive. Perhaps most unsettling in the eyes of residents were the charcoal burners who plied their trade in the innermost reaches of the forest. Tending their cone-shaped furnaces (see Figures 1 and 2) in the coolness of night and cutting wood by day, *charbonniers* were nomadic and nocturnal by necessity. With their smoke-blackened faces, itinerant lifestyle, and intimate association with fire, it was little wonder that villagers viewed them with suspicion.[71]

68 *Cahier de doléances*, Pierrefontaine-les-Varans, 15 March 1789, in Jouvenot, *Le bailliage de Baume-les-Dames*, 378. On the prices, see Vion-Delphin, "Forêts et cahiers de doléances," 19.

69 *Cahier de doléances*, Cubry-les-Soing, in Godard and Abensour, *Cahiers de doléances du bailliage d'Amont*, 1:439.

70 Arlette Brosselin et al., "Les doléances contre l'industrie," in *Forges et forêts: recherches sur la consommation proto-industrielle de bois*, Denis Woronoff, ed. (Paris: Éditions EHESS, 1990), 23.

71 Bertrand, "Les hommes et la forêt dans les communautés villageoises," 204–6.

Together, these unspoken cultural anxieties underscored Comtois perceptions of industry as a threatening interloper.

RESOURCE COMPETITION AND INTERNAL FRICTION

Railing against industry, as did 30 percent of the *cahiers* of the *bailliage* of Vesoul, obscured inhabitants' own role in the woodland crisis.[72] Many communities profited from forge masters' and towns' rising fuelwood demands by selling them portions of their *affouage*, a practice that violated the custom in spirit and letter.[73] Villagers also regularly pastured their animals in the woods, which, if done to excess, interfered with trees' regeneration. In both cases, rural elites had the advantage. Formulated according to property ownership, tax contributions, proof of residence, building size, and family status, the customs governing common grazing and communal wood and timber allocations favored landed over landless, old residents over new, and patriarchal households over widows and orphans. Even when modified to include the poor, these resource distribution arrangements still guaranteed the most propertied inhabitants a disproportionate share of communal benefits. In the case of pasturage, large landowners maximized their claims by sending greater numbers of beasts – and bigger ones, too, like horses and cows, as opposed to sheep and pigs – to dominate grazeable wastes and woodland. The same held true for *affouage*. In Franche-Comté, most communities distributed firewood according to *le marc la livre*, a system through which inhabitants received an allotment based on their place on the commune's tax rolls, and timber was allocated *par toisé*, proportional to the dimensions of homes and outbuildings. Practices varied from village to village. Some distributed half of wood according to *le marc la livre* and the remainder equally; others followed a two-thirds/one-third ratio. As a result, some affluent households received a surplus of wood, which they then sold at a profit, while smallholders and the landless poor received reduced portions or nothing at all.[74]

Affouage disparities were already in place prior to the French conquest, but a combination of factors caused them to grow more acrimonious over the course of the eighteenth century.[75] First was the

[72] Brosselin et al., "Les doléances contre l'industrie," 16–17.

[73] Corvol, "L'affouage au XVIII siècle," 395–96.

[74] Bertrand, "Les hommes et la forêt dans les communautés villageoises," 191–92; Corvol, "L'affouage au XVIII siècle," 390–407.

[75] On the great variety in *affouage* customs in the north and east, see Victor Alexis Désiré Dalloz et al., *Répertoire méthodique et alphabétique de législation, de doctrine et de*

crown's reformations of the province's forests from the 1710s forward. Seeking to limit the number of claimants wherever possible, royal *réformateurs* produced a stringent interpretation of beneficiaries, based on the original territorial extent of a village and the age and location of homes within its perimeter.[76] A second factor was demographic change. Even as the population of Franche-Comté mounted and the number of rights holders grew, the size of communal woodlots and crown and seigneurial allocations remained constant, resulting in ever-smaller shares of wood overall. Under these circumstances, legally recognized claimants became increasingly possessive of their franchise. Not only did they resist legislation intended to make *affouage* more equitable, but they also sought to decrease or eliminate the modest portions granted to newcomers, widows, orphans, and the landless. Making the beneficiaries of *affouage* more defensive still was the growing insistence by seigneurs-turned-industrialists that they be included in common rights, even as they worked to refute long-standing communal claims in their own woods.

Relatively few Comtois *cahiers de doléances* addressed the issue of *affouage*, perhaps because the men who figured most prominently in drafting them were the ones most favored by the current system.[77] Those that did tended to focus on rebuffing the claims of seigneurs and other pretenders. "May the right of *affouage* enjoyed by [the seigneur] M. de Saint Mauris in the communal woods of Raddon et Chapendu as well as that which is similarly enjoyed by the barrister Reigney both be abolished," began the complaint of one Haute-Saône community.[78] In the *bailliage* of Baume-les-Dames, where the hardships of mortmain remained common, eight villages denounced the exorbitant annual fees that they paid for *affouage* in seigneurial woods. "May [these charges] and other payments of this sort be entirely suppressed," insisted the

jurisprudence en matière de droit civil, commercial, criminel, administratif, de droit de gens et de droit public, new ed., vol. 25 (Paris: Bureau de la Jurisprudence générale du royaume, 1849), 558, §1810.

76 The 1669 Ordinance mandated that *affouage* customs be upheld, but sought to pare the roster of recipients back to pre-1669 levels. See the 1669 Ordinance in Isambert, vol. 18, title 25, art. 11, and title 21, art. 1, 2, 5, 10.

77 On the social dimensions of writing of the *cahiers*, see McPhee, "'The Misguided Greed of Peasants'?," 253; also Shapiro and Markoff, *Revolutionary Demands*, 136–40.

78 *Cahier de doléances*, Raddon et Chapendu (Haute-Saône), 16 March 1789, in Godard and Abensour, *Cahiers de doléances du bailliage d'Amont*, 2:311.

inhabitants of Villers-Saint-Martin, adding that in any case, "the communities supply the wood that is needed."[79]

Notably absent in these *cahiers'* complaints is any acknowledgment of the strife that *affouage* disputes caused within their communities, though demands for fuelwood and timber were clearly on the rise: over the course of the century, the province's population boomed, reaching nearly 700,000 inhabitants by 1789.[80] One of the few communities to call attention to the issue was Granges-la-Ville, a village near the Swiss border, whose petitioners urgently highlighted the gross unfairness of *le marc la livre*. "The custom," they argued in their *cahier*, "brings opulence to the rich and horribly crushes the indigent, the widow, and the orphan, by taking from them ... their own property that divine Providence saw fit to create[,] without any manual labor by men[,] for the equal use of the poor as well as the rich."[81] Over the course of the Revolution, when even the humblest members of the community gained the opportunity to be heard, conflict over *affouage* in Franche-Comté would grow louder and more rancorous. In the meantime, the rural poor grabbed what wood they could and where, without regard to felling methods or season, and local authorities, unwilling or unable to enact a fairer system, turned a blind eye to the ensuing deterioration. As the signatories of one Comtois *cahier* noted, "When there are infractions, one is often obliged to stay quiet in order not to ruin a poor unfortunate."[82] Barred a fair share in this most vital of common rights, the landless and other ineligible inhabitants had little incentive to practice conservation. Forest edges, clearings, and other areas where supervision was limited were picked bare, and coordinated woodland incursions multiplied.

THE FAILINGS OF THE FOREST ADMINISTRATION

If rural inhabitants resented their seigneurs and had cause to be bitter about each other, they forged remarkable unity in their antipathy for the

[79] *Cahier de doléances*, Villers-Saint-Martin-le-Sec (Haute-Saône), 16 March 1789, in Jouvenot, *Le bailliage de Baume-les-Dames*, 473.

[80] Boichard uses Necker's figures, which calculated 678,000 Francs-Comtois in 1783. Vion-Delphin's calculation of 775,000 Francs-Comtois in 1790 suggests a higher number of inhabitants before the Revolution. Vion-Delphin et al., *Les hommes et la forêt en Franche-Comté*, 112, 121.

[81] *Cahier de doléances*, Granges-la-Ville, 17 March 1789, in Godard and Abensour, *Cahiers de doléances du bailliage d'Amont*, 2:44.

[82] *Cahier de doléances*, Herimoncourt, 18 March 1789, in Jouvenot, *Le bailliage de Baume-les-Dames*, 92.

Eaux et Forêts, the royal agency responsible for the nation's woods and waters. Even more than iron manufactures, the Eaux et Forêts was regarded as an intruder by Comtois peasants and seigneurs alike. Starting in 1700, the crown's *réformateurs* worked their way through the royal and communal forests of the province, enacting the methods of regulation and enforcement laid out in the 1669 Ordinance. In addition to installing a strict hierarchy of guards, sergeants, officers, surveyors, clerks, *maîtres particuliers*, and, at the top, a regional *grand maître*, the reformers established a conspicuous physical presence for the Eaux et Forêts, signaling its authority in the cornerstones, roads, and boundary ditches they carved from the soil. Marking out woodlots, creating coppices and timber reserves, the reformers and woodland officers who came after them emphasized naval, industrial, and urban uses over traditional agropastoral needs. In so doing, they altered both the aesthetics and the economy of Franche-Comté's forests.

The implications of these changes were still becoming apparent in Franche-Comté in 1789. Once rich in oak, beech, hornbeam, willow, poplar, pine, and fir, Comtois forests had noticeably declined in diversity since the early part of the century under the specialized demands of wood-burning industry and naval contractors. Complaining that "the wood of oaks is becoming very rare and is indispensably necessary for the inhabitants of the countryside for buildings and repairs ... and for the construction of their wagons, carts and other plowing instruments," the inhabitants of Betoncourt-lès-Brotte blamed "the brokers of the navy" who "carry off all the wood proper for the aforesaid usages, and restrict them for all other normal usages except for the service of the king."[83]

Such restrictions inevitably frustrated inhabitants. To gain the crown's approval to obtain timber from their own communal woods, municipal authorities had to undertake a costly and time-consuming rigmarole of surveys, correspondence, and petitions. These measures would be defensible if they had actually protected the forest as they were intended. Yet as copious criticisms attested, the Ordinance's highly centralized approach largely fell short. The crown "without doubt believed it had seen to [the administration of the woods] by the establishment of the Eaux et Forêts," observed the *cahier* of Bourbévelle,

[83] *Cahier de doléances*, Betoncourt-lès-Brotte, 17 March 1789, in Godard and Abensour, *Cahiers de doléances du bailliage d'Amont*, 1:182–83.

but it was "precisely since that time that the forests began to fall into the ruin we see them into today."[84]

The experience of Mont-de-Laval, a small parish on the Doubs plateau, was typical. Hoping to procure enough timber to roof their church, fix their public fountains, and complete other municipal projects, the community wrote to the king in June 1783 for permission to harvest their *quart en réserve*, a mandatory timber reserve comprising one-quarter of their communal woods.[85] Off-limits to inhabitants without the express consent of the crown, these reserves were integral to the Ordinance's mission of safeguarding the nation's timber supply and assuring its regeneration. In practice, however, the reserves created economic and environmental predicaments of their own. As the letter from Mont-de-Laval explained, the community's woods contained 200 arpents (103 hectares), of which 50 arpents had been designated for protection when the crown's *réformateurs* had surveyed the village's woods in 1737.[86] Since that time, the lots assigned for inhabitants' use had been felled five times, to the point that they were "no longer planted with anything but young pines of two or three feet around which are useless for timber."[87] The *quart en réserve*, by contrast, was "filled with decaying Pines that long ago reached their value and maturity," including three hundred that were "entirely dry without leaves nor Branches, rotten, hollow and which can serve only as Firewood."[88] In the damp and frosty climate of the Haut-Doubs, trees were more susceptible to the ravages of insects, fungi, and weather. If they were to be harvested as timber, they could not be left to grow indefinitely – a fact local residents knew well.[89] Though polite, the parish was clearly exasperated at seeing desirable trees disintegrate while their buildings fell into disrepair. Mont-de-Laval's woodlots, their request noted, occupied "steep, craggy, and nearly inaccessible" slopes, far from any manufacture or town. Extracting the timber would be too costly for either naval or fuelwood merchants,

[84] *Cahier de doléances*, Bourbévelle, 17 March 1789, in Godard and Abensour, *Cahiers de doléances du bailliage d'Amont*, 1: 218.

[85] Title XXIV, art. 2, of the 1669 Ordinance required all religious and lay communities to set aside one-quarter of their forests for *futaie*.

[86] AN Q1 183, *Projet d'arrêt du conseil et procès verbaux d'aménagement et de reconnaissance le tout relatif aux Bois des habitants et communauté de Mont-de-Laval.*

[87] AN Q1 183. [88] AN Q1 183.

[89] On the problems of applying the 1669 Ordinance to conifers, see Schaeffer, "La forêt résineuse comtoise," 243–44; and Vion-Delphin, "L'homme et la forêt dans le Haut-Doubs," 274.

whereas the inhabitants had "an indispensable need of the trees for themselves." Thus, the letter continued, putting forth a local vision of conservation, since the wood in the reserve had already decayed, permitting it to be cut could only lead to "taller and more beautiful" pines and "procure the Greatest Benefits for the Supplicants' Descendants."[90]

In the end, the crown granted Mont-de-Laval's request. Yet it is easy to see how the community, like hundreds of others with similar needs across France, would have found the Eaux et Forêts' transformation of the forest and its curtailment of customary rights maddening. In forests where the administration established its presence early on, inhabitants occasionally reacted with violence, most spectacularly in the demoiselles' uprising in Chaux in 1765.[91] In woods not regulated until the 1760s, coppice sprouts were undergoing only their first or second harvest when the Revolution began. Regardless of when the 1669 Ordinance and forest administration were introduced to their communities, however, inhabitants were nearly universally dismayed at their impact. The villagers of Germéfontaine expressed a common sentiment when they observed, "The establishment of the *maîtrises* is not at all conducive to the best exploitation of [our] woods."[92] Indeed, criticism of the forest administration and the 1669 decree abounded in the *cahiers* of all three orders of Francs-Comtois, from the petition of the Dole nobility to the clerical petitions of the districts of Dole, Aval, and Besançon.[93]

The problem went beyond bureaucratic formalities and ecologically unsound constraints. Rather, other enduring difficulties hampered the crown's woodland policies. The worst of these involved royal finances. Despite its avowals to safeguard the forests, the crown itself set the standard for negligent exploitation, repeatedly using sales of timber, fuelwood, and administrative offices as stopgap sources of income. This practice, which Colbert had suppressed as part of his reforms in the 1660s, was revived by Louis XIV to pay for military campaigns toward the end of his reign. It continued thereafter, with occasional interruptions, until the Revolution. The reconstruction and maintenance of the French naval fleet, virtually wiped out during the Seven Years' War, also consumed enormous amounts of timber. In under a decade, France's navy assembled nearly one hundred new frigates and ships of

[90] AN Q1 183. [91] Vion-Delphin, "Forêt, frontière et commerce du bois," 203.

[92] *Cahier de doléances*, Germéfontaine (Doubs), 19 March 1789, in Jouvenot, *Le bailliage de Baume-les-Dames*, 400.

[93] See discussion in Vion-Delphin, "Les Forêts du Nord de la Franche-Comté," 48.

the line.[94] At a per-vessel rate of two to three thousand full-grown oaks plus six to eight thousand pines and firs, each ship represented a massive investment of forest capital.[95] Much of it came from Franche-Comté, which together with Burgundy contributed more than half the domestic timber to the main naval yard at Toulon.[96]

The tension between exploiting the forest and enforcing the Ordinance also carried over to everyday operations. As amply documented in their *cahiers de doléances*, many Francs-Comtois viewed the guards and officers of the Eaux et Forêts as parasites whose primary occupation was extorting fees from the inhabitants in exchange for access to their most vital assets. The parish of Chouzelot, on the banks of the Loue River, complained that the fees their *maîtrise* charged "sometimes absorbed more than half the value" of the annual wood harvest.[97] The inhabitants of Amance-et-Leugney echoed these sentiments. "Tears would better express their misery than their words," their *cahier* noted, explaining that annual expenses for their woods amounted to "at least 20 livres per year" for marking, felling, and policing their lots, plus "more than 60 livres each time" the officers of the Eaux et Forêts visited.[98] The *cahier* of Briaucourt in the Haute-Saône reported that the service fees the *maîtrise* charged were so high that the village had to buy its wood from other communities.[99]

Parishes were so overwhelmed by these "ruinous formalities, procedures and monstrous fees," wrote the comte d'Essuile, an influential agronomist, that they "wanted to be rid of [their communal woods] and

[94] Acerra, "Marine militaire et bois de construction," 113; also James C. Riley, *The Seven Years War and the Old Regime in France: The Economic and Financial Toll* (Princeton, NJ: Princeton University Press, 1986), 80–82.

[95] On average, a hundred-year-old oak yielded 28 cubic feet of wood. According to the 1783 *Encyclopédie méthodique*, one 74-gun vessel required at least 80,000 cubic feet of wood, while a 116-gun ship needed 127,000 cubic feet. See discussion in Acerra, "Marine militaire et bois de construction," 114; and Devèze, "Les forêts françaises à la veille de la Révolution," 194. Ships grew larger in the eighteenth century. Glete, *Navies and Nations*; and John Charnock, *A History of Marine Architecture* (London: R. Faulden, 1800–1802): 3:140.

[96] Vion-Delphin, "La fourniture des bois de marine en Franche-Comté," 459; also Bamford, *Forests and French Sea Power, 1660–1789*, chaps. 8 and 9.

[97] *Cahier de doléances*, Chouzelot, in Vion-Delphin and Lassus, *Le bailliage de Quingey*, 123–24.

[98] *Cahier de doléances*, Amance-et-Leugney, 18 March 1789, in Jouvenot, *Le bailliage de Baume-les-Dames*, 411–14. The *cahier* of Ivrey and Combelle enumerates the price of each fee. See ADJ Lp1618.

[99] *Cahier de doléances*, Briaucourt, 17 March 1789, in Godard and Abensour, *Cahiers de doléances du bailliage d'Amont*, 1:262.

thought they could recoup their costs by destroying them in every way possible."[100] In reality, most parishes envisioned less drastic solutions. Rillans, a tiny hamlet near Baume-les-Dames, reflected the views expressed in at least a third of Comtois *cahiers* in demanding that forest responsibility be transferred to local authorities, who could do the work more cheaply and efficiently. "The *bailliages* should handle the sale, marking, and harvest of the *quarts en réserves*, to the exclusion of the *grands maîtres*, the royal council, and everyone else," insisted the Rillans residents in their *cahier*. "Up to this point," the petitioners added, "it must be forcefully said that half the profit from the sale of the reserves is a pure loss for the communities, because of the immense fees [the administration] incurs."[101]

The fees were all the more odious because officers personally benefited from them. As with the judiciary and many other royal functions, venality governed the higher offices of the Eaux et Forêts. Sold by the crown to raise cash, *maîtrises* were owned and operated by wealthy nobles (and occasionally nonnobles) who aimed to become wealthier still from their office's emoluments. As the Champagne forge master Pierre-Clément Grignon observed, "Forest offices like nearly all others are the products of finance[;] they have devolved to all who have money and who wish to invest their funds advantageously."[102]

The crown's forest guards were also compensated mostly by the fines they levied. In the *maîtrise* of Dole, for instance, the *garde général* received one-quarter of all the penalties collected.[103] The arrangement encouraged corruption and carelessness. The parish of Byans-sur-Doubs, for instance, accused their *maîtrise* of fabricating crimes and pocketing the fines. "For some time," the Byannais charged in their *cahier*, "[forest officers] have had the habit of making frequent visits to communal woods in their jurisdiction and of condemning the communities' foresters to considerable fines for alleged wrongdoing." "Even the woodcutters appointed annually to cut the lots [were fined]," the cahier added, "often for imaginary offenses." The villagers were especially

100 Essuile quoted in Sée, "Les forêts et la question du déboisement," 18.
101 *Cahier de doléances*, Rillans, 20 March 1789, in Jouvenot, *Le bailliage de Baume-les-Dames*, 554–57.
102 AN F[10] 403, Le Chevalier Grignon, *Observations sur la dégradation des forêts, particulièrement de celles de la Bourgogne, du Lyonnais, du Beaujolais et du Dauphiné et sur les précautions à prendre pour leur conservation*, 1778.
103 C.A.R.A.N., *Histoire de forêts*, 110.

annoyed because they had only hired the woodcutters to comply with the 1669 Ordinance.[104]

Occasionally guards and officers went beyond coercion and actively despoiled the forests in their care. Inhabitants of Chouzelot protested that while their own woodsmen "responded personally to crimes and abuse," the guards of the seigneur and *maîtrise*, "instead of conducting a better watch and administration of the woods ... degrade the bulk of it themselves, so that the communities dare not even complain, fearing recrimination."[105]

Maîtres and *grands maîtres* did not escape reproach either. Lamenting their negligent attitude, Martène, a *subdélégué* in Burgundy, remarked that *grands maîtres* "only appear in the provinces like streaks of lightning, scarcely do they stay one or two nights in the towns where they [are directing] the deliveries of harvested wood; they never see the forests nor rivers if it isn't from afar, from their carriages. They earn huge profits for making a tour of two or three weeks each year."[106] The industrialist Grignon expressed an analogous view in his *Observations on the Deterioration of the Forests* of 1778. "Some *maîtrises*," he asserted, "and it is the greater part, look upon the duties of their appointment as *corvées*, to the point that they make no visits to the forests to look after their conservation and regeneration."[107]

To be sure, there were administrators who, by their own indications, worked diligently to uphold the Ordinance. François Joseph Legrand de Marizy, the *grand maître* of Franche-Comté from 1754 until the Revolution, left ample records of his efforts over his thirty-five years of service. In carrying out exceptional fellings, like the village of Mont-de-Laval's request to cut its *quart en réserve*, Legrand de Marizy followed the Ordinance to the letter: first he visited the site to assess its timber, then he drafted a management plan that maintained a set number of seed trees, and lastly he returned to the village to supervise

[104] *Cahier de doléances*, Byans-sur-le-Doubs, in Vion-Delphin and Lassus, *Le bailliage de Quingey*, 131.

[105] The *cahier* of neighboring Lavans echoed this observation. *Cahiers de doléances* of Chouzelot and Lavans, in Vion-Delphin and Lassus, *Le bailliage de Quingey*, 120, 125.

[106] Martène, *subdélégué* of Saint-Jean-de-Losne, to Feydeau de Brou, *intendant* of Bourgogne, 1780, as cited in L. Blin, "Sur trois maîtrises des eaux et forêts et la navigation de la Saône," *Annales de Bourgogne* 42 (1970): 177.

[107] AN F^{10} 403, Grignon, *Observations sur la dégradation des forêts*, 6.

the trees' harvest and auction.[108] Legrand de Marizy's assiduousness, though in stark contrast to the carelessness of his peers, may still have been motivated by self-interest. He received a handsome commission for his trouble, just as he did with other tasks. Moreover, his frequent endorsements of parishes' request to cut their timber effectively contributed to the forests' diminution and intensified pressures they were supposed to alleviate.[109]

Practical hurdles also hindered forest protection. For one, the *grands maîtres*' jurisdictions were too large and the woods in their care too numerous to enable anything more than a cursory knowledge of them. Besides Franche-Comté, Legrand de Marizy's command extended over all of Alsace and Burgundy. He was expected to tour the entire region each year to inspect operations and investigate disputes, then return to Paris for the winter to draft his recommendations.[110] The workload made it hard for communities to get his attention. The Chozelois complained that the *maîtrise* at Besançon was "too distant from their village, obliging [inhabitants] to make many expensive voyages each year."[111]

Officers' limited competence also posed problems. Until 1826, forest personnel received no formal training. The qualifications for guards were minimal: one had to be at least twenty-five years old, male, Catholic, and of good moral character, and know how to read and write.[112] For higher-level officers, financial standing was more crucial than moral character, but still no special knowledge of woodland management was required. Officers and guards simply arrived on the job unaware of logging, pruning, and planting techniques, and more ignorant still of forest ecosystems. Those who were so inclined could consult Duhamel du Monceau's primers for woodland proprietors, which exhaustively detailed every aspect of forest management. Some industrious *maîtres*, like Dominique Antoine Telles d'Acosta, the

108 On average, there were sixteen *baliveaux*, or seed trees, per arpent (about thirty-two per hectare). Vion-Delphin et al., *Les hommes et la forêt en Franche-Comté*, 92.

109 Grignon observed that *maîtres* were all too willing to approve communities' or inhabitants' felling requests because of the fees they earned in the process. AN F[10] 403, Grignon, *Observations sur la dégradation des forêts*, 6.

110 On Marizy and the duties of the *grands maîtres*, see Waquet, *Les grands maîtres des Eaux et Forêts*, 384, 168, 175.

111 *Cahier de doléances*, Chouzelot, in Vion-Delphin and Lassus, *Le bailliage de Quingey*, 124.

112 As per title II, art. 1, of the 1669 Ordinance. See also R. Blais, "Contribution à une histoire des gardes forestiers," *Revue forestière française* 38, no. 1 (1986): 18.

grand maître of Champagne from 1752 to 1784, even produced their own manuals.[113] But there was little incentive for such diligence. The venal system prevented officers from being promoted. Thus the majority learned their trade on the job, if at all, with no other guidelines to go on than the Ordinance's limited instructions. Grignon groused that "merit, knowledge, and experience in the matter are not necessary traits for holding a post; ... young men with money acquire the right to preside over the administration and regulation of woods, that is to say the real riches of the state."[114] Lacking basic silvicultural skills, officers at the highest levels of the forest administration were ill equipped to create sound management plans, much less modify the Ordinance to suit local conditions.[115]

These failings by the Eaux et Forêts vexed the Francs-Comtois beyond all others, for in their view, the forest administration's methods were neither economically tenable nor ecologically wise. "One knows from experience," the petitioners of Solemont explained, "that forests exploited not [all at once] but by letting the small trees grow and felling the old, sustain themselves better."[116] Solemont's villagers echoed the sentiments of hundreds of other Comtois communities when they demanded that "the *maîtrise des eaux et forêts* be suppressed entirely and forever in all the province."[117] In expressing this wish, the peasantry were not necessarily opposing the goals of forest supervision and management. Rather, they objected to the means by which these goals were carried out, in some cases arguing that the crown should be even more attentive.[118] "Wood is a necessity as vital as bread," the inhabitants of Bourbévelle in the Saône river basin warned. "It is high time that the administration seriously concern itself with such an important subject."[119]

113 Dominique Antoine Telles d'Acosta, *Instruction sur les bois de marine et autres: Contenant des détails relatifs à la physique & à l'analyse du chêne* (Paris: Duchesne [&] Clousier, 1782; *Supplément*, 1784; *Second supplément*, 1786).

114 AN F^{10} 403, Grignon, *Observations sur la dégradation des forêts*, 5.

115 Both Grignon and Poncelin asserted this. Poncelin, "Réponse à la question: 'Quel est le meilleur aménagement possible des bois de la province de Franche-Comté?'" *Mémoires de l'Académie de Besançon* (Besançon: Académie de Besançon, 1771).

116 *Cahier de doléances*, Solemont (Velle-sous-le-Mont, Doubs), 16 March 1789, in Jouvenot, *Le bailliage de Baume-les-Dames*, 293.

117 *Cahier de doléances*, Solemont (Velle-sous-le-Mont, Doubs), 16 March 1789, in Jouvenot, *Le bailliage de Baume-les-Dames*, 293 and 227. Out of 212 *cahiers* in the district of Baume-les-Dames, 123 demanded the *maîtrises'* suppression.

118 Bertrand, "Les hommes et la forêt dans les communautés villageoises," 171.

119 *Cahier de doléances*, Bourbévelle (Haute-Saône), 17 March 1789, in Godard and Abensour, *Cahiers de doléances du bailliage d'Amont*, 1:218.

Among alternatives the parishes proposed, two concerns dominated: who should handle marking and felling the forests, and who should guard them. Many communities felt that since their seigneurs' forest officer was already familiar with the area, *he* should handle forest operations, not the officers of the Eaux et Forêts. "The seigneurs' officers[,] ... being on the scene, know the woods better and the communities' need," argued inhabitants of Vaucluse.[120] This approach especially made sense where the *maîtrise*'s and the seigneur's woodland jurisdictions overlapped, as in Chaux-les-Châtillon, just a few kilometers to the north of Vaucluse. Eliminating redundancy would be an "advantage to the nation," the *cahier* declared, because the fees of the Eaux et Forêts were so exorbitant.[121] When it came to guards, however, most parishes rejected both seigneurial and royal oversight. The petitioners of Vernois-lès-Belvoir explained,

> We have woodsmen and foresters in our community, two mayors who are sworn to uphold the law, *échevins* [municipal magistrates] to oversee the guarding of the woods and forests, meadows and other holdings of our community, so that one can make no damage there and to uphold order in the area. In seems to us that these communal guards, mayors, [and] magistrates are sufficient to see to the policing of the exterior as well as interior of the village.[122]

Most *cahiers* also insisted that the parish itself should benefit from the fines collected.[123]

In sum, in calling for the abolition of the *maîtrises* in favor of greater communal control and communal benefit, the rural inhabitants of Franche-Comté promised to be thriftier, more rigorous, and more effective stewards than the Eaux et Forêts. The *cahier* of Hérimoncourt explained, "By transferring [the process] to our judges, it will cost less. By diminishing the price of fines, more reports will be made, the law will be in greater force and consequently there will not be as many abuses."[124]

120 *Cahier de doléances*, Vaucluse, 16 March 1789, in Jouvenot, *Le bailliage de Baume-les-Dames*, 269.

121 *Cahier de doléances*, Chaux-lès-Chatillon [*sic*], 18 March 1789. On "exorbitant fees," see the *cahier* of Hérimontcourt; on seigneurial officers working more effectively and cheaply, see the *cahier* of Froidevaux. All three in Jouvenot, *Le bailliage de Baume-les-Dames*, 299, 92, and 292.

122 *Cahier de doléances*, Vernois-lès-Belvoir, 17 March 1789, in Jouvenot, *Le bailliage de Baume-les-Dames*, 330.

123 *Cahier de doléances*, Charquemont, 19 March 1789, also the *cahiers* of Adam-lès-Passavant and Lanans, in Jouvenot, *Le bailliage de Baume-les-Dames*, 248, 430, 465.

124 *Cahier de doléances*, Hérimoncourt, 18 March 1789, in Jouvenot, *Le bailliage de Baume-les-Dames*, 92.

Evidence suggests that particularly in the uplands, rural communities did indeed manage their woodlands effectively prior to the establishment of the *maîtrises*. In his preliminary report on Franche-Comté in 1688, prior to enacting reforms in the area, the *grand maître* and *réformateur* Jean Sanguinière observed that though the parcels formerly belonging to the Spanish Habsburgs were badly degraded, mountain parishes had maintained their forests by protecting felled areas until they grew large enough to withstand cutting and pasturing again.[125] These "conservative communities," forest historian Réné Schaeffer has argued, apparently did a good job; mid-eighteenth-century timber assessments from the Haut-Doubs attested that trees more than six feet around were common, and trees eight to twelve feet around were not unusual.[126]

That big timber should persist longer in the uplands compared to Franche-Comté's plains stands to reason. Upland villages were farther removed from population centers along the Doubs River as well as from the manufacturers multiplying in the Saône and Ognon valleys. Unless they were near a good-sized waterway, their isolation also shielded them from the avidity of naval procurers. The mountain forests were reformed relatively late – not until the 1740s in the Haut-Jura – so the Ordinance's impact was limited.[127] Lastly and most importantly, even after their woods had been restructured in accordance with the 1669 decree, many communities refused to abide by the new policies and instead maintained their traditional practices. As the abbot of Montbenoît, in the Haut-Doubs, protested in a letter of 1719, the coniferous forests of the region were completely unsuited to the shelterwood felling system stipulated by the Ordinance. Founded in the eleventh century, the monastery at Montbenoît had been an important early initiator of settlement in Franche-Comté. From experience, its residents favored the more moderate approach known as *jardinage*, described by the abbot as "cutting one tree out of twenty."[128] Years after their woods' reorganization, parishes around Montbenoît continued to resist the crown's procedures. As a local

[125] Schaeffer, "La forêt résineuse comtoise," 165.

[126] AD Doubs B 17 100–550; as cited in ibid., 166. One foot, or *pied*, was roughly equivalent to an English foot: 32.5 centimeters or 12.8 inches. Ronald Edward Zupko, *French Weights and Measures before the Revolution: A Dictionary of Provincial and Local Units* (Bloomington: Indiana University Press, 1978).

[127] AN Q1 184 (Doubs); Q1 406 (Jura); also Q1 183 on the *réformation* conducted in 1737 by the *maître particulier* at Besançon.

[128] Schaeffer, "La forêt résineuse comtoise," 243; and Turc, "L'aménagement des bois de sapin de Franche-Comté," 445.

judge noted in 1728, "one cannot cut by *tire et aire* without exposing the woods to total ruin."[129] The survival of large trees in the region even after midcentury, combined with the resolve of these communities to uphold their time-honored forestry practices, lends substance to Comtois claims of superior woodland custodianship.[130]

Of course, proposals to proscribe or even abolish the Eaux et Forêts in favor of local control were not without their own pitfalls. The selection method of *jardinage* led to abuse if conducted without care. When villagers along the Doubs River realized profit could be made by supplying timber to naval procureurs, big trees declined in their areas. Community-appointed foresters and woodsmen, furthermore, were vulnerable to social pressures that the crown's *garde général*, usually an outsider, was better able to ignore. Several *cahiers* alluded to this problem, which they blamed on penalties demanded by the law. The "excessive expense" of fines for woodland infractions, noted the *cahier* of Lanans, "often silences the forest guards, perpetuating by this means the abuses and damage."[131] As for the suggestion that seigneurial agents should handle the management of communal woods, this, too, was a risky strategy, for the seigneurs' behavior in the past half century showed little indication that they would act in the best interest of inhabitants. On the contrary, many seigneurs had shown their willingness to muscle in on communal woods and even revoke customary rights when it suited their interests. Yet if these proposals had considerable shortcomings, the crown's authoritarian policies and fumbling practices had proven little better at assuring the nation's social and ecological well-being. For the benefit of rural communities as well as the environment, the forest administration and the 1669 Ordinance would require substantive revision, if not outright elimination.

A SALTY STRUGGLE

The Saline Royale at Arc-et-Senans, site of a radical effort to reconfigure the geography of raw materials and fuel supplies, offers a particularly

[129] Schaeffer, "La forêt résineuse comtoise," 244.

[130] *Jardinage* remains important in the Vosges and Jura for the same reasons the monks cited: though not as lucrative as *tire et aire*, it reduces risk of erosion. Andrew Liston and Arne Rumbold, "A Forest Study Tour in North East France," *Quarterly Journal of Forestry* 83, no. 2 (1989): 107–13.

[131] *Cahier de doléances*, Lanans, 23 March 1789, in Jouvenot, *Le bailliage de Baume-les-Dames*, 465.

revealing example of the crown's woodland authoritarianism. As noted in Chapter 1, saltworks had been important sources of seigneurial power in Franche-Comté for centuries. After the French conquest, these lucrative manufactures became the property of the crown. The most important among them was Salins, where an especially rich *saumure*, or brine, bubbled naturally from the earth.[132] Here the crown rapidly expanded production from the 1680s forward, sending its agents to exploit the forests of surrounding communities for salt production. By the 1770s, with the hillsides around Salins denuded and distant villages complaining about mounting wood prices, the crown realized it needed a new solution. It found one in the person of Claude-Nicolas Ledoux, one of the most popular and accomplished architects of the age. Appointed saltworks commissioner of Lorraine and Franche-Comté in 1771, Ledoux would bring a visionary zeal to his post.[133]

Unlike the beleaguered administrators before him, who had tried in vain to eke more and more wood from increasingly barren forests, Ledoux realized that for saltmaking to continue at Salins, the fuel problem would have to be approached from another angle. As an architect, he approached the dilemma spatially, looking first at the manufacture's physical characteristics and surroundings. His findings were not promising. The salt spring and surrounding community of Salins were wedged deep in a narrow gorge surrounded by rising plateaus, all of which slowed the transport of fuel and increased its expense. Overcoming these obstacles would require a substantial reconfiguration of the saltworks, which, indeed, is what Ledoux undertook to do. Realizing that water moved more readily than wood, he opted to shift salt production some thirteen miles to the north, to a spot outside the hamlets of Arc and Senans where open space and ample woodlands converged near the Forêt de Chaux. To transport the *saumure* to the site, he installed a gently sloping, twenty-one-kilometer brine aquaduct, or "*saumoduc*," of interlocking fir logs.[134] Once the brine arrived, the boiling pans of Ledoux's newly constructed Saline Royale would transform it into "white gold."

[132] The other Franche-Comté saltworks at Montmorot apparently had lower output or a less rich *saumure* than Salins. The *cahiers* refer to *saumure* as *muire*. See, for example, ADJ Lp1618, *cahier de doléances* of Port-Lesnay.

[133] Ledoux was an architect for the Eaux et Forêts prior to obtaining his saltworks position. Michel Gallet, *Claude-Nicolas Ledoux, 1736–1806* (Paris: Picard, 1980).

[134] Claude-Nicolas Ledoux, *L'architecture considérée sous le rapport de l'art, des moeurs et de la législation* (repr., Paris: Hermann, 1997).

In its architectural emphasis on harmony, efficiency, and order, the Saline Royale was, and remains, a monument of Enlightenment thought. Drawing on elements of Palladian classicism, Newtonian logic, and Masonic fraternalism, Ledoux's design integrated the material and social elements of production in a rigid semicircle of lodgings and workshops arrayed around a director's office and residence, the saltworks' functional and visual hub. The striking, panoptic structure, along with the never-realized utopian "city of Chaux" that Ledoux later dreamed of building around it, is today widely recognized as a masterpiece, a precursor of the planned industrial communities of the nineteenth and early twentieth centuries. At the same time, however, the saltworks' alien incongruity amid the lush fields of the Val d'Amour and its obsessively ordered, barren arrangement – three lone oaks soften its stony severity today – speak volumes about relations between eighteenth-century rational improvers and the objects of their ambition. Ledoux and his reformist colleagues wished not so much to enhance the natural workings of the environment as to impose a new, supranatural order upon it. As with the implementation of the 1669 Ordinance in the rangy woods of Franche-Comté, the Saline Royale reflected the desire to exploit resources efficiently through uncompromising control. Understandably displeased at this new force in their forest, the villagers around the Forêt de Chaux proved uncooperative at giving way.

Ever since the crown's reformation of Chaux in 1724 and 1731, the forest's *usagère* communities had found themselves competing with forge masters and naval procurers for wood from their own lots. Although the sandy Doubs and Loue river basin in which the Forêt de Chaux stood lacked ore deposits on the scale of those in northern Franche-Comté, its sweeping expanse of oaks, beech, and hornbeam nonetheless suffered impoverishment at the hands of charcoal burners and woodcutters, who devoured nearly half the yearly wood harvest fueling six ironworks north and east of the woods. The thirty-odd *usagère* communities who relied on the forest for their domestic needs also constituted a significant strain, as did their potteries, tanneries, tileworks, limekilns, sawmills, and basket weavers. The glassworks at La Vieille-Loye consumed another 4 percent of the forest's fuelwood.[135] Together, these demands generated ongoing conflict. At least twice, revolt, or the threat of revolt, forced the crown to adjust its apportionment of Chaux – once in response to the demoiselles'

[135] See consumption figures in Husson, "L'histoire du paysage forestier français," 136.

uprising of 1765, which won the parishes' inhabitants a slightly increased allocation of fuelwood, and again in 1768 when the controller general reduced the minimum size of wood harvested in Chaux from eight *pouces* around (about eight and a half inches) to six. The latter move was intended to appease the "wails and complaints of all the orders of the city" of Dole regarding the high price and unavailability of fuel.[136]

Despite the evidence of heavy socioeconomic and environmental pressure on the forest, the crown granted the new saltworks at Arc-et-Senans exclusive use of 22,000 arpents (about 11,400 hectares), more than half of Chaux's total area. In addition, it entrusted the *maître de la saline* with the task of felling and delivering *usagère* communities' annual wood supply.[137] The inhabitants of these villages were less than pleased with the arrangement. According to the irate inhabitants of Arc-et-Senans, the foot guards entrusted with reporting on and seizing livestock in the saltworks' woodlots were "men at loose ends and leech[es] of the people," who "look out for their own interest" and "ruin the people ... under the veil of conserving the forest."[138] The villagers of Liesle, four kilometers northeast of the Saline Royale, devoted some seven pages of their *cahier* to condemning salt contractors and the system that favored them. Since 1766, the Lieslois noted, they had been deprived of their former grazing, gathering, and fuelwood rights, though their village was barely a stone's throw from Chaux. Now more than half their communal lots were allocated for the saltworks, "which only [paid] for a third of their actual value."[139] The result, exclaimed the petitioners, was misery:

Not having sufficient wood for their own consumption relative to their numbers, [the inhabitants] are in the sad necessity of lacking it nearly always, such that during the harsh winters their situation is most pitiful, being obliged to heat themselves and cook their food by digging up the most meager roots,

[136] AN C 128, *Mémoire sur les difficultés qu'éprouve la ville de Dole de s'approvisionner de bois de chauffage, et sur les moyens d'y remédier*; letters of 26 April, 3 May, and 27 May 1768 between de Beaumont, *contrôleur des finances*, and Lacoré, *intendant* of Franche-Comté. See also discussion of 21 June 1768 ruling regarding the reduction in fuelwood size in Vuillier, *Mémoire concernant les droits d'usage indéfini de la ville de Dole*, 14–16.

[137] *Arrêts du Conseil d'état* of 29 April 1773, 8 June 1774, 12 June 1774, and 11 July 1775, in Vuillier, *Mémoire concernant les droits d'usage indéfini de la ville de Dole*, 18–20.

[138] *Cahier de doléances*, Arc-et-Senans, 19 March 1789, in Vion-Delphin and Lassus, *Le bailliage de Quingey*, 198. In Lorraine, communities also suffered shortages despite an abundance of woods because they were barred access to the woods by saltworks. *Les Eaux et Forêts*, 176.

[139] *Cahier de doléances*, Liesle, 17 March 1789, in Vion-Delphin and Lassus, *Le bailliage de Quingey*, 171.

even cutting fruit trees to cover this vital need and [avoid] risking ruinous fines by the salt-farmers.[140]

Their paltry allotments of wood were "all the more distressing," the Lieslois continued, because of the salt contractors' cupidity:

Each day we see strangers taking our timber[,] which they often obtain under false pretenses, [and] not having need of it themselves, they resell it to others; while in our parcel of the same woods, we are scarcely able to obtain one foot of wood for every hundred *toises* of houses, which shows the radical vice of the administration that distributes the wood.[141]

Even parishes whose woods were spared these demands had reason to grumble about the saltworks because, as *cahiers* throughout Franche-Comté attested, the price of salt was too high and the quantities they received too low. The villagers of Orsans, on the Doubs plateau, blamed it on corruption: "although [salt] is abundant in this province ... the price has been considerably increased to provide for foolish expenditures, like administrative offices and comedies," they complained.[142] The inhabitants of Thervay-lès-Balançon noted that salt prices were so high that they had had to resort to smuggling to obtain the precious commodity.[143] Although many *cahiers* claimed that Comtois salt was being "conducted to foreigners," at least part of the shortage was due to the mediocre quality of the local product, which at times was barely fit for consumption. "Since 1775," noted the *cahier général* of the district of Arbois, inhabitants had been receiving only "salt of the worst quality, mixed with two-thirds nonsalt matter, destructive to the health of the people and leading to the ruin of livestock."[144] The residents of Boujailles, in the marshy uplands east of the Forêt de la Joux, likewise protested that the salt sold to them was "useless for salting the cheeses of the mountains."[145]

As with ironworks and the Eaux et Forêts, some *cahiers* proposed closing down saltworks and transferring control of salt distribution to

140 Ibid. 141 Ibid. One *toise* was six *pieds*, or roughly six and a half feet (1.9 meters).

142 *Cahier de doléances*, Orsans, 16 March 1789, in Jouvenot, *Le bailliage de Baume-les-Dames*, 409.

143 AD Lp1618, *Cayer de doléances pour servir aux habitans de la communauté de Tervay les Balancon* [*sic*], 16 March 1789.

144 ADJ Lp1619, *Cayer general ... des Remontrances, plaintes et doléances des Bourgeois, habitans et manans tant de cette ville* [Arbois], *que des communautés ressortissantes par appel en ce Baillage*, 19 March 1789.

145 ADJ Lp1618, *Cahier de doléances*, Boujeailles [Boujailles], n.d. [March 1789?], art. 17. See also *cahier general* of Salins, which demands that the saltworks at Arc-et-Senans be closed and that the residents receive only the good salt from Salins.

the communities. Vaucluse, in the Dessoubre Valley, envisioned this as the "liberty of each parish to take salt wherever it finds it convenient."[146] The majority of *cahiers*, however, asked simply for an increased allocation of salt, "in grains, not loaves," and at lower prices.[147] In addition, the communities most affected by the saltworks' fuelwood allocations, in the corridor between Arc-et-Senans and the "Petit Montagne" south of Salins, insisted they at least be properly remunerated. The petitioners of Arc-et-Senans reasoned, "The saltworks neighboring [the village] are the cause of the scarcity of wood. It would seem that the communities should be compensated by a sufficient quantity of salt."[148] The inhabitants of Champagne-sur-Loue preferred their compensation in the form of wood. Invoking the notion of a fair exchange of values, they demanded, "If it is necessary for the exploitation of the Saltworks of Franche-Comté that the King continue to withhold Woods that belong to the Communities; he should Expressly order the ... Managers of the said Saltworks to Sell and deliver at a Just price to Each and all inhabitants of the region ... all Wood necessary for Heating, Building and the Rest."[149]

Considering its inspired architecture and cut-rate fuelwood concessions, the Saline Royale ought to have flourished. Instead, it never turned a profit during its entire time in operation.[150] Nor could Ledoux's engineering, a triumph of rational design, prevail over the inexorable power of nature. Thirty percent of the brine flowing along the twenty-one-kilometer aquaduct from Salins spilled, evaporated, or was siphoned off by thieves before reaching the boiling pans at Arc-et-Senans.[151] Outside its stern walls, moreover, the Saline Royale gave rise to discord and disarray. Its contractors strong-armed wood from local communities, rode roughshod over customary rights, drove up prices, degraded the ecology of the surrounding area, and provoked resentment and resistance among inhabitants. These troubles were all the more dire because the Forêt de Chaux had already been fraught with conflict before the saltworks'

[146] *Cahier de doléances*, Vaucluse, 16 March 1789, in Jouvenot, *Le bailliage de Baume-les-Dames*, 269. The *cahier* of Vauclusotte is more explicit on this point; Jouvenot, *Le bailliage de Baume-les-Dames*, 277.

[147] ADJ Lp1618, *Cahier de doléances*, Communailles, 19 March 1789.

[148] *Cahier de doléances*, Arc-et-Senans, 19 March 1789, in Vion-Delphin and Lassus, *Le bailliage de Quingey*, 197.

[149] ADJ Lp1618, *Cayer de doléances*, Champagne [Champagne-sur-Loue].

[150] It closed in 1895.

[151] Anthony Vidler, *Claude-Nicholas Ledoux: Architecture and Social Reform at the End of the Ancien Regime* (Cambridge, MA: MIT Press, 1990), 126–33.

construction. Like the stubborn *saumure*, which ran its own course even when channeled, the communities surrounding Chaux refused to be confined to smaller and smaller shares of the woods. In the long run, their defiance would wear down administrators and outlast industries.

As the *cahiers* consistently demonstrated, conflicts over woodland resources ranked high among the concerns of rural inhabitants at the end of the eighteenth century. If the tension surrounding the Saline Royale was especially bitter, it is because the crown's ambitions there seemed so out of step with the basic demands of the populace. But the aggravations of increased competition for diminished resources were everywhere in Franche-Comté, from the Saône river valley to the Haut-Jura peaks. By constricting the silvopastoralism so essential to Franche-Comté's economy and by replacing the region's famous *haute futaie* with ironworks-oriented coppices, state, seigneurial, and industrial encroachment transformed communities' relationship to the forest as well as the forest itself.[152]

Some historians have interpreted the *cahiers*' widespread demands for the suppression of industries, seigneurial guards, and the Eaux et Forêts as simply an attempt to "eliminate a rival." Yet a closer look at the petitioners' language suggests that something more was at work.[153] Time and again, the *cahiers* recalled the entitlements they had enjoyed under the "former sovereigns" and insisted on their reinstatement. The *cahier* of Liesle, for example, asserted, "[We] demand an ancient right of use, pasturage, and access that we had in the Forêt de Chaux, which we have been deprived of since 1766."[154] If access was not an option, the Lieslois called for wood to be sold at a "just price." In underscoring the historical legitimacy and economic modesty of their demands, the Francs-Comtois were articulating a specifically local interpretation of liberty, property, and fraternity, one in which the viability of the community took precedence over commercial production and in which the allocation of resources accommodated shared interests in a limited environment. The Francs-Comtois hoped to realize this vision when they welcomed the Revolution in 1789.

[152] On grazing in the margins and being fined for it, see *cahier de doléances*, Arc-et-Senans, 19 March 1789, in Vion-Delphin and Lassus, *Le bailliage de Quingey*, 198.

[153] Brosselin et al., "Les doléances contre l'industrie," 27–28.

[154] *Cahier de doléances*, Liesle, 17 March 1789, in Vion-Delphin and Lassus, *Le bailliage de Quingey*, 172.

4

"Seduced by the word liberty"*

Woodland Crisis and the Failure of Revolutionary Reform

The national interest is armed with a spade; self-interest is armed with an axe.

Barère de Vieuzac, 1790[1]

It is an evil not to have subjected to the Police générale the forests that were sold by the Republic. The purchasers have razed them entirely, and timber, which is becoming rare, will one day be lacking.

Quirot, Commissioner of the Doubs, *Compte de la situation politique du département du Doubs pendant le mois de Messidor an 6* (June–July 1798)[2]

Armed with sheaves of their districts' *cahiers de doléances*, deputies from every corner of France converged on Versailles in the first week of May 1789 to take part in the Estates General. Expectations ran high among the participants, as well as the thousands of peasants, bourgeois, clerics, and nobles who had contributed to the grievance petitions. The first gathering of its kind in 175 years, the meeting portended far-reaching political and economic change.

Within only a short time, however, the sense of anticipation and excitement that characterized the outset of the Estates General would be

* The quotation in the chapter title is from Delisle de Moncel, *Mémoire sur le repeuplement*, 5.

1 Bertrand Barère de Vieuzac, *Rapport des comités réunis des domaines, des finances, de l'aliénation des biens nationaux, de la marine, du commerce et d'agriculture: Sur les bois et forêts nationales*, 6 August 1790, in *Archives parlementaires de 1787 à 1860: Recueil complet des débats législatifs et politiques des chambres françaises, imprimé par ordre du corps législatif sous la direction de mm. J. Mavidal et E. Laurent. Première série (1787 à 1789)*, 17:630–36 (hereafter *AP 1ère série*).

2 AN F^{1cIII} Doubs 6.

tempered by dissent and unease about the nature and course of reform. In some areas, anxiety over the possible repercussions of the deputies' actions, combined with intense economic hardship, gave rise to a wave of panic and popular violence that came to be known as the Great Fear.[3]

Franche-Comté numbered among the regions racked by unrest in the spring and summer of 1789. Focused above all on seigneurial privilege, the turmoil was particularly prevalent in the Haute-Saône and Doubs. Seeking to redress grievances they suspected might not be resolved otherwise, artisans and peasants, often encouraged by more affluent members of the community, converged on abbeys and estates to demand that the occupants renounce their privileges in writing in the presence of a notary.[4] The inhabitants of Menoux, for example, threatened their octogenarian countess with bodily harm and the destruction of her chateau unless she signed over legal title to all her woodland holdings in the village. The countess complied.[5] The revolts did not always involve such legal niceties, however. In many cases, irate inhabitants destroyed dovecotes, pursued game, grabbed timber, and generally seized the opportunity to destroy the hated symbols of seigneurial prerogative. Fueled by long-standing resentment (as well as by the consumption of copious amounts of wine, according to many reports), the participants' bids for restitution and parity spoke to a new political climate. As the throng of four hundred that strong-armed the seigneur of Jasney into relinquishing his privileges insisted, "he has to be just like us and not more than us."[6]

Remarks like these made clear that the violence sweeping through Franche-Comté was not merely a settling of scores; rather, the participants sought to equalize what had become an increasingly inequitable relationship and push back seigneurs' iniquitous greed, including their expropriations of communal forests. Not everyone saw it that way, of course. Touring Franche-Comté at the height of the upheaval in late July 1789, the English agronomist Arthur Young dismissed the political purposefulness of the events, citing them instead as evidence of the region's "backwardness." "The mob plunder, burn, and destroy, in complete ignorance," he declared. Nor did he set much store by the achievements

[3] For a general discussion, see Alan Forrest, *Paris, the Provinces and the French Revolution* (London: Arnold, 2004), 60–62.

[4] On the Great Fear in Franche-Comté, see Girardot, *Le département de la Haute-Saône pendant la Révolution*, 1:129–48. and Mayaud, *La Franche-Comté au XIX siècle*, 7, 34–35, 22.

[5] Girardot, *Le département de la Haute-Saône pendant la Révolution*, 1:142.

[6] Quoted in ibid.

of the National Assembly, which he accused of being out of touch. "[Amid] this universal mass of ignorance," he added, "there are men every day in the Estates, who are puffing themselves off for the FIRST NATION IN EUROPE! the GREATEST PEOPLE IN THE UNIVERSE! as if the political juntos, or literary circles of a capital, constituted a people."[7]

Young gave the concerns of provincial inhabitants and the reformist ambitions of the new lawmakers too little credit, but in identifying the gulf of comprehension, expectation, and communication that separated them, he put his finger on a problem that would plague the Revolution throughout. From 1789 forward, legislators would endeavor to renovate the nation through an astonishingly bold series of measures: abolishing feudalism and slavery, reorganizing the church, overhauling the geography of administration by dividing the country into departments (see Map 1), even transforming the measure of time through the creation of a new, decimal-based calendar. Agricultural and woodland reform were central to this vision of a rationalized France, yet time and again, innovators' schemes would run aground on local resistance because of their incompatibility with inhabitants' needs and interests or incongruity with the social and physical landscape.

This was especially true of revolutionary efforts at forest reform. The task of devising uniform laws for a nation as topographically diverse as France was daunting enough, but unraveling the tangle of local practices, each rooted in history, culture, and the constraints of climate and terrain, and each fraught with their own rivalries and entrenched interests, proved more challenging than legislators anticipated. In almost every instance, from decrees encouraging clearing and enclosure to proposals for abolishing communal wood-gathering and grazing, what was championed in one region was vehemently rejected in another. Within communities, moreover, the reform initiatives exposed and sometimes exacerbated tensions between landowners and the landless, grazers and growers, industry and agriculture, and individuals and associations. Coming amid these conflicts, legislative attempts to overhaul the forest administration and rationalize woodland management would prove both ill considered and unenforceable.

This chapter examines the halting efforts of revolutionary lawmakers to develop a new and comprehensive forest policy between 1789 and 18 Brumaire, year VIII (9 November 1799). Focusing on the debates over proposed woodland legislation, as well as related initiatives like the auction of the *biens nationaux* and the division of the commons, it

[7] Young, *Travels in France*, 156–59, 26–27 July 1789.

highlights a period of contingency and conflicting hopes. For more than a century, the 1669 Ordinance had governed the nation's forest realm, but during the tumultuous revolutionary decade, individuals and groups representing a welter of interests would attempt to wipe the slate clean and fundamentally reshape the allocation and management of France's most vital natural resource. Although few of their schemes, which ranged from complete privatization to reinforced communal control, successfully moved from legislation to implementation, the ideas and debates fomented during this time nonetheless had a lasting and ultimately deleterious effect on the social and natural ecology of French woodlands.

Even as the Revolution nurtured hopes of change for the better, a convergence of forces would end up depleting the nation's forests on an unprecedented scale and foreclosing the aspirations of communities that had petitioned in their *cahiers* for the restoration and fortification of their woodland use rights. Mounting pressures on the forest included the material exigencies of war, widespread land sales to pay off swelling government debts, unchecked clearing and felling by landed and landless alike, and, not least, the abrupt dissolution of the Eaux et Forêts without a workable replacement in place. At the same time, lawmakers' efforts to transform property relations in the name of individual liberty – a key tenet of the Revolution – undercut the long-standing and comparably sustainable forms of communal forest stewardship championed by the peasantry in their *cahiers*. By insisting upon proprietary individualism as the key to citizenship, revolutionary legislators dealt a mortal blow to collective identity and interests and, in the process, fostered predatory and competitive behavior among inhabitants faced with the loss of woodland access. Together, these forces would give rise to an erroneous, but politically useful *légende noire* of environmental destruction at the hands of an anarchic peasantry during the Revolution.[8]

"LEND A HAND TO THE OFFICERS CHARGED WITH ENFORCING THE LAWS"

After the National Assembly formed amid high drama and tumult in June 1789, one of its members' first priorities was to devise a replacement for

[8] On the *légende noire*, see G. Buttoud, "Les projets forestiers de la Révolution (1789–1796)," *Revue forestière française* 1 (1983): 9–20; Plack, "Agrarian Reform and Ecological Change"; McPhee, "'The Misguided Greed of Peasants'?"; and Peter McPhee, *Revolution and Environment in Southern France, 1780–1830: Peasants, Lords, and Murder in the Corbières* (Oxford: Oxford University Press, 1999), 141–47.

France's widely despised forest administration, the Eaux et Forêts. Special commissions of the crown had already presented forest reorganization schemes in 1768, 1776, and 1787, but few of their suggestions had been implemented.[9] Now legislators hoped to formulate and execute a much grander reform, one that involved overhauling not only France's woodland management but also the national economy writ large. They had their work cut out for them. From midsummer onward, members of the Committee of Agriculture, Industry, and Commerce met three evenings a week for hours at a time to discuss hundreds of proposals, letters, and complaints that poured in from around the country, each with their suggestions for improvement. As the letters demonstrated, the stakes in any sort of reform were high, as was the potential for acrimony.

The committee's work was soon complicated by increasing turbulence in the capital and around the country. In addition to widespread tax evasion, spurred by the expectation that a more equitable economic system was imminent, communities across France pushed the limits of their newfound freedoms, in the pursuit of local interpretations of woodland liberty.[10] Inhabitants of the Haute-Saône village of Fédry, for example, gathered in late October 1789 to craft an agreement to "fell high timber, fruit trees, and other dead wood … located throughout the forests."[11] Recognizing that their plan might run afoul of the law, they included a clause acquitting their forest guards "of whatever might arise [as a result of] this felling."[12] That left enforcement up to the district *garde general*, who arrived for an inspection in December and found "various offenses committed in the woods of the community of Fédry." When he tried to issue citations, the disgruntled guard later reported, he became the target of "rebellion, violence, and maltreatment … on the part of many inhabitants of the village."[13]

Hoping to rein in such behavior, the National Assembly issued a string of forest-related decrees, starting in November 1789 with a proclamation accusing rural inhabitants of "taking the liberty of entering [the forest] daily, usually in armed mobs," to steal fuelwood and sell it in nearby towns. Invoking the king's authority, the directive warned that "His Majesty expressly forbids all persons from entering the woods,

[9] Andrée Corvol, *L'homme aux bois: Histoire des relations de l'homme et de la forêt XVIIe–XXe siècle* (Paris: Fayard, 1987), 233–34.

[10] On taxes and rural tensions see McPhee, *Revolution and Environment in Southern France*, 79–90, also Peter McPhee, *The French Revolution, 1789–1799* (Oxford: Clarendon Press, 2002), 67–70.

[11] ADHS B 9255. [12] ADHS B 9255.

[13] ADHS B 9255. Legal proceedings stemming from the case continued into August 1790.

individually or in groups, to commit infractions." In the same breath, however, the proclamation also confirmed the legitimacy of use rights for gathering "dry and fallen wood" – a proviso that presaged the difficulties legislators faced in discriminating between licit and illicit woodland acts and in regulating the myriad overlapping claims to the forest.[14]

Adamant though this and subsequent pronouncements were, they remained for the most part utterly toothless. Other than pressuring municipal leaders to "lend a hand to the officers charged with enforcing the laws" and to "print, publish, and post" the November decree everywhere, the National Assembly was powerless to ensure compliance.[15] Instead, the general state of disorder and organizational disarray that descended upon the forests in the summer of 1789 would continue for most of the Revolution.

Part of the difficulty in restraining rural inhabitants lay in the mixed messages that lawmakers themselves, torn between liberalization and regulation, were transmitting. On the one hand, the National Assembly declared that it was committed to replacing the privileged rule of the Old Regime with a constitution founded on the "moral and collective" will of the people. On the other hand, when communities like Fédry acted on their moral and collective will, they were accused of insubordination and abuse. The National Assembly's second woodland proclamation, the 11 December 1789 Decree Concerning the Repression of Offenses Committed in the Wood and Forests, reflects these competing impulses. In part a reiteration of its ineffectual injunction a month earlier, the decree called on "all the inhabitants of the realm to attend to the conservation of and maintain the respect due to all properties." At the same time, it placed the "forests, woods and trees ... under the safeguard of the nation, the law, the king, the courts, the administrative and municipal assemblies, communes, and national guards."[16] In sanctioning the authority of just about every supervisory body to pursue and punish delinquents except the Eaux et Forêts, the decree undercut both the legitimacy and authority of the existing woodland administration.

[14] 3 November 1789, *Proclamation du Roi pour la conservation des forêts et bois*, in J. B. Duvergier, ed., *Collection complète des lois, décrets, ordonnances, réglemens, avis du Conseil-d'Etat, publiée sur les éditions officielles du Louvre, de l'Imprimerie nationale, par Baudouin; et du Bulletin des Lois; de 1788 à 1830 inclusivement, par ordre chronologique*, 2nd ed., 30 vols. (Paris: Chez A. Guyot et Scribe, 1834–38), 1:55. (Hereafter cited as Duvergier, *Collection complète des lois*.)

[15] Ibid.

[16] 11 December 1789, *Décret concernant la repression des délits qui se commettent dans les forêts et bois*, in Duvergier, *Collection complète des lois*, 1:61.

Two weeks later, the National Assembly took another step in this direction by formally charging the local and regional administrations with the conservation of the woods and forests. Though the measure was buried deep within a larger catalog of the duties of the newly constituted primary and administrative assemblies, formed via law of May 1790, it marked a decisive break from the regulatory philosophy that had dominated France's woodlands since the twelfth century.[17] Thus emboldened, some municipal and regional officials now went further, dismissing unpopular forest officers and replacing them with their own appointees.[18] By undermining the *maîtrises* and assigning local authorities guardianship of the forest, the National Assembly was repudiating a historically troubled structure of woodland governance. But what would take its place?

The Committee of Agriculture was inundated with opinions, proposals, and insights – most of them unsolicited – on this question. Views tended to fall into three categories: those who advocated restructuring the Eaux et Forêts and wanted to maintain woodland control in the hands of the state, those who pushed for drastically decentralizing forest oversight and vesting it in the control of communities, and those who favored wholly privatizing the forests and eliminating communal access altogether.

One of the earliest to speak out in favor of administrative reorganization was Dominique Antoine Telles d'Acosta, a former *grand maître* of Champagne. In his *Project of Reform and Plan for a New Administration of the King's Woods and Forests*, published in pamphlet form in December 1789, Telles d'Acosta advocated the creation of an autonomous forest bureau, separate from the vested interests of community leaders and law enforcers and staffed by guards and officers rigorously trained in the art of silviculture.[19] Sympathetic to the *grands maîtres*' loss of earnings and status, Telles d'Acosta revived an idea that had previously been proposed by the physiocrats: renting communal forests out to private investors and using the profits to buy Eaux et Forêts officers out of their titles.[20]

[17] 22 December 1789, *Décret portant constitution des assemblées primaries et des assemblées administratives*, art. 6, section III. In Georges Bourgin, ed., *La révolution, l'agriculture, la forêt: Lois et règlements* (Paris: Éditions du Comité des Travaux historiques et scientifiques, 1989), 194.

[18] G. Buttoud, "Les propriétaires forestiers privés et l'État," in "Société et forêts," special issue, *Revue forestière française* (1980): 11.

[19] Dominique Antoine Telles d'Acosta, *Projet de réforme et plan d'une nouvelle administration des forêts et bois du Roi* (1789).

[20] See discussion in Devèze, "Les forêts françaises à la veille de la Révolution," 207–8.

The General Assembly of *grands maîtres* soon followed with a statement in which they sought to counter proposals to transform their offices or even abolish the Eaux et Forêts outright. "Far from being harmful, even burdensome to the State, as has been imputed," the statement's eight signatories averred, "the *Grands-Maîtres* have, at all times, denounced abuses, protesting above all the [woodland] concessions that have in such a disastrous way turned to the disadvantage of public property."[21] Turning the tables on their critics, they argued that schemes to suppress the *maîtrises* were "for the most part the work of people who desire to be freed of a supervision [that is] irksome for them" and who had seized "this moment of general effervescence to try to mislead the masses."[22] "If there is abuse in the administration," insisted the *grands maîtres* plaintively, "it will be reformed; if there are reprehensible Officers, they will be punished, but above all the execution of the Ordinance must be assured and the powers of the Officers in charge of carrying it out must be preserved."[23]

Such promises fell on deaf ears. As the *cahiers*' voluminous complaints against the forest administration demonstrated, dislike of the Eaux et Forêts was pervasive. *Grands maîtres* could insist that they were motivated by altruistic concern, but the venality of their offices alone discredited them. After a century and a half of unwelcome and sometimes unjustifiable interventions in the nation's forests, the Eaux et Forêts had far more skeptics than supporters. As Marc-François Bonguiot, deputy from the Jura, observed, "although the Ordinance of Louis XIV was ostensibly created to see to the forest's conservation in a special way, it actually had the creation of venal offices as its goal."[24]

In calling for the *maîtrises*' suppression and the expansion of communal oversight, Bonguiot reflected the long-standing Comtois desire for autonomy.[25] In this respect, his views fell into the second category of

[21] *Observations sur le régime des forêts, et sur la nécessité de conserver la forme actuelle de leur administration. Arrêt à l'assemblée générale des grands-maîtres des Eaux et Forêts de France, le 29 janvier 1790* (Paris: Imprimerie L. Jorry, 1790), 16. Also Waquet, *Les grands maîtres des Eaux et Forêts*, 317; and Buttoud, "Les projets forestiers," 14.

[22] *Observations sur le régime des forêts*, 17, 34–35. Previous attempts to suppress the *maîtrises* took place in 1576 and 1614.

[23] *Observations sur le régime des forêts*, 34–35.

[24] Marc-François Bonguiot [Bonguyod], *Opinion de Marc-François Bonguiot (du Jura) sur l'administration forestière* (Paris: Imprimerie nationale, 1795 [year III]), 4.

[25] Bonguiot [Bonguyod], *Opinion de Marc-François Bonguiot (du Jura)*, 19–36, titles I, II, VII, VIII, IX.

woodland reformers, the advocates of administrative decentralization. Bonguiot was joined in his views by François Louis Jérôme Baron, a deputy from the Marne and member of the *comité des Domaines* (the committee on state properties), who believed that the supervisory authority of the district and departmental assemblies should take the place of the *grands maîtres* while at the same time retaining the administrative structure of the Eaux et Forêts. The local assemblies, Baron argued, should "protect with all their power the agents of the Eaux et Forêts."[26] Others took a more extreme view. "The Nation cannot administer the national forests itself at all; that is a fact," asserted one anonymous member of the *Domaines* committee in a pamphlet of 1790. Rather than creating a new administration that would only replicate the failures of the old, the author proposed contracting the work out to lessees and stewards in a system akin to tax farming.[27]

Yet a third group of commentators favored privatization, arguing that the state could save on the high costs of management if forests were in private hands and that personal interest encouraged better stewardship.[28] Louis Simon Martineau, a deputy from Paris, insisted that "all woods under 500 arpents [640 acres] in size should be liquidated" because "woods produce more in the hands of individuals than public regimes."[29] This was not the first time such a proposal had been made. As early as 1768, the *contrôleur général* Étienne Maynon d'Invault had suggested dividing up and selling the royal forests, on the basis of the physiocratic claim that individual proprietors made better conservationists than the state.[30] Though the idea had never gained much traction previously, legislators now began giving it a much more serious look because of the pressing problem of the national debt.

[26] François Louis Jérôme Baron, *Projet de l'organisation de l'administration des Eaux et Forêts, proposé au Comité des Domaines de l'Assemblée Nationale, par M. Baron, Membre de ce Comité, & Député du département de la Marne* (Paris: Imprimerie nationale, 1790), 3.

[27] *Plan d'administration des forêts nationales et de décret par un membre du comité des domaines de l'Assemblée nationale*, 7 October 1790, in *AP 1ère série*, 19:496–501. See discussion of other viewpoints in Buttoud, "Les projets forestiers," 14–15.

[28] See the discussion and dismissal of the cost-saving argument in Étienne Cunin, *Observations sur le projet d'aliénation des forêts nationales; par Étienne Cunin, Membre du Comité des domaines, député du département de la Meurthe* (Paris: Imprimerie nationale, n.d. [1790?]), 4.

[29] Louis Simon Martineau, 10 May 1790, in *AP 1ère série* 15:454.

[30] Étienne Maynon d'Invault, *contrôleur général*, 1768, cited in *Les Eaux et Forêts*, 171.

"AT THE DISPOSAL OF THE NATION"

Already enormous when the crown first summoned the Estates General in the winter of 1788, France's debt grew to unprecedented proportions within the first few months of the Revolution. Tax evasion, exacerbated by the National Assembly's decision to have the peasantry compensate seigneurs for the privileges that had been abolished on 4 August 1789, contributed significantly to the increase.[31] Casting about for a source of revenue sizeable enough to offset the mushrooming deficit and desperate for even short-term profits, some legislators looked covetously to the church's expansive holdings.[32] Though the proposal was hugely contentious – debate continued for a month and ranged from questions of ecclesiastical duty to the logistics of clerical compensation – in the end, the National Assembly voted by a margin of 568 to 346 to place "all ecclesiastical properties at the disposal of the nation."[33] The meaning of "disposal" became clearer six weeks later, when legislators decreed that clerical lands, together with crown lands, would be sold to pay off the nation's debts.[34]

For advocates of proprietary individualism, the ruling represented a major victory. Here, at last, was the opportunity to put into practice the theory they had long espoused – that creating a nation of landowners would encourage investment in agricultural improvements and result in increased prosperity for all of France.[35] The country's financial desperation won over the remainder of the legislators.[36] When it came to selling

[31] Doyle, *Oxford History of the French Revolution*, 117; McPhee, *French Revolution*, 70–71.

[32] Charles Maurice de Talleyrand, bishop of Autun, put the initial proposal before the National Assembly in a speech on 10 October 1789. *AP 1ère série* 9:398–404. For more on Talleyrand (1754–1838), see Louis S. Greenbaum, *Talleyrand, Statesman-Priest: The Agent-General of the Clergy and the Church of France at the End of the Old Regime* (Washington, DC: Catholic University of America Press, 1970).

[33] For the debates over the motion, see *Discussion sur les biens ecclésiastiques*, 13 October 1789 forward, in *AP 1ère série* 9:415–40.

[34] The National Assembly decreed on 19 December 1789 that the lands of the crown and clergy would be sold to pay off the debt and help underwrite new projects. Details followed in the *Décret relatif à une instruction pour l'aliénation des biens nationaux*, 1 June 1790, in *AP 1ère série* 16:26.

[35] Clerical properties were put up for sale from 9 July 1790 forward and reached their peak in 1791. See discussion in Annie Gay, "Les acquéreurs de biens nationaux à Dole," *Société d'émulation du Jura: Travaux* (1990): 229–31.

[36] Forty legislators abstained from the final vote. *Adoption du décret sur les biens ecclésiastiques*, 2 November 1789, *AP 1ère série* 9:648–49.

the forested areas of these holdings, however, consensus was harder to achieve. The struggle that ensued in the National Assembly over whether to liquidate all or part of the country's woodlands or keep them in national control would draw lawmakers on both sides of the issue into protracted deliberation of the value of forest resources and what role, if any, the state should play in their care.[37]

The question was one of potentially enormous significance as well as complexity. Former ecclesiastical woodland holdings constituted more than 800,000 hectares, and royal forests covered another 500,000 hectares.[38] Selling them meant disentangling customary rights, reconfiguring leases and concessions, and assigning monetary values to parcels. Recognizing the challenges that such a project would involve, the National Assembly decided on 21 December 1789 to temporarily exempt woods and forests from the sales of *biens nationaux* that they had decreed three days earlier.[39] It further instructed the committees of clerical affairs and *Domaines* to determine what of this enormous acreage could and should be sold.[40]

By spring, members of the *Domaines* committee were making their views plain. Expanding on Colbert's focus on naval timber, they emphasized forests' broader importance as a critical national resource – a resource that, in turn, required national oversight. The expropriation of ecclesiastical properties and their aggregation with the state's woodland holdings represented an enormous opportunity, they argued. Far from being carved up and auctioned, these resources should be kept under national control, closely regulated, and ameliorated for the betterment of all France. As Barère de Vieuzac, the spokesman for the committee and deputy from the Hautes-Pyrénées, stressed in his April 1790 report to the

[37] On the debates over the alienation of ecclesiastical property, the auction of state forests, and the conflict between advocates of public utility and supporters of private ownership, see Frédéric Ogé, "Appropriation communautaire et/ou appropriation étatique de la forêt sous la Révolution," in *Propriété et révolution: Actes du colloque de Toulouse, 12–14 octobre 1989*, ed. Geneviève Koubi (Paris: Éditions du Centre national de la recherche scientifique, 1990).

[38] The figure for the crown (domainial) woods and forests is from Barère de Vieuzac, 10 April 1790, *Rapport du comité des domains sur la vente et l'aliénation des domains de la couronne*, in *AP 1ère série* 12:637–38. See also Devèze, "Les forêts françaises à la veille de la Révolution," 181; and C.A.R.A.N., *Histoire de forêts*, 80.

[39] Decree of 21 December 1789, art. 2, as cited by Barère de Vieuzac, 10 April 1790, in *AP 1ère série* 12:638.

[40] The report of the Comité pour l'aliénation des biens domainiaux et ecclésiastiques was presented to the National Assembly on 9 May 1790, by Delley d'Agier, the committee spokesperson. *AP 1ère série* 15:446.

National Assembly, woods constituted a "precious portion" of the former crown lands, both in terms of their profit-making potential and their "great utility" for the navy, construction, industry, and urban fuel.[41] In light of their unique value, Barère told the Assembly, forests should be the focus of a "separate administration" and "not be included in the plans for division that you have wisely decreed for the other [former crown] lands."[42] Barère's remarks were echoed by the abbé Gouttes, who noted that personal interest would be likely to tempt purchasers to "take the best part of the woods for himself," rather than waiting until the trees were big enough for the needs of the navy, thus depriving the nation of its first line of defense.[43]

For the most part, these commentators underscored the critical importance of fuelwood and timber to the nation's economic and strategic health. A few, however, looked beyond the commodity value of trees to highlight forests' broader socioeconomic value. Étienne Cunin, a deputy from the Meurthe and member of the *Domaines* committee, warned that woodland auctions would have a grave impact on the provinces of Lorraine, Trois-Évêchés, and Franche-Comté because they would lead to the consolidation of forests in the hands of a few speculators, thereby cutting communities off from their use rights, depriving rural flocks of forest pasturage, driving up the price of fuel, harming mining and manufacturing, and ultimately causing the general depopulation of the region.[44]

Voicing a compromise between keeping and selling the forests was the deputy Pierre-Victor Malouet, a lifelong administrator in colonial and naval affairs, who suggested auctioning only small-to-medium-sized woods and retaining those "exceeding 50 arpents" (64 acres), "so as to conserve these woods for the navy."[45] Sales of larger parcels could be considered on a case-by-case basis by local departmental assemblies.[46] Cunin, however, doubted whether such sales would even make a difference in the national debt. Flooding the market with land would "diminish competition and lower the overall price," he noted, adding, "what guarantee could the [property] transfers' partisans give that this enormous mass of land will find buyers?"[47]

41 Barère de Vieuzac, 10 April 1790, in *AP 1ère série* 12:638.
42 Ibid.
43 Debate 10 May 1790, in *AP 1ère série* 15:454.
44 Cunin, *Observations sur le projet d'aliénation*, 21, 18–20.
45 Debate 10 May 1790, in *AP 1ère série* 15:454.
46 Debate 10 May 1790, in *AP 1ère série* 15:454.
47 Cunin, *Observations sur le projet d'aliénation*, 17.

Barère de Vieuzac, for his part, focused his energies on dissuading legislators from permitting large-scale auctions of the *bois nationaux*, lest France lose the naval and industrial advantages it currently enjoyed. "To sell the forests, to entrust them to the property of individuals, is to give them up to a destructive use; it is to expose the realm to seeing the dearth and costliness of wood increase considerably in a short time; it is to risk losing in twenty-five years one of the resources most necessary to France," he fulminated.[48] Addressing the National Assembly in August 1790 on behalf of the *réunis* – a subcommittee comprising members of the *Domaines*, Navy, Finances, *Aliénation des biens nationaux*, Agriculture and Commerce committees – Barère exhorted his fellow deputies to "take a wise and enlightened role for the conservation and regeneration of the woods throughout the realm."[49]

France's national forests were at a critical juncture; yet in the end, the woodlands' fate remained contingent and uncertain. Torn between the urgency of preserving France's forest assets and the desire to liquidate them in the interest of private property (and fiscal survival), the National Assembly ultimately chose to waffle. On the heels of Barère de Vieuzac's speech, legislators issued an initial decree protecting large forests from auction. Little more than two weeks later, however, they announced the sale of woodland parcels smaller than fifty hectares.[50] Founded on the idea of the public good, these arguments set the stage two years later, when the nationalization of émigré properties led to a fresh round of debates over whether to retain the wooded parcels or sell them alongside émigrés' arable.

Such contradictory maneuvers – on the one hand acknowledging the need to conserve woodlands, and on the other hand permitting fragmentary sales – were typical of the equivocal path revolutionary lawmakers would take in their protracted pursuit of forest reform. Nonetheless, Barère de Vieuzac and his fellow committee members did convince the National Assembly of the merits of state oversight. On 6 August 1790, following the presentation of Barère de Vieuzac's report, the Assembly voted overwhelmingly in favor of the principle of an autonomous forest administration.[51] Advocates of more extreme woodland liberalization were thwarted, at least temporarily.

[48] Barère de Vieuzac, *Rapport des comités réunis*, 7. [49] Ibid., 4.

[50] *Décret pour excepter de la vente des biens nationaux les grandes masses … des forêts nationales*, 6 August 1790, in Bourgin, *La Révolution, l'agriculture, la forêt*, 194; also decree of 23 August 1790 permitting sales of *bois nationaux* of less than 50 hectares two kilometers distant from large forests. Buttoud, "Les projets forestiers," 12.

[51] See 6 August 1790 discussion in *AP 1ère série* 17:630–36; also *AP 1ère série*, 6 September 1790, 18:623.

"SUCH DESIRABLE BENEFITS"

From this point forward, legislators turned their attentions back to forest reform. Many believed there was no time to lose. For close to a year, the forests had had little coordinated oversight. Months of administrative ambiguity and regional authorities' expectation that the Eaux et Forêts would soon be transformed had given rise to confusion, abuse, and neglect in some areas and dubious assertions of power in others. In the Somme, for example, distressed landowners noted that inhabitants were clearing forests, having "interpreted the decrees in their favor." Their actions, the correspondents warned, would "only do harm" because the land was "very steep and arid" and "not capable of producing anything."[52] Voicing similar complaints, the *procureur du roi* at the *maîtrise* of the Eaux et Forêts in Besançon fretted over the "abuses that many municipalities have authorized in misconstruing the laws."[53]

Not all woodland misconduct skewed toward destruction. In a series of apparently widespread incidents in 1789–90, rural inhabitants deliberately impeded scheduled timber harvests in nearby clerical forests, even though the fellings had already been approved prior to the nationalization of ecclesiastical properties. Timber harvests generated opposition for many reasons: they disrupted grazing rights for years at a time, reduced the availability of the dead and fallen branches that were used for fuel, and increased competition within the community for the woodland resources that remained. For lawmakers, however, the land and timber revenues that the sales were expected to generate were far more important. Chiding local authorities in March 1790 for "giving certain decrees of the National Assembly an interpretation contrary to their true meaning," legislators declared that timber harvests could "neither be stopped nor disturbed by any group nor individual, under the pretext of the decree ... of 11 December 1789."[54]

52 Memo from growers in the village of Grouches (Grouches-Luchuel), department of the Somme, discussed by the Committee of Agriculture on 3 December 1790, in Fernand Gerbaux and Charles Schmidt, eds., *Procès-verbaux des Comités d'agriculture et de commerce de la Constituante, de la Législative et de la Convention*, Collection de documents inédits sur l'histoire économique de la Révolution Française, Ministère de l'Instruction Publique (Paris: Imprimerie nationale, 1906), 1:686; also AN F^{10} 332.

53 Among the abuses, he cited pasturage where "shoots and saplings ... were not yet able to withstand it." Letter from M. Bevalet, *procureur du roi*, from the *maîtrise* of Besançon, discussed by the Committee of Agriculture on 30 August 1790 in ibid., 1:488.

54 *Décret concernant les oppositions mises par les municipalités à la coupe de bois dépendant d'établissemens ecclésiastiques*, 11 March 1790, in Duvergier, *Collection complète des lois*, 1:114.

The frequency with which such decrees were issued attests to their ineffectiveness. Edicts reminding officials of their duties and clarifying the chain of command went unheeded or simply added to the sense of uncertainty surrounding the forest's oversight. For example, an announcement of August 1790 stated that "the jurisdiction of the Eaux et Forêts still exists: it has only lost its influence over hunting violations." Spelling out the matter further, the edict observed, "The municipalities have no role other than supervising the implementation of decrees and informing on infractions. The administrative assemblies must prevent the municipalities from exaggerating this role."[55]

In addition to municipal rule-bending, more spontaneous and destructive forms of woodland resistance continued apace, particularly illegal wood gathering and clearing. Observers usually cited economic desperation as the primary motivator. "The situation is critical," noted Jacques François Laurent de Vismes, the correspondent for the *comité des Domaines*, observing that "amid the needs which are born of difficult occasions, and which are growing during a harsh season, offenses are multiplying in the woods, and all the vigilance of the public authority can scarcely protect the forest from total devastation."[56]

Delisle de Moncel, author of a 1791 memorandum on reforestation and conservation, blamed opportunism and misguided ideology. "Individuals from the indigent sector of the populace" were responsible for the revolts, he insisted, arguing that they had been "seduced by the word liberty."[57] Nonetheless, plenty of indications suggested that wrongdoing was occurring at all levels of the community, among them landowners and industrialists relishing the decline of the Eaux et Forêts' oversight. As the Conseil général of the Haute-Saône noted in a December 1790 report on the causes of forest deterioration in the region, the "current freedom of every individual to construct private forges" was contributing greatly to the scarcity of wood, as was the "egoism with which exploitations are conducted in the woods of individuals, who, today, think less of the means of perpetuating their revenues than of augmenting them for the

[55] *Teneur de l'instruction de l'Assemblée nationale sur les fonctions des assemblées administratives*, 12 August 1790, chap. 4, "Domaines et bois," excerpted in Bourgin, *La Révolution, l'agriculture, la forêt*, 194.

[56] Jacques François Laurent de Vismes [Devisme], *Rapport du comité des Domaines sur les moyens de pourvoir provisoirement à la conservation des bois*, 19 December 1790, in *AP1ère série* 18:564.

[57] Delisle de Moncel, *Mémoire sur le repeuplement*, 5.

moment with too-frequent fellings."[58] Other commentators echoed these sentiments, decrying in missives to the Committee of Agriculture the "property transfers suffered in the privatizations of former crown land."[59]

As members of the rural community grew more confident in expressing and pursuing their self-interest, they also became more confrontational. Fearing for their safety, forest guards and officers began abandoning their posts in droves. As a former master wolf hunter, Delisle de Moncel summoned the image of feral predators in the forests' midst in claiming that "hordes of brigands, too often weapons in hand, have infested [the forests]: guards, principally those of the national woods, have been threatened, chased off, and even shot."[60] Baron, the deputy from the Marne, similarly observed that guards who stayed on were "subjected to insults, mistreatment," and even murder.[61] Yet in the greater number of cases, the problem may have been rather more prosaic: the paltriness or even nonpayment of their salaries gave forest personnel little incentive to stay on when they were not welcome.[62]

As unpopular as the 1669 Ordinance and the Eaux et Forêts were, it was increasingly clear that from the standpoint of resource protection, having neither a coherent woodland policy nor agents to enforce it was even more problematic. Legislators agreed on this point, but determining a course of action remained difficult.[63] Members of the *comités réunis*, which had been designated by the National Assembly in June 1790 to take a comprehensive approach to the issue, could scarcely agree themselves on the best approach to woodland reform, and the clamor from concerned citizens, petitioners, and other interested parties outside the

58 AN F^{1cIII} Haute-Saône 5 (dossier 1), *Extrait des registres des déliberations du Conseil général du département de la Haute-Saône, Vesoul*, 8 December 1790, "Rapport du Comité d'agriculture et de commerce concernant les Bois et forêts," 77–84.

59 See memo by M. Decaux, a former land steward, discussed by the Committee of Agriculture, 15 November 1790, in Gerbaux and Schmidt, *Procès-verbaux des Comités d'agriculture*, 1:643.

60 Delisle de Moncel, *Mémoire sur le repeuplement*, 5. On his wolf-hunting exploits, see Delisle de Moncel, *Méthodes et projets pour parvenir à la destruction des loups dans le royaume* (Paris: Imprimerie royale, 1768).

61 Baron, *Projet de l'organisation de l'administration des Eaux et Forêts*, 3.

62 On the conditions of forest guards, see Iain Cameron, "The Policing of Forests in Eighteenth-Century France," in *Police and Policing: Past and Present Society Colloquium* (Oxford: Oxford University Press, 1983), 1–27; Graham, "Policing the Forests of Pre-industrial France," 157–82; and Blais, "Contribution à une histoire des gardes forestiers," 17–26.

63 de Vismes, *Rapport du comité des Domaines*, in *AP 1ère série* 18:564.

committee had only grown louder. Some of the correspondence focused on the need for technical improvements, like a 1791 memorandum by Philibert Charles Varenne de Fenille, a pioneer of French forestry and member of the Royal Society of Agriculture, in which he urged legislators to rectify the "unfortunate consequences of the scientific errors into which the authors of 1669 Ordinance had fallen" and to implement "the most profitable route, be it for the Public, be it for Proprietors, to follow in managing coppices and timber."[64] Delisle de Moncel similarly pushed for a "forest regulation adapted to local conditions." "What degree of esteem can the famous Ordinance of 1669 enjoy, its many inconveniences aside," he asked, "if it makes no distinction between [the] climates of the north and south?"[65] Still others emphasized the need to fund reforestation projects, increase guards' wages, or even use the new forest administration as a job creation program to train "idle men" to improve the "measurement and cultivation" of the forests.[66]

What is perhaps most striking about these recommendations, beyond their sheer number, is how few of them defended the interests of customary usage. Cunin, the deputy from the Meurthe, noted the importance of forest pasturage and firewood rights to the regional economy, but far more commentators held up forests' deterioration as proof of the carelessness with which inhabitants treated resources they used or possessed in common.[67] To these critics, customary claims would have to be restricted, eliminated, or banished to the far fringes of the woods for genuine reform to be achieved.[68] Delisle de Moncel, for example, proposed banning dairy and cattle herds from young coppices for five to six years until the shoots were no longer succulent enough to tempt the beasts' taste buds. Acknowledging that such a move would be unpopular with rural inhabitants "since the prohibition would be real and the future benefit uncertain," Delisle recommended compensating those who complained.[69]

64 Philibert Charles Varenne de Fenille, *Observations sur l'aménagement des forêts, et particulièrement des forêts nationales. Présentées à l'Assemblée Nationale, par la Société Royale d'Agriculture, le 9 juin 1791* (Imprimerie de la Feuille du Cultivateur, 1791), 4, 7.

65 Delisle de Moncel, *Mémoire sur le repeuplement*, 11.

66 AN F^{1cIII} Haute-Saône 5; M. Decaux, memo, in Gerbaux and Schmidt, *Procès-verbaux des Comités d'agriculture*, 1:643.

67 AN F^{10} 403, *Mémoire sur l'abus des usages dans les bois*, May 1789; also 30 August 1790 letter from M. Bevalet, *procureur du roi*, from the *maîtrise* of Besançon, in Gerbaux and Schmidt, *Procès-verbaux des Comités d'agriculture*, 1:488.

68 *Plan d'administration des forêts nationales et de décret*, 1790, 16–17.

69 Delisle de Moncel, *Mémoire sur le repeuplement*, 13.

One of the few to speak up in favor of customary and communal usage was Pierre-Joseph Balland, a forest administrator from Dole and member of the *Cercle Social*, a radical political club. Addressing *cultivateurs* – comparatively affluent landowning peasants – in a 1791 pamphlet, Balland argued that laws suppressing use rights "hastened the forests' ruin sooner than their conservation, because wood is an absolutely necessity, more than bread."[70] If deprived of wood, he warned, citing the case of the 1765 demoiselles' revolt in the Forêt de Chaux, the peasantry would be "forced to steal it."[71] In that instance, Balland reminded his readers, the crown had denied the villages bordering the Chaux the wood to which they were entitled, leading inhabitants to obtain it through force and reducing the forest to a "most disastrous state." As soon as the *grand maître* delivered "at a modest price" the timber and fuelwood they were owed, however, "order was reestablished, farming increased, and the forest prospered."[72] "Our legislators know that a good people exists, worthy of careful attention," Balland declared, concluding that the government could "bring happiness" to the realm by "meeting the need for wood everywhere."[73]

As it turned out, the decree that the National Assembly finally ratified on 29 September 1791, announcing the formation of a new forest administration, would bring little happiness or improvement.[74] One of the National Assembly's last acts before the October 1791 installation of the new Legislative Assembly, the law was initially hailed with relief. The departmental assembly of the Haute-Saône, for example, welcomed "such desirable benefits of [the new] administration" and promised to assist future forest commissioners in "stopping as much as possible the devastations that until now have been only too harmful to the nation and individuals."[75]

Moderate in nearly every respect, the decree rebuffed more extreme reform suggestions, like selling off the forests entirely or entrusting their care to private contractors, in favor of establishing a new administrative

70 Balland, *Observations sur l'administration des forêts*, 20.

71 Such thefts, usually committed under cover of night, were "infinitely disastrous and ruined the forests" because of the clumsiness the darkness entailed. Ibid., 22.

72 Ibid., 23. 73 Ibid., 24.

74 *Décret sur l'administration forestière*, 15–29 September 1791, in Duvergier, *Collection complète des lois*, 3:271–86; and *Décret concernant le nombre, la repartition et le traitement des agents de la Conservation générale*, 15 September 1791, in Bourgin, *La Révolution, l'agriculture, la forêt*, 217–20.

75 AN F^{1cIII} Haute-Saône 5, November 1791.

superstructure, the Conservation Générale des Forêts. In place of the former *maîtrises*, the Conservation Générale would be divided into districts and subdistricts, operated by trained professionals in the manner that Telles d'Acosta and others had recommended.[76]

Yet despite the care that went into these formulations – extremely specific in its details, the plan spelled out particulars like the number of holidays that guards and inspectors could take and the time frame in which they had to file reports – the administrative reorganization nonetheless fell far short of expectations. For one thing, the September 1791 decree put off discussing the technical aspects of management like marking, felling, allocating, and reforesting, promising instead that a "law on management" would be devised "without delay."[77] For another, the appointment of new commissioners, *conservateurs*, and inspectors, stipulated by the law, was suspended after less than six months in March 1792. In their place, the Legislative Assembly temporarily restored former officers of the Eaux et Forêts.[78] Discussion of the promised "law on management" also ground to a halt.

Only one element of the September 1791 decree took hold with any real force, and it had both grave and enduring consequences for the nation's forests. Breaking with the 1669 Ordinance's stance on private woodlands, policymakers ruled that henceforth, the state's regulatory authority would be limited to national and communal forests. "Woods belonging to individuals will cease to be subject" to the forest administration, the decree announced, "and each proprietor will be free to administer and dispose of them in the future as he sees fit."[79] As in the case of the auction of the *biens nationaux*, the move reflected legislators' effort to compromise between advocates of privatization and boosters of centralized state oversight. In so doing, they appeased landowners and avoided addressing the questions that had arisen in that earlier debate: how the state can justify regulating natural resources in a society based on individual liberty, and whose interests those regulations should serve.

[76] Telles d'Acosta, *Projet de réforme*; and Telles d'Acosta, *Projet d'une nouvelle administration pour les forêts*.

[77] *Décret sur l'administration forestière*, title XV, art. 4, in Duvergier, *Collection complète des lois*, 3:284. Until then, the regulations of the 1669 Ordinance would remain in force.

[78] The law of 11 March 1792 suspended the personnel already in place. Buttoud, "Les projets forestiers," 16.

[79] *Décret sur l'administration forestière*, title I, art. 6, in Duvergier, *Collection complète des lois*, 3:272.

It was a rash move. Suddenly free to exploit their forests in whatever manner they pleased, landowners began felling, clearing, and fragmenting their holdings in a bid to reap as much profit as they could, as quickly as possible. This result was particularly the case with the former ecclesiastical properties put up for sale by the state from August 1790 forward and the émigré holdings that would be auctioned off in similar fashion two years later.

"WHAT MAKES THE POOR INTO SLAVES"

If private landowners obtained the woodland liberty they had hoped for from the September 1791 decree on the forest administration, *usagère* communities had the opposite experience. The reforms granted municipal authorities greater oversight than they had had under the *maîtrises*, but they did not necessarily bring about greater equity in the way that woodland resources were distributed. Rather, the weakening of state as well as seigneurial authority allowed previously veiled disparities within communities to emerge and intensify. Likewise, legislators' delay in developing a new law on forest management, together with their decision to maintain, at least for now, the dictates of the 1669 Ordinance, perpetuated controls that rural inhabitants had railed against in their *cahiers*. On top of these disappointments, the auction of national, ecclesiastical, and, later, émigré woodlands brought increased hardship to many rural communities as the new forest proprietors – notably bourgeois entrepreneurs, merchants, and industrialists – sought to unfetter their acquisitions of customary and collective claims.[80]

Prior to the Revolution, the crown's need for naval timber and, to a lesser extent, industrial fuel had dictated woodland policy. The crown had now been toppled, but in a clever hat trick, legislators kept the same approach in place, emphasizing timber and fuelwood production on the grounds that it was vital to the "Nation" and the nebulously defined *bien public*. The only drastic deviation from the Old Regime – the exemption of private landowners from state oversight – was likewise justified by the assertion that individual, unrestricted proprietorship enhanced the public good through investment and improvement. Throughout these contortions – for contortion is indeed what legislators went through to maintain the illusion that individuation and exclusion conserved forests

[80] On the purchasers of *biens nationaux*, see Forrest, *Paris, the Provinces and the French Revolution*, 133–34.

and served the public interest better than various collective forms of resource governance – the entreaties of rural inhabitants concerning their customary rights were met with indifference or with policies that were neither practical nor beneficial.

This was the case with rural appeals concerning *affouage*, the right to firewood and timber in state, communal, and seigneurial woods. The question was of particular interest in Franche-Comté, where woodland resources were comparatively abundant but their distribution discriminatory. Legislation to produce a more equitable form of *affouage* had already been enacted sixty years prior to the Revolution, but it was largely disregarded by municipal elites, who benefited from the status quo.[81] For similar reasons, the *cahiers de doléances* were largely silent on the subject. Nonetheless, by the end of the century, the magnitude and pervasiveness of woodland pilfering could not be overlooked, and its root causes – exclusion, frustration, and acute necessity among those barred access to *affouage* – became the focus of clamorous demands for reform.[82] Calls for an overhaul of the system, particularly the *le marc la livre* custom of apportioning wood relative to the amount recipients paid in taxes, came from administrators and the rural poor alike. "This custom, overly favorable to rich landowners," observed a report of the Haute-Saône Conseil général, "has legitimated the conduct of the poor who are forced to scour and pluck the communal woods to obtain for themselves this object of basic necessity, which they have been deprived of by [*le marc la livre*] distribution." "One knows only too well the pernicious effects of this daily damage," the report's authors added.[83]

The National Assembly responded first by digging in its heels and next by muddying the waters. Despite the panoply of other agricultural and woodland issues under review, legislators ruled on 31 May 1790 to preserve existing *affouage* customs and make no alterations.[84]

81 Dalloz et al., *Répertoire méthodique*, 25, 558, §1809; Bertrand, "Les hommes et la forêt dans les communautés villageoises," 192.

82 See discussion in Jean-Louis Clade, *Mutations, permanences, ruptures dans le canton de Rougemont, Doubs: 1789–1815*, Collection du bicentenaire de la Révolution française 4 (Paris: Les Belles Lettres, 1986), 4, 140–41; and T. Jolas, "Bois communaux à Minot (Côte-d'Or)," in "Société et forêts," special issue, *Revue forestière française* (1980): 219.

83 AN F^{1cIII} Haute-Saône, *Extrait des registres des déliberations du Conseil général du département de la Haute-Saône*, "Rapport du Comité d'agriculture et de commerce concernant les Bois et forêts," Vesoul, 8 December 1790.

84 *Loi, sur le décret de l'Assemblée nationale, du 21 du même mois, concernant la distribution des bois communaux en usance*, 31 May 1790, in Baudrillart, *Recueil chronologique*, tome 1, 493.

Referencing the apparent "misinterpretations" that had taken place in response to its decrees on taxation the previous autumn, the Assembly emphasized that it "did not intend to bring about any change in the manner in which communal woods under exploitation should be distributed among its beneficiaries." Rather, it insisted that the *marc la livre* system of apportionment be upheld, and that absentee landowners and their tenant farmers should still be accorded shares of wood and timber, even if they were no longer paying taxes "as they had previously" – taxes that, according to custom, were the basis for their inclusion.[85]

The 1790 ruling only fanned the flames of discontent. With so much at stake and the injustice of the situation so stark, rural inhabitants pressed lawmakers harder than ever to uphold the Revolution's stance on the rights of the individual and replace *le marc la livre* with *partage par feu* (allotments per household) or even *partage par tête* (per person).[86] By 1794, with the radicalization of the new regime intensifying, lawmakers did just that, ruling on 26 Nivôse, year II (15 January 1794), that wood issuing from communal lands should henceforth be allocated equally per person.[87] This decree served as an extension of the National Convention's earlier law of 10 June 1793 authorizing the division of the commons, yet like *partage*, discussed below, it ultimately failed to receive wide implementation.[88] As with so many revolutionary reforms, the decree overreached in its attempts to impose uniformity. Among other flaws, it did not differentiate between firewood and timber – meaning that landless individuals who owned no buildings would receive construction wood and be required to pay the fees associated with it, when what they really wanted was more heating fuel. Landowning elites, likewise, stood to lose their lucrative surplus and be forced to buy timber if their equal allocation fell short. Lawmakers hoped to offer a solution to the *affouage* imbalance, but this was not it. In the absence of a workable approach, most

[85] *Loi ... concernant la distribution des bois communaux en usance*, 31 May 1790, in Baudrillart, *Recueil chronologique*, tome 1, 493.

[86] See discussion in AN F^{1cIII} Haute-Saône 5; and AN F^{1cIII} Haute-Saône 6, "État décadaire pendant les mois de Brumaire, Vendémiaire, Messidor, Frimaire An V." See also Vion-Delphin et al., *Les hommes et la forêt en Franche-Comté*, 113–15; and Pierre Goujon, "Legislation et droits d'usage en France dans la premiere moitié du XIXe siècle," in *La Coutume et la lois: Etudes d'un conflit*, ed. Claude Journès et al. (Lyon: Presses universitaires de Lyon, 1986), 15–25.

[87] *Décret de la Convention nationale, rélatif au partage de la coupe des bois communaux*, 26 Nivôse, year II (15 January 1794), in Baudrillart, *Recueil chronologique*, tome 1, 522.

[88] On the decree's uneven interpretation, see Dalloz et al., *Répertoire méthodique*, 25, 546.

communities simply carried on as usual, with little relief for residents who received insufficient portions or were barred from the benefits of *affouage* altogether. Petitioners from Courchaton, in the Haute-Saône, minced no words about their frustration. They asserted in March 1794:

> As for our community, the Rich want to be able to annihilate the poor and deprive them of what rightly comes to them, to the point that there are those who have only a quarter of a cord at most and others who are rich who have from 12 to 100 cords, which leads to the ruin of the woods by the unfortunates, [their] having nothing to heat themselves and bake their bread.[89]

Affouage would continue to be a source of festering resentment in Franche-Comté, as well as a cause of sporadic upheaval, throughout the Revolution.[90] As a summary of departmental affairs in the Doubs in 1800 reported, "Repeated controversies over the mode of distribution, be it coppices and tree branches, be it large timber, have forced the administration to draft a memo to administrators and municipal officers reminding them of the conclusions of the law of 21 May 1790."[91] When revolution again erupted in February 1848, the unresolved struggle over *affouage* would emerge as a rallying point for rural inhabitants across the region.

Revolutionary legislators' unwillingness or inability to put an end to the manifestly inequitable and contentious *le marc la livre* system of wood distribution stands in marked contrast to their determined support for the *partage des biens communaux*, or division of the commons. A look at *partage*'s origins and eventual outcome is instructive, both in comparison with lawmakers' stance on *affouage* and as an example of how land reform served as an arena in which notions of citizenship, national identity, and the public good were tested and contested in an endless double helix of ambition and discontent.[92]

The push to parcel the commons into arable plots dated to at least the mid-eighteenth century, the time when the physiocratic movement was scoring success through legislative incentives to landowners to clear and cultivate their *friches*. Persuaded by physiocrats' claim that only

89 AN F^{10} 329, 6 March 1794.

90 See discussion in Roger Marlin, "Un problème d'affouage en Franche-Comté au milieu du XIX siècle," *Revue d'histoire économique et sociale* 2 (1957): 167–97.

91 AN F^{1cIII} Doubs 6. The document cites 21 May 1790, which was the date of the decree; the law was not issued until 31 May 1790.

92 For a definitive study, see Nadine Vivier, *Propriété collective et identité communale: Les biens communaux en France 1750–1914* (Paris: Publications de la Sorbonne, 1998).

individuated property ownership could bring about the surpluses necessary for stimulating population growth, commerce, and industry, some communities, notably in the Auvergne, carried it out.[93] Nevertheless, proposals to individuate the commons, whether wooded or fallow, were rebuffed in the first several years of the Revolution. Just as Barère de Vieuzac and his fellow members of the *comité des Domaines* fended off suggestions to alienate the national woodlands completely, lawmakers excluded clauses endorsing enclosure, partition, and the abolition of communal pasturing from the final draft of the Rural Code that was promulgated in September 1791.[94]

Less than a year later, however, the Legislative Assembly hastily issued a decree on the heels of the king's removal in August 1792, announcing that "this year, immediately after the harvests, all communal lands and usages other than the woods will be divided among the citizens of each community."[95] Proponents of the measure, among them landowning Girondins in the Committee of Agriculture, expressed the by-now-familiar assertion that communal usage impeded productivity and gave rise to abuse.[96] "That which belongs to all belongs to no one," insisted Jean-Baptiste Aveline, president of the Committee of Agriculture, in a preamble to the first draft of the *partage* decree, "and a collective despoilment is only a continual exercise in devastation and pillage."[97] By giving "men without occupation" a propertied stake in the nation, Aveline continued, the division of the commons would transform a "loose, lazy, obdurate, and sullen" populace into self-sufficient and industrious citizens.[98]

The argument sounded persuasive in theory, but achieving *partage* in practice would turn out to be hampered by social, ecological, and

[93] See discussion in Gerbaux and Schmidt, *Procès-verbaux des Comités d'agriculture*, 3:9; and Weulersse, *La physiocratie*, 4, 92–93.

[94] Barère de Vieuzac, *Rapport des comités réunis*, in *AP 1ère série* 17:630–36. On the Rural Code, see *Décret concernant les biens et usages ruraux et la police rurale*, 28 September 1791, in Bourgin, *La Révolution, l'agriculture, la forêt*, 49–62.

[95] *Décret relatif au partage des biens et usages communaux*, 14 August 1792, in Duvergier, *Collection complète des lois*, 4:306.

[96] On the composition of the committee and its members, see Gerbaux and Schmidt, *Procès-verbaux des Comités d'agriculture*, 1:v–vi.

[97] Jean-Baptiste Aveline, *Rapport et projet de décret sur le partage & le défrichement des biens communaux, faits et présentés, au nom du Comité d'Agriculture, par M. Aveline, Député du Département du Calvados, Président du Comité d'Agriculture* (1792), 8.

[98] Property would be "a link that binds all citizens to the State," Aveline, *Rapport et projet de décret*, 12, 8.

economic complexity. As written, the 14 August 1792 *Décret relatif au partage des biens et usages communaux* gave no details on how land that had been utilized in common for centuries should be carved up and allocated. Instead, it promised that "the Committee of Agriculture will present the design of the decree in three days."[99] This promise, like legislators' earlier pledge to issue new woodland management guidelines "without delay," would prove wildly optimistic.[100] Not only did the Committee of Agriculture have countless other concerns competing for its attention – from peat extraction to pin factories to the proposed "maximum" price on grain and bread – but the Legislative Assembly and, after it, the Convention were deeply preoccupied with the challenges of mounting a war when more than half the nation's army officers had left, with propping up an economy in which luxury goods manufacturing and trading had collapsed, and with handling the discord over the Civil Constitution of the Clergy and the trial and execution of the king.

In the meantime, debate over the *partage* decree continued. Critics argued that once parceled up, terrain that had been used in common for a multitude of purposes would be susceptible to reconsolidation in the hands of the few.[101] "Those who urge the division of the commons have no other goal than to monopolize the whole," warned one petitioner to the Convention in March 1793, "for I assure you that three years after the division two-thirds of the portions will have been sold to the solicitors of *partage*."[102] Robespierre took a stronger stance, avowing, "the equality

[99] *Décret relatif au partage des biens et usages communaux*, section 4.

[100] See discussion of *partage* in vols. 2 and 3 of Gerbaux and Schmidt, *Procès-verbaux des Comités d'agriculture*; also Georges Bourgin, ed., *Le partage des biens communaux: Documents sur la préparation de la loi du 10 juin 1793*, Collection de documents inédits sur l'histoire économique de la Révolution française (Paris: Imprimerie Nationale, 1908).

[101] Aveline dismisses this issue, noting that the possibility that the poor will sell their lots is "to take the exception for the rule." Aveline, in Bourgin, *Le partage des biens communaux*, 347.

[102] Letter to the Convention from "le Citoien Dumas," 21 March 1793, in AN F^{10} 329. The author identifies himself as a "past pupil of the Ponts et Chaussées and former geographer for the émigré d'Artois." On the arguments for and against *partage*, see Daniel Solakian, "Territoires naturels et communautés: La loi du 14 août 1792–10 juin 1793," in *La nature en révolution 1750–1800*, ed. Andrée Corvol (Paris: Éditions L'Harmattan, 1993), 145–55; John Markoff, *The Abolition of Feudalism: Peasants, Lords, and Legislators in the French Revolution* (University Park: Pennsylvania State University Press, 1996), 485–86; and McPhee, *Revolution and Environment in Southern France*, 130–31.

of property is a chimera."[103] Those who would advocate property rights over individuals' right of existence, he declared, were no less than "*assassins du peuple.*"[104]

These admonishments notwithstanding, the August 1792 *partage* announcement was warmly received in some regions. Officials from the commune of Mello, in the Oise, for example, urged the Convention

> to decree the method of division of the commons; rural people, eager to benefit, are excited in a thousand ways on this subject; it is no longer possible to contain them; they transact with the nobles or even threaten the municipal officers, while [the latter] seek to make them understand the necessity of waiting for the law's dictates on the method of *partage*.[105]

Petitioners from the village of Louhans, in the Saône-et-Loire, likewise urged,

> Legislators! Our happiness and that of our children is in your hands. The health and prosperity of the public depend on you. Hasten yourselves to produce the decree on the method and implementation of the division of the commons; you will make happy ten million citizens who, in fertilizing these uncultivated lands, will forever bestow abundance on the state and will heap blessings upon you.[106]

In Franche-Comté, views on *partage* ranged from hostility to approval among village bigwigs and the landless poor alike, depending on the local economy, land customs, and relative suitability of the terrain to division. In the Haute-Saône, municipal officials embraced *partage* with enthusiasm, not least because they stood to benefit personally. This was the case in Amage, located at the foot of a large forest in the department's lake district, where well before the *partage* decree was even announced in August 1792, unhappy villagers reported that their recently established municipal officers were "so exalted by their positions that they have divided on their own accord and following their whims two sections of the communal lands."[107] The new leaders had apparently had their eyes

[103] Maximilien Robespierre, "Discours sur la propriété," 24 April 1793, in Christian Biet, ed., *Les droits de l'homme* (Paris: Imprimerie nationale, 1989), 486–87. Robespierre was a vocal opponent of *partage*. See Florence Gauthier, "De Mably à Robespierre: De la critique de l'économique à la critique du politique, 1775–1793," in *La guerre du blé au XVIIIe siècle*, Florence Gauthier and Guy-Robert Ikni, eds. (Paris: Les Editions de la Passion, 1988), 111–44.

[104] Gauthier, "De Mably à Robespierre," 132. On Jacobins and land reform in general, see Jones, *Peasantry in the French Revolution*, 164–66.

[105] AN F^{10} 330, 11 December 1792, in Bourgin, *Le partage des biens communaux*, 546–47.

[106] AN F^{10} 333, commune of Louhans, Saône-et-Loire, in Bourgin, *Le partage des biens communaux*, 591.

[107] ADHS B 255, March 1790.

on the hamlet's limited lowland commons for some time, and they used their new authority to carry out the partition they had coveted. The resulting allocation, the petitioners protested, was carried out "without justice nor equity."[108]

Similar abuses of power were observed in March 1793 by the attorney general of the Haute-Saône, who remarked that "the Conseil-Général of Montigny-lès-Cherlieux [*sic*] has made himself guilty of the most shocking misconduct. ... he himself decided the division of thirty-two arpents of woods and prompted the other inhabitants to consent to this division."[109] A buyer's agent subsequently bought out residents' portions. Nestled in wooded valley of the Ougeoutte River, Montigny-lès-Cherlieu had abundant woods but little arable, leading the community's Conseil général to parcel up the forest. Such divisions were especially lucrative for their acquirers because they yielded timber profits prior to being put to the plow.

In the upland Doubs community of Indevillers, by contrast, the poorest inhabitants of the community pushed hardest for *partage*, and the village's self-described "middling landowners" furiously resisted. In this high-elevation district close to the Swiss border, winters were long and the growing season short. Animal husbandry and the cooperative cheese dairies known as *fruitières* constituted the primary sources of income. Though the village was endowed with comparatively expansive communal holdings, rights to use those lands were enjoyed by a select few – in this case, landowners whose payment of an annual fee granted them access to the best pasturage. Mistakenly interpreting the 14 August 1792 *partage* announcement as obligatory, Indevillers landowners protested to the Convention in April 1793 that the commons were actually not common at all. "The lands called commons," asserted proprietors from the village, "have always been attached to individual property and formally transmitted in contracts."[110] Such arrangements often dated back centuries, the result of charters and other agreements with noble families.[111]

Arguing against the claims of *partage* advocates, Indevillers' most affluent inhabitants insisted that collective grazing was the most

108 ADHS B 255, March 1790. 109 AN F[1cIII] Haute-Saône 6.

110 AN F[10] 329 Partage des communaux, 1792-an III, *Mémoires et pétitions addressés au Comité d'Agriculture, Doubs.*

111 See discussion in David Hunt, "Peasant Movements and Communal Property during the French Revolution," *Theory and Society* 17 (1988): 273.

cost-effective and productive use of the village's resources, and that dividing them would destroy the local economy: "A single fountain suffices for all; the animals have more terrain in which to go about[,] thereby becoming bigger and more robust; the owners are free from guarding the animals and from [building] fences whose construction and maintenance would be very costly and difficult."[112] Given these conveniences, the fees they paid to secure them, and the effort they had put into clearing and cultivation, they did not intend to give up their commons to the "first person who comes along who has never possessed anything nor acquired anything by his own industry, who has often been the plague of his place of residence because of his laziness and indolence."[113]

In the climate of anticipation that the announcement of *partage* created, resentments once shared by inhabitants against the crown and seigneurs were now shifting to reveal long-simmering antagonisms within the community itself. Enraged at the complacent self-interest of the village's well-off, Indevillers' poor responded with invective of their own, accusing the beneficiaries of the commons of taunting them with their privilege. "What makes the poor into slaves," they protested, "is to hear the rich constantly saying: 'You still do not possess our pastures for cultivation, you will never have them!'" Petitioners supported their demands for democratic *partage* by citing their disproportionate contributions to the war. "Only a few" of Indevillers' rich had enlisted, they observed, whereas great numbers of the parish's "poor citizens" had "abandoned their fathers, mothers, wives, and children in order to support the Republic."[114]

Two planting seasons came and went before the National Convention finally issued procedures for dividing the commons. Making its debut on 10 June 1793, the *Décret concernant le mode de partage des biens communaux* was far bolder than either its detractors or its supporters could have imagined. In a blow to *partage*'s physiocratic advocates, who had envisioned a mode of distribution weighted in favor of large landowners, the June decree dictated that municipally owned commons should be divided equally among inhabitants, regardless of age, status, or – most astonishingly of all – gender.[115] By encouraging individual freeholding on a mass scale, the decree also caused widespread consternation and

[112] AN F[10] 329. [113] AN F[10] 329.

[114] Reprinted in Bourgin, *Le partage des biens communaux*, 444.

[115] On the allocation of lots to men and women, see *Décret concernant le mode de partage des biens communaux*, section II, art. 1, reprinted in Bourgin, *Le partage des biens communaux*, 728–39. Aveline's initial draft argued against allocation per person and proposed instead a multilevel allocation based on household size, status (landowner,

confusion among rural inhabitants, rich and poor, who relied on customary and collective usage to survive. Seeking clarification, the Saint-Julien *Société des amis de la Liberté et de l'Égalité* inquired of the Committee on Agriculture, "The division being made per person, will children count? A bastard, nourished in a community for one year, will he be included?"[116] Most distressing of all for proprietors like those of Indevillers, *partage* proposals required the approval of only one-third of the community's voters to be enacted.[117]

At a time when skepticism about the Revolution was on the rise and the support of the rural populace was urgently needed, legislators of the Convention took a leap of faith in *partage*'s ideological premise – that landownership promoted participatory, responsible citizenship – and opened access to the divided commons as widely as possible. For all its grand intentions, however, the much-anticipated decree failed to create an engaged citizenry, much less improve the citizens' economic well-being. During the three years the law would be in force, only a quarter of eligible communities implemented it, and of these, many did so only under direct pressure from national and departmental authorities.[118]

A number of factors contributed to *partage*'s failure. Obtaining eligibility was one. To petition for a legal division of their commons, communities had to owe no taxes or other debts to the state.[119] This clause alone created an insuperable obstacle for many communities. In addition, only certain types of *biens communaux* could be divided; public thoroughfares, town squares, large woods, marshes, and mineral-rich sites were off-limits. Some wetlands qualified for *partage*, but only after inhabitants carried out drainage and other improvements.[120] As for the communal

sharecropper), and place on the municipal tax rolls. See Aveline's report in Bourgin, *Le partage des biens communaux*, 354ff.

116 AN F[10] 329.

117 The decree instructed municipalities to call an "assembly of inhabitants " to approve *partage*, but the one-third clause allowed a minority to gain sway. *Décret concernant le mode de partage des biens communaux*, section III, nos. 9, 10, 16. See also Vivier, *Propriété collective et identité communale*, 124–28.

118 On sites of *partage* and legislators' efforts to push the decree, see Jones, *Peasantry in the French Revolution*, 47–54, 146–49; also Peter Jones, "Agrarian Radicalism during the French Revolution," in *Reshaping France: Town, Country, and Region during the French Revolution*, ed. Peter Jones and Alan Forrest (Manchester: Manchester University Press, 1991), 137–51; and Hunt, "Peasant Movements and Communal Property," 255–83. On communities that did not respond, see AN F[10] 329.

119 *Décret concernant le mode de partage des biens communaux*, section I, art. 8.

120 *Décret concernant le mode de partage des biens communaux*, section I, art. 8. The decree also allowed easements for riparian and pathway rights.

forests, the law permitted their partition, but in failing to stipulate how yields from fellings that were already underway should be distributed, the decree "[gave] rise to difficulties, lawsuits, and even brawls" in "many departments."[121] The exclusion of all other terrain where inhabitants exercised use rights – including national and private holdings – left many hamlets with little or no acreage to divide.[122]

In Franche-Comté, geography played a decisive role in communities' decisions on *partage*. Communal holdings in the Doubs and the Jura often encompassed steep slopes and wind-blasted plateaus, making them unsuitable to agriculture. Nonetheless, they provided critical grazing, firewood, and foraging, and in this respect they were already being put to the best use possible. As the departmental administration of the Doubs observed in July 1795, "The soil of the department being mountainous, the commons started off as nothing more than nondivided lands, because covered with one or two inches of soil, they could serve neither for crops nor for meadow, but for excellent pasturage."[123] Since livestock constituted the principal wealth of the Doubs, the report continued, *partage*, if enacted, would "necessitate the reduction of animals and the suppression of more than one hundred *fruitières*."[124]

Even in the well-watered river basins of the Haute-Saône, where in some cases municipal authorities had enacted *partage* without inhabitants' consent, the department's Conseil général came out firmly against it. Deliberating the question in November 1793, council members worried that dividing up the commons would lead inevitably to their sale and in turn "strip the community of these resources in which they have the greatest interest of conserving, be it for the annual yields from which they can draw a significant advantage for assuaging their habitual needs, be it for providing a last reserve in extraordinary and unforeseen circumstances."[125]

Recognizing the threat posed by the parcels changing hands, the law stipulated that partitioned lots could not be divided or sold for ten years. Thereafter, however, the lots would be fully alienable, thereby increasing the likelihood that they would change hands and be consolidated in private ownership.[126] As a result, inhabitants balked when the sought-after law became a reality. As the Doubs departmental administration

[121] AN F[10] 329, Louis-Jérome Gohier, Ministre de la Justice, au Citoyen Président de la Convention nationale, 23 August 1793.
[122] *Décret concernant le mode de partage des biens communaux*, section IV, art. 12.
[123] AN F[1cIII] Doubs 6. [124] AN F[1cIII] Doubs 6. [125] AN F[1cIII] Haute-Saône 5.
[126] *Décret concernant le mode de partage des biens communaux*.

observed, the "indigent class has seen in this law only a momentary benefit and its future ruin; there *partage* has not been deliberated, only possession in common."[127]

Communities that did implement *partage* often found, to their dismay, that the size and configuration of the parcels rendered them inadequate or even unfeasible for cultivation.[128] In such cases, recipients ended up renting out or selling their allocations after only a few years, most often to the same ex-nobles and *cultivateurs* who were already the largest landowners in the district.[129] The popular society of the Jurassien villages of Maynal and Augea denounced the trend, noting that

> rich egotists, natural enemies of equality . . . have monopolized the properties; they have rented all the lands that belonged to the non-*cultivateur* landowners and, calculating with greed the product of the fields and the value of the suppression of the unjust rights with which it was gained, they have found the means, in subletting to the poor *laboureur*, to sell him the benefits of the revolution that is your work.[130]

Jacobin egalitarianism may have influenced the law's final form, but at the end of the day it could not alter its emphasis on the interest of the individual over the welfare of the whole. As a result, the division of the commons, like the auction of ecclesiastical, national, and émigré properties, for the most part fortified the power of rural elites. Moreover, in sundering the cooperation, no matter how limited or contentious, that communal usage entailed, the land distribution initiatives of the Revolution promoted indifference toward one's fellow citizens and disrupted inhabitants' sense of shared accountability for the natural resources they used in common. No longer beholden to the needs of the collective, individual proprietors began exploiting their possessions as they pleased. Those left without access to resources, in turn, were reduced to grabbing what they could. In this way, the Revolution's endorsement of atomization in the ostensible service of the nation would prove to have a profound effect on the environment.[131] Now, far more than ever before, a

[127] AN F^{1cIII} Doubs 6. [128] See discussion in Jones, "Agrarian Radicalism," 147.

[129] See discussion in Jones, *Peasantry in the French Revolution*, 163; also Gay, "Les acquéreurs de biens nationaux," 229–42.

[130] AN F^{10} 264.

[131] The Directory suspended egalitarian *partage* in 1796 and replaced it with auction to the highest bidder. This trend toward freeholder entitlement became even more pronounced from Napoleon forward. *Loi portant qu'il sera provisoirement sursis aux poursuites résultant de l'éxecution de la loi du 10 juin 1793 sur le partage des biens communaux*, 21 Prairial, year IV (9 June 1796), in *Bulletin des lois de la République française*, ser. II, no. 52, 12–13.

national policy on resource conservation was needed. The old models of communal oversight had been dealt a grievous blow.

FEDERALIST REVOLT AND THE RÉBELLION DES MONTAGNES

As the reactions to *partage* in Franche-Comté suggest, communities and individuals responded to revolutionary land reforms in ways as complex as their particular circumstances. Nonetheless, the overall mood in the region in the summer of 1793 was one of increasing disillusionment. Among the poorer peasantry, dissatisfaction flowed from their unsuccessful efforts to resolve *affouage* injustices, from *partage*'s predictable failure to be a panacea, and from the consolidation of *biens nationaux* in the hands of land speculators. Among the administrative elite, many felt mounting discomfort over decisions being made at the national level that they believed infringed too severely upon Franche-Comté's historical autonomy. The decisions included the resolution, decided in January, to execute the king without recourse to the primary assemblies; the passage in the spring of a series of war-related measures establishing a "maximum" on grain prices; the formation of departmental surveillance committees to report on émigrés, foreigners, and other suspects; and the dispatching of "representatives on mission" to monitor the supply of men and materials to the front. As the summer progressed, additional measures would only increase Comtois ire.

Spurred by these discontents, tensions in Franche-Comté boiled over into revolt. Two particularly dramatic instances, quite different in character, stand out both for what they reveal about Comtois willingness to take up arms against the state and as a reflection of how the Revolution magnified and ruptured internal rifts in the region.[132]

The first incident unfolded in the Jura in June. Reacting against the National Convention's expulsion and arrest of twenty-nine Girondin legislators on 2 June, administrators from nearly 60 percent of France's eighty-three departments condemned the action in writing. Letters from the Doubs, the Jura, and the Haute-Saône numbered among the protest.[133] Nonetheless, the Jura alone in Franche-Comté took its grievance

[132] On uprisings against the *levée en masse* in the Haute-Saône, see Girardot, *Le département de la Haute-Saône pendant la Révolution*, 3:18–21, also 3:332.

[133] Girardot, *Le département de la Haute-Saône pendant la Révolution*, 2:223–35. On the response of the Doubs, see Jean Ferrez, "La Convention face au mouvement Jurassien, janvier–août 1793," in *Le Jura contre Paris: Le mouvement fédéraliste jurassien de 1793. Actes de la table-ronde du 3 avril 1993*, ed. Henri Hours (Lons-le-Saunier: Société d'émulation du Jura, 1994), 94.

a step further. Joining administrators from a handful of other areas, among them the neighboring department of the Ain, plus Normandy, Brittany, and Picardy, the Jura's "Federalist" administrators – the term of rebuke by which the movement came to be known – pushed for a military as well as political response.[134]

Their recruitment effort was astonishingly successful. Jurassiens had already demonstrated their willingness to defend the Republic before. In contrast to the foot-dragging shown toward conscription elsewhere, the men of the department had exceeded in record time the military quota set by the Convention in February.[135] Like their compatriots in the Doubs and the Haute-Saône, the majority of Jura inhabitants had never fully accepted French rule and welcomed a regime promising representative governance and an end to *seigneurial* privilege. For this reason, Jurassiens, particularly municipal and departmental leaders who had gained power through the Revolution, were all the more dismayed to see a new form of authoritarian centralization emerging in Paris. Rallying support with the accusation that "the Convention no longer exists except in fiction," the Jura's Lons-le-Saunier–based Conseil général organized a force of twelve to fifteen thousand volunteers in a matter of weeks, with the goal of marching on Paris to restore the Convention's independence.[136]

Not all Jurassiens endorsed the Federalist cause, however. Many of the department's popular societies vehemently opposed it. By 1793, political fault lines in Franche-Comté, as in Paris, were no longer the divisions between republicans and monarchists – the province had few of the latter to begin with and fewer still after noble emigration – but between educated and affluent bourgeois who had gained power and wanted to see it secured in law, and the "bourgeoisie of the boutique" – shopkeepers,

[134] On Federalist responses in general, see Paul R. Hanson, *The Jacobin Republic under Fire: The Federalist Revolt in the French Revolution* (University Park: Pennsylvania State University Press, 2003).

[135] Ferrez, "La Convention face au mouvement Jurassien," 83–84. The Haute-Saône and Doubs were similarly prompt in meeting their conscript quota. Jean-Paul Bertaud, *The Army of the French Revolution: From Citizen-Soldiers to Instrument of Power*, R. R. Palmer, trans. (Princeton, NJ: Princeton University Press, 1988), 94; Robert Monnet, "Les voluntaires comtois, 'soldats de l'an II,'" *Revue historique des armées* 5, no. 2 (1978): 158.

[136] ADJ 1 L 74, folio 32, 7 June 1793. See also Henri Hours, "L'administration du Jura pendant le mouvement fédéraliste: Rôle et enjeu des institutions," in *Le Jura contre Paris: Le mouvement fédéraliste jurassien de 1793. Actes de la table-ronde du 3 avril 1793* (Lons-le-Saunier: Société d'émulation du Jura, 1994), 31.

clerks, industrial agents, and poorer *laboureurs* who dominated the region's popular societies and sought a more inclusive democracy.[137] Dole's popular societies were especially active, in large part because of long-running rivalries between Dole, Besançon, and Lons-le-Saunier. At the time of the French conquest a century and a half before, Dole had been the province's administrative seat, but the city had suffered a humiliating decline after Louis XIV transferred the Franche-Comté *parlement* to the more populous and pliant Besançon in punishment of the Dolois' resistance to his invasion. Dole was passed over again in 1790 when the National Assembly selected Lons-le-Saunier as the Jura capital.[138] The rise of Jacobin popular societies gave Dole's politically engaged community an opportunity to regain some of the influence it had lost.[139] When the Convention's representatives on mission arrived in the Jura in the spring of 1793, popular society members helped ferret out potential "traitors" and assisted in the arrest of elected officials.

Enmity between the Jura's popular society members and its bureaucratic elite eventually played a key role in the Federalist defeat. Urged on by René-François Dumas, the former head of Lons-le-Saunier's popular society, the Convention issued a series of increasingly punitive decrees against the Jura throughout July and early August.[140] One, dated 27 July, demanded that the insubordinate officials "cease all functions under pain of death" and report before the Revolutionary Tribunal in Paris. Though the decree initially strengthened their resolve, a small but demoralizing skirmish a few days later with a corps of National Guardsmen from Dole fatally weakened the administrative insurgents' stomach for violence.[141] Threatening to lay siege to Lons-le-Saunier the way it was presently besieging Lyons, the Convention forced the department's Conseil général

137 Lawyers comprised the majority of the administrators, followed by landowning farmers, merchants, doctors, and ex-nobles. Claude-Isabelle Brélot, "Les 'fédéralistes' jurassiens: Républicanisme, contre-révolutionnaires et réseaux d'amitié." In *Le Jura contre Paris: Le mouvement fédéraliste jurassien de 1793. Actes de la table-ronde du 3 avril 1993*, Henri Hours, ed. (Lons-le-Saunier: Société d'émulation du Jura, 1994), 61–74.

138 On the clash between Dole and Besançon, see Ted W. Margadant, *Urban Rivalries in the French Revolution* (Princeton, NJ: Princeton University Press, 1992), 253–54.

139 On relations between political clubs in Franche-Comté and Paris, see Danièle Pingué, "Les Jacobins de province et Paris: Le cas franc-comtois," *Annales historiques de la Révolution française*, no. 330 (2002): 101–13.

140 Ferrez, "La Convention face au mouvement Jurassien," 93–97.

141 Several Lons-le-Saunier Federalists died in the clash, which occurred at Tasseniėres on 2 August 1793. Rousset, *Dictionnaire géographique*, 3:584.

to renounce its rebellion on 9 August.[142] As punishment, it also stripped Lons-le-Saunier of its administrative oversight and transferred all authority, courts, and financial offices to Dole, the lone Jurassien town that remained loyal to Paris throughout the conflict.

This might have been the end of the affair – certainly the insurgent administrators who resumed their regular duties thought so – were it not for the onset of the Terror a month later. René-François Dumas, now president of the Revolutionary Tribunal, made it his mission to punish the participants for their insubordination.[143] After a year, he finally succeeded in bringing thirteen Jurassiens, some of them only tangentially related to the uprising, before the court. Twelve were guillotined the next day.[144] By enforcing the department's subordination in blood, the Revolutionary Tribunal sowed bitterness where there had been hope and assured that the uprising would continue to resonate for decades, both in the region's hostility toward state rule and in the ongoing antagonism between the department's popular and elite political factions.

The second uprising to erupt in Franche-Comté occurred in the Doubs in late August 1793. In contrast to the revolt in the Jura, which involved departmental administrators, the "*rébellion des montagnes*" was a markedly rural and grassroots affair.[145] Dubbed the "petite Vendée" by contemporary Montagnard critics, who likened it to the counterrevolutionary resistance in the west, the rebellion began in the upland hamlet of Sancey, on the Maîche plateau, and spread within a day to the nearby

[142] Bruno Benoît, "Le Jura, les jurassiens et Lyon à l'époque du soulèvement lyonnais (mai–octobre 1793)," in *Le Jura contre Paris: Le mouvement fédéraliste jurassien de 1793. Actes de la table-ronde du 3 avril 1993*, ed. Henri Hours (Lons-le-Saunier: Société d'émulation du Jura, 1994), 143–58.

[143] A close friend of Robespierre, Dumas became vice president of the Revolutionary Tribunal in September 1793 and president in April 1794. See "Délibérations de la Société populaire [de Lons-le-Saunier]. Appendix 41: Les deux Dumas," *Société d'émulation du Jura*, 6th ser. (1896): 335–40.

[144] Ferrez, "Les jurassiens devant le Tribunal révolutionnaire," in *Le Jura contre Paris: Le mouvement fédéraliste jurassien de 1793. Actes de la table-ronde du 3 avril 1993*, Henri Hours, ed. (Lons-le-Saunier: Société d'émulation du Jura, 1994), 173–208.

[145] See correspondence between departmental administrators of the Doubs and the Convention, AN F^{1cIII} Doubs 6, 1794–95. On the villages involved, see Jean-Luc Mayaud, "Pour une généalogie catholique de la mémoire contre-révolutionnaire: La Petite Vendée du Doubs," in *Religion-Révolution: Actes du colloque de Saint-Florent-Le-Vieil, 13–15 mai 1993*, ed. Jean-Clément Martin (Paris: Anthropos/Economica, 1994), 217.

cantons of Vercel, Pierrefontaine-les-Varans, and Le Russey.[146] In all, some 1,200 men from more than thirty communities were involved. Their tactical aim was clear, even if their ultimate objective was not: breaking into village halls across the region, they intended to seize weapons that had been confiscated since May in accordance with the law on the disarmament of suspects. When these efforts garnered them only a meager cache of arms, the insurgents shifted their sights to the district capitals, which had more sizeable arsenals. By this point, however, the department's National Guard troops had assembled and were in full pursuit. Managing only a partial attack on one of the arsenals, the insurgents found themselves on the run by 4 September and utterly routed the day after that.[147]

Of the 1,000 men gathered on 5 September at Plaimbois-du-Miroir, high above the valley of the Dessoubre, half fled that night under the cover of darkness. The rest made a dash for Switzerland the next morning. Ambushed near Bonnétage, five miles from the frontier, some were shot, some surrendered, and the rest raced on to swim to freedom across the Doubs River, which marked the French-Swiss border. Nearly three hundred made it to the other side. Yet even for them, escape remained elusive. Under pressure from France, Neuchâtel authorities repatriated half the fugitives before the month's end.[148]

Fueled in part by rumor – rebels believed that a prominent local noble and antirevolutionary conspirator, the Chevalier de Malseigne, was arriving any minute from Switzerland with an army of émigrés and that Prussian prisoners of war liberated from nearby Baume-les-Dames were about to join them – the revolt ignited quickly and burned out just as fast.[149] No reinforcements arrived to aid the insurgents. Rather, beyond the cluster of villages involved, Doubs inhabitants showed little support for the uprising. Unlike the Federalist revolt in the Jura, in which

[146] Roger Marlin, "Essai d'interprètation de l'insurrection paysanne du Haut-Doubs, dite Petite Vendée, en septembre 93," *Mémoires de la Société d'émulation du Doubs*, new ser., no. 31 (1989): 33.

[147] Mayaud, "Pour une généalogie catholique," 216. On the disarmament of suspects in the region, see AD Doubs L 56, L 170, L 838, L 867.

[148] Mayaud, "Pour une généalogie catholique," 216.

[149] On the insurrectionary role of Malseigne and other nobles, see Mayaud, "Pour une généalogie catholique," 223–24; and Jean Egret, "La Révolution aristocratique en Franche-Comté et son échec, 1788–1789," *Revue d'histoire moderne et contemporaine* 1, no. 4 (1954): 245–71. On the role of rumor, see Maurice Carrez, "L'opinion publique aux environs de Sancey et Belvoir (Doubs) sous la Révolution française: Essai d'interprétation," *Annales historiques de la Révolution française* (October–December 1992): 542.

townspeople welcomed insurrectionary leaders back into their posts and reprisals were meted out primarily at the level of the Revolutionary Tribunal, the other communities around Maîche, Pierrefontaine-les-Varans, and Russey not only participated in quelling the insurrection as civilians and members of the National Guard but also informed on insurgents after the fact.[150]

This difference played out in the rebellion's comparatively fierce repression. Nearly 500 suspects were arrested, twenty deported, forty-six imprisoned, and forty-three executed for participating in the uprising.[151] The great majority of the condemned – as many as two-thirds – were *cultivateurs*. Day laborers and domestic servants, some of whom may have been coerced to participate by their *cultivateur* employers, made up the second largest group, while an assortment of municipal officers, artisans, and minor employees of the state comprised the rest.[152] That fellow Doubiens were so willing to turn on them suggests that they rejected the rebels' motives or regarded them as threats to their own well-being.

No clear statement of their motives survives in the records, but most historians agree that the catalyst was the 24 August 1793 announcement of a new *levée en masse*. Preoccupied with the late-summer harvest and already shorthanded because of the mass mobilization in the spring, the peasantry – particularly landowning *cultivateurs* who needed day laborers at this time of year – may have viewed the further loss of men as disastrous.

Other discontents also likely contributed. As the rebels' concern with recovering confiscated weapons suggests, the May 1793 law on the disarmament of suspects may have been one of them. Guns were a prized commodity everywhere, but they were especially valued in the perilous Doubs uplands, where "wolves and all wild beasts who devour the countryside" were a constant threat.[153] Having gained wider freedom to bear arms in 1789, flock owners were reluctant to give them up.[154]

Indignation about the Civil Constitution of the Clergy also played a role. Throughout much of Franche-Comté, but particularly in the Doubs,

[150] Towns most sympathetic to the Revolution, like Pierrefontaine-les-Varans, energetically resisted the revolt. Abbé Louis Besson, "La Petite Vendée," *Mémoires de l'Académie des sciences, belle-lettres et arts de Besançon* (1861): 61, 57.

[151] AD Doubs L 215, L 213. [152] Mayaud, "Pour une généalogie catholique," 219.

[153] Quoted from the *cahier de doléances* of Bretonvillers, March 1789, in Jouvenot, *Le bailliage de Baume-les-Dames*, 369–70. Bretonvillers was among the communities that participated in the revolt.

[154] Marlin, "Essai d'interprètation de l'insurrection paysanne," 40–45.

the revolutionary reorganization of the Catholic clergy that had been enacted in July 1790 had foundered upon the region's ingrained suspicion of Gallican Catholicism.[155] Dechristianization efforts further antagonized Franche-Comté's faithful, pushing some into active opposition. Of twenty-eight priests in the district, only two took the state-mandated oath of loyalty to the Constitution.[156] The nonjuring parish priest of Sancey, where the "petite Vendée" rebellion originated, was among those forced to step down.[157]

The Civil Constitution of the Clergy was not only disruptive from the standpoint of spiritual practice but also a source of socioeconomic hardship, particularly for the better-off sector of the peasantry that accounted for the majority of the rebels. For these families, the priesthood provided an essential avenue of social advancement. The imposition of the ecclesiastical oath, and the refusal of the majority of Doubs clergy to swear it, threatened to close this avenue. Once they were removed from their posts and went back to their natal villages, moreover, many nonjuring priests from the Doubs found themselves being pursued by Bernard "the Iron Pick" de Saintes, the department's aggressively antipapal representative on mission in the Doubs. De Saintes' excessive zeal fueled further dismay among the priests' supporters.[158]

Interestingly, inhabitants of Sancey – or at least the tax-paying, landowning *coqs du village* involved in drafting the village's *cahier de doléances* – had also expressed exceptional devotion to the crown in their March 1789 petition.[159] In a passage that stands out among the Comtois *cahiers*, most of which repeated the same perfunctory phrases of fidelity (if any), Sancéens gushingly assured Louis XVI of their "love and faithfulness, their absolute submission to sacrifice their fortunes and their lives to maintain the crown of France on his head," even as they also called on the Estates General to suppress the Eaux et Forêts.[160] Such sentiments help to explain the insurgents' link to the Chevalier de

[155] On Comtois religiosity, see Frank Tallett, "Religion and Revolution: The Rural Clergy and Parishioners of the Doubs, 1780–1797" (PhD thesis, University of Reading, 1981).

[156] Carrez, "L'opinion publique," 543.

[157] When priest's replacement arrived in Sancey, a crowd of fifty women or men dressed as women attacked him. Carrez, "L'opinion publique," 544.

[158] Girardot, *Le département de la Haute-Saône pendant la Révolution*, 3:4–26.

[159] Mayaud, "Pour une généalogie catholique," 223–24; Carrez, "L'opinion publique," 542.

[160] *Cahier de doléances*, Sancey-le-Grand, in Jouvenot, *Le bailliage de Baume-les-Dames*, 359–60.

Malseigne, who not only fought on the side of the Prussians and the English but also tried repeatedly to launch a royalist uprising in Franche-Comté.[161]

Among *cultivateurs*, the Convention's ham-fisted land reforms also fostered dissatisfaction. Like the village of Indevillers, just a few miles to the north, the centers of the revolt were located in a rugged and deeply carved terrain of recessed valleys, high peaks, and plunging cliffs. Most were dairying communities in which access to pasturage was hard to come by and jealously guarded.[162] Landowners in these villages, as in Indevillers, may have viewed the June law on *partage* with alarm or aversion.[163] This was the case in Sancey-le-Grand, where inhabitants resoundingly rejected a proposal to partition their commons in August 1793.[164]

Cultivateurs similarly viewed the April 1793 decree of a "maximum" price for foodstuffs as a threat to their livelihood. Four years earlier, many of the villages involved in the revolt had called for the "free circulation of grain" in their *cahiers de doléances*. Now they were being asked to accept price controls that, while favorable for purchasers, posed problems for sellers. In Belvoir, three kilometers from Sancey, the municipal authorities flatly refused to enforce the decree.[165]

Together, these measures and their unwelcome effects may have been enough to convince *cultivateurs* and other like-minded citizens around Sancey that the Convention had become out of touch with rural concerns and that taking up arms was the only choice to defend the Revolution that they had envisioned. Though poorly conceived and chaotic, the *rébellion des montagnes*, like other rebellions that punctuated Franche-Comté's past, expressed inhabitants' desire for self-determination and protested against its betrayal. From the *loups de bois* who resisted the French conquest, to the dogged demoiselles of the Forest of Chaux, to the insurrections in the Jura and the Doubs in the summer of 1793, Francs-Comtois repeatedly showed their willingness to fight for their autonomy

161 Mayaud, "Pour une généalogie catholique," 223–24.

162 For demands for greater freedom to graze in communal woodlands, see the *cahiers de doléances* of Laviron, Vyt-lès-Belvoir, Vellerot-les-Belvoir, and Vernois-lès-Belvoir (all active in the revolt), in Jouvenot, *Le bailliage de Baume-les-Dames*, 314–26. On pasturage among upland communities, see Boichard, "L'espace agricole comtois," 322–25.

163 On the role of *partage* in the revolt, see Carrez, "L'opinion publique," 554 and Mayaud, "Pour une généalogie catholique," 225. On *fruitière* communities, see Mayaud, *Les secondes républiques du Doubs*, 101–22.

164 Carrez, "L'opinion publique," 554.

165 Ibid., 561.

even when faced with a much more powerful foe.[166] Threats and the use of force squelched the revolts this time around, but the internal and external struggles that precipitated them would continue to flare in the region through 1848.

"NEVER WAS THERE A MORE FAVORABLE MOMENT"

The execution in July 1794 of Robespierre and his supporters – among them the Jura's own René-François Dumas – shifted the balance of power back to the Convention and marked the start of a period of legislative reaction and reappraisal. Administrators anxious about the continued deterioration of the nation's forests hoped that they might finally be able to create a new forest administration and accompanying woodland policy that covered not only policing but also management. Although revolutionary legislators had issued an astonishing forty-five laws and fifty decrees relating to the forest between 1790 and 1794, their efforts at comprehensive reform, much less enforcement, had fallen short.[167] Jean-Pierre Couturier, a deputy to the Convention from the Moselle, voiced the concerns of many when he observed, "The irreparable dilapidation that has taken place in all the forests of the Republic since the Revolution reflects the most urgent and direst necessity of organizing a financially viable administration."[168] Yet arriving at a solution would not be easy. Despite evidence that the administrative upheavals of the past five years had greatly accelerated overuse, the Thermidorian Convention as well as the Directory would continue to wrangle over the same unresolved issues that had undone the forest's first administrative overhaul in 1791: what and whose interests the forest should serve, and how it should be managed accordingly. Their inability to reconcile this debate assured that reaching an agreement on reform would be a long time coming.

Couturier, for example, remained adamantly opposed to any form of state oversight.[169] As a deputy to the Convention and, under the

166 Maurice Gresset, "Les Francs-Comtois entre la France et l'Empire," in *Régions et régionalisme en France du XVIIIe siècle à nos jours* (Paris: Presses Universitaires de France, 1977), 106–13.

167 Figures cited in Louis Badré, *Histoire de la forêt française* (Paris: Éditions Arthaud, 1983), 118. See also McPhee, *French Revolution*, 167.

168 Jean-Pierre Couturier, *Project de décret sur les véritables principes d'une administration forestière, constitutionnelle et paternelle, par Couturier, Député à la Convention nationale, par le Département de la Moselle* (Paris: Imprimerie nationale, n.d. [1794/5?]), 1.

169 See discussion of Couturier's earlier 1792 proposal in Buttoud, "Les projets forestiers," 16; also Badré, *Histoire de la forêt française*, 117. On Couturier and his political

Directory, as a member of the Council of Five Hundred, he repeatedly pushed for woodland decentralization and the sale of *biens nationaux*, especially nationalized clerical properties, in small-sized lots. "Landowners are interested in the conservation and improvement of their properties," he observed in a proposal submitted to the Convention, "one sees the proof of this fact in the remarkable difference that existed at the start of the Revolution, between the forests of the former seigneurs ... and those that the administration had confided to the blood-sucking officers of the maîtrises." "The first," he added, "offer the eye the most beautifully kept and most charming spectacle of nature; the second by contrast offer only a spectacle of hideous and heart-rending devastation."[170]

Marc-François Bonguiot, a deputy from the Jura, shared Couturier's view. In April 1795 he asserted, "I see no reason to establish a specific commission for the forest sector; one will be convinced of this [by observing] that the lands of the nation, by means of [their] ongoing sales, will soon be limited to large forests."[171] Condemning the regulatory structure that had been decreed in September 1791 as an "attack on the sovereignty of the people," Bonguiot asserted that forest oversight should be "simple and close to its subjects" and emphasize active and disciplined agents and "prompt and exemplary" punishments.[172] Only by incorporating forest administration into the fabric of local and regional governance would these goals be possible, he insisted.

Increasingly, however, other voices advocated a more centralized approach. In his *Réflexion sur les forêts de la République*, C. J. B. L. Hébert, a former *mâitre particulier* who had been imprisoned during the Terror, proposed that the *Conservation générale* decreed in 1791 be revived, with only minor adjustments to the responsibilities of midlevel guards and inspectors. More strikingly, he also called for private forests to be regulated by the state.[173]

Growing favor for a national-level oversight was also apparent in the writings of Jean-Baptiste Rougier de la Bergerie, an agronomist and former deputy from the Yonne who had been a forceful advocate of

positions, see Émile August Nicolase Bégin, *Biographie de la Moselle* (Verronnais, 1829), 317–19.

170 Couturier, *Project de décret*, 3.

171 Bonguiot [Bonguyod], *Opinion de Marc-François Bonguiot (du Jura)*, 9.

172 Ibid., 7.

173 C. J. B. L. Hébert, *Réflexion sur les forêts de la République*. Prairial, year III (May 1795); see discussion in Buttoud, "Les projets forestiers," 17.

woodland privatization in 1792. At the time, Rougier de la Bergerie had claimed that the power of private interest made for better management than administrative salaries. Forest guards only encouraged infractions by "waging a perpetual war on woodland citizens," he contended.[174] By 1795, however, Rougier de la Bergerie was beginning to have doubts. Chiding the government for inattention, he exclaimed in his *Traité d'agriculture pratique*, "Would that the *Convention nationale* would soon take efficient measures against the devastation of the woods and forests, [in] which, for a long time, there has also been vandalism!"[175]

As it happened, the work of crafting "efficient measures" would fall to the Directory, which proved uninterested.[176] The new government set the tone for its stance on forest reform on its very first day by transferring responsibility for woodland management to Administration de l'Enregistrement et des Domaines, the agency responsible for collecting dues on state-owned properties.[177] The move was aimed at collecting higher profits.

Continuing in this vein two months later, the Directory loosened restrictions on the auction of national woodlands. The law dated 2 Nivôse, year IV (23 December 1795), authorized the division and sale of woods smaller than 150 hectares, as long as they were one kilometer from the *massifs forestiers*.[178] This was triple the surface area that had been deemed auctionable in 1790, and twice as close to the *massifs*.[179] The decision not only contracted the arboreal buffer ringing the nation's major forests but also glutted the market further, thereby driving down woodlands' price and counteracting the benefit of selling them.

The Council of Five Hundred's decision in Prairial, year IV (June 1796), to suspend egalitarian *partage* had the similar effect of benefiting

174 *Opinion, non prononcée, de Rougier de la Bergerie sur l'administration des forêts*, 2 March 1792, in *AP 1ère série* 39:313.

175 J. B. Rougier de la Bergerie, *Traité d'agriculture pratique, ou Annuaire des cultivateurs du département de la Creuse et pays circonvoisins, avec des vues générales sur l'économie rurale* ... (Paris, year III [1795]), 385.

176 The Convention referred to the anticipated forest reforms in its *Code des délits et des peines*, of 3 Brumaire, year IV (25 October 1795), in *Bulletin des lois*, year IV, no. 204, law. no. 1221, title III, des gardes champêtres et des gardes forestiers (art. 38–47).

177 *Arrêt* of 4 Brumaire, year IV (26 October 1795), by the Comité des Finances placing the forest administration under the direction of the Enregistrement et des Domaines.

178 *Loi qui ordonne la vente des bois d'une contenance de moins de 300 arpens*, 2 Nivôse, year IV (23 December 1795), in Baudrillart, *Recueil chronologique*, tome 1, 524.

179 *Loi, sur le Décret de l'Assemblée nationale, du 6 du même mois, qui excepte les grandes masses de bois et forêts nationales, de l'aliénation des biens nationaux*, 23 August 1790, in Baudrillart, *Recueil chronologique*, tome 1, 495.

land speculators while handing over communal acreage at bargain rates. Instead of partitioning the commons equally among inhabitants, the Directory authorized auctioning them, a mode of allocation that favored the affluent and intensified disparities in access to natural resources.[180]

While the Directory did take up the issue of forest reform, the debate between the merits of centralization and sale again proved to be the sticking point. The project began in December 1795, when the Council of Five Hundred appointed Alexandre Besson, a former deputy to the Convention and now member of the Five Hundred from the Doubs, to lead a committee to create a new resolution on the forest administration. Under the Convention, Besson had pushed to speed the sale of émigré properties, in part as a reaction against the long persistence of mortmain in the Doubs. Yet he staunchly opposed the alienation of national forests. Making his committee's case before the Five Hundred in Fructidor, year IV (September 1796), Besson asserted, "Never was there a more favorable moment for creating a general management of the forests of the Republic than [now,] when the majority of the woods and all the largest forests are in the hands of the government." "If these same forests were sold," he added, "what authority would be powerful enough, and what supervision would be active enough to force private landowners to deprive themselves of the product of their land for three or four generations?"[181] "On the contrary," Besson warned, "have we not seen ... timber forests sold by the government fall immediately under the axe of greed?"[182] Noting that other measures were also needed to combat France's wood crisis, like increasing the use of coal and suppressing illegal appropriations, Besson underscored the importance of establishing a "distinct and centralized administration that, in encompassing all the branches of this vital element, will manage them with knowledge as to the general interest."[183]

The proposal never had a chance with the Five Hundred. Though Besson presented it with passion and detail, his fellow deputies castigated it as "useless, infeasible, [and] ruinous." Jean-Pierre Couturier cheekily offered a 152-article version of his own. Even Jacques-Antoine Boudin, one of Besson's own committee members, dissented. Invoking the same

[180] *Loi portant qu'il sera provisoirement sursis aux poursuites résultant de l'exécution de la loi du 10 juin 1793 sur le partage des biens communaux*, 21 Prairial, year IV (9 June 1796), in *Bulletin des lois de la République française*, ser. II, no. 52, 12–13.

[181] Alexandre Besson, presentation to the Council of Five Hundred, 27 Fructidor, year IV, in *Le moniteur universel*, no. 362, 2me jour complémentaire de l'an 4 (18 September 1796), 1439.

[182] Ibid. [183] Ibid., 1440.

tired saw that the forests would be better off "in the hands of individuals," Boudin suggested postponing "all further discussion of the national forests until peacetime."[184] The rest of the Council agreed. It would be another three years before the Directory considered woodland reform again.

"VIOLATIONS EVERYWHERE"

In the meantime, France's forests degenerated more rapidly than ever. Troubles ranged from the predictable, like the encroachment of grazing flocks on unguarded woodlands, to the brazen, like the theft of twenty-two cartloads of timber from state forests by a Haute-Saône forge master.[185] Reporting on the situation during the month of Vendémiaire, year VI (October 1797), Grégoire Piguet, the commissioner of the Directoire exécutif for the Haute-Saône, painted a picture as bleak as it was reproachful.[186] "On one side," he noted, "the communes sell their timber reserves without authorization, on the pretext of needing to pay the [military] volunteers and transport goods to the army."[187] "On the other side," he continued, "the pretext of serving the navy favors speculation by stave merchants who filch considerably and daily from forests designated for felling whether they have been marked for harvest or not."[188]

Piguet and other Comtois administrators blamed inadequate oversight. "The variation that has taken place in the administration of woodland policing, indeed the absence of policing in the first part of the Revolution, has seen infractions excessively increase," he lamented.[189] The commissioner of the Doubs village of Vuillafans similarly despaired, "The indifference shown toward their duties by many communes' forest and pasture guards renders their surveillance fruitless; in other communes there are no guards at all, with the result that in nearly all the communes of the canton the forests are considerably degraded, the commons are invaded and private property is ... pillaged."[190]

[184] Jean-Etienne-Robert Pons-Saint-Martin, deputy from the Aveyron; Jean-Pierre Couturier; and Jacques-Antoine Boudin, deputy from the Indre, in response to Besson's presentation to the Council of Five Hundred, 27 Fructidor, year IV, in *Le moniteur universel*, 1451.

[185] ADHS 17 P 358 Tribunal de Vesoul, "Dossiers d'instruction pour les délits forestiers, Messidor An V" (June/July 1797).

[186] On Piguet, see Girardot, *Le département de la Haute-Saône pendant la Révolution*, 3:162, 3:282.

[187] AN F1cIII Haute-Saône 6, October 1797. [188] Ibid. [189] Ibid.

[190] AN F1cIII Doubs 11, 20 Vendémiaire, year VI (October 1797).

Given their working conditions, it is not surprising that guards neglected their duties and abandoned their posts. According to Alexandre Besson, guards who had remained in place through the Revolution "nearly everywhere suffered harassment from rural municipalities, which ought to protect them [but], on the contrary, refuse them civic spirit certificates when they will not close their eyes to infractions."[191] Piguet, the Haute-Saône administrator, further noted that the devaluation of paper currency had "rendered guards' salaries nearly null."[192] Already low during the Old Regime, guards and officers' wages now hovered around zero. "This lack of payment," Piguet warned, "has not only caused conservation to be neglected but has given rise to infractions by guards themselves."[193]

Beyond the issue of nonpayment, the war was making it nearly impossible to fill guard vacancies. "Hands are so rare that only men who are errant or unworthy of the public confidence present themselves," the municipal council of Vuillafans lamented in March 1797, "all the more so since communes that are for the most part small or poor cannot pay wages high enough to fully employ one or more honest men."[194] Indeed, most of Vuillafans' forest guards had already resigned, and the few remaining municipal officers who had not done so themselves had had to seek help from the National Guard.[195]

With their services in high demand and little oversight to speak of, some unscrupulous forest officers took the opportunity to make money for themselves. As Paturot, a district commissioner in the Doubs, reported, state forest officers were demanding to be "paid in advance, in silver" before they would agree to mark communities' woodlots for felling.[196] Only one village was able to meet this condition. "The others have not felled [their] lots at all, which has given rise to violations everywhere," the frustrated commissioner noted.[197]

Some cash-strapped communities found ways to circumvent forest officers altogether. Quirot, the Doubs' careworn departmental commissioner, described the phenomenon in December 1797. "I have been

[191] Besson, presentation to the Council of Five Hundred, 27 Fructidor, year IV (13 September 1796), printed in *Le moniteur universel*, 18 September 1796, 1438.

[192] AN F^{1cIII} Haute-Saône 6, October 1797. Metal currency was reintroduced in 1798.

[193] AN F^{1cIII} Haute-Saône 6, October 1797.

[194] AN F^{1cIII} Doubs 11, 27 Germinal, year V (16 April 1797). The letter responded to the accusations by the cantonal commissioner, Regnaud.

[195] AN F^{1cIII} Doubs 11.

[196] AN BB 18 267, 16 Ventôse, year IV (6 March 1796).

[197] Ibid.

informed that in a very great number of communities the delinquents make arrangements with the injured parties, often in the presence of the municipal agents," he noted, adding, "after the settlement has been made, silence is imposed on the guards."[198] "By forcing the communes and others to go around it," Quirot fumed, "the forest administration's excessive cost is contributing to the ruin of the forests."[199]

"THE NEED FOR A NEW FORESTRY ORGANIZATION IS FELT EACH PASSING DAY"

By 1798, three years after the Council of Five Hundred had rejected the reforms proposed by Besson's committee, the situation had become too dire to ignore. "The need for a new forestry organization is felt each passing day," reported Claude-François Balivet, the new commissioner of the Haute-Saône.[200] In response to pressure not only from administrators but also from other rural notables, the deputies of the Five Hundred at last launched another attempt at woodland reform, commissioning eight of their members to draft a new Forest Code.[201]

Led by Joseph Clément Poullain de Grandprey, deputy from the Vosges and son of a *maître particulier*, the new committee had an easier assignment in some respects than Besson and his colleagues. While sales of nationalized properties continued, the push for complete woodland privatization no longer loomed as large as it once had.[202] Many of the nation's largest and most desirable lots had already been auctioned off, and those that remained were netting ever lower prices.[203]

Poullain de Grandprey's report, presented on 16 Ventôse, year VII (6 March 1799), offered no great surprises. Rather, like Besson's rejected schema four years earlier and Poullain de Grandprey's own, largely overlooked recommendations of 1793, the new proposal outlined a compromise between an overarching state administration and one fragmented among the municipal and departmental councils. It also recommended that the forest administration be returned to the Ministry of Interior's

198 AN F[1cIII] Doubs 6. 199 Ibid.

200 Ibid. On Balivet, see Girardot, *Le département de la Haute-Saône pendant la Révolution*, 2:264.

201 See discussion in Buttoud, "Les projets forestiers," 18.

202 Discussions of sales of the *domaines nationaux* are common in the 1799 *Moniteur universel*. See for example AN F[10]15 (1799–1800) concerning the auction of futaie in Loir-et-Cher in Year VIII.

203 Vion-Delphin et al., *Les hommes et la forêt en Franche-Comté*, 115.

oversight and that the state revert to the stricter limits on woodland auctions laid out in 1790.[204] Though relatively meek on the subject of private landowners, it nonetheless acknowledged the need for controls. As Poullain de Grandprey noted, "Properties are sacred; the constitution guarantees their unfettered enjoyment; but if their conservation is essentially linked to the public interest, this must prevail, for the constitution also intends that nothing can be done that harms another."[205]

For the first time since 1791, the government seemed finally poised to act. Local and regional administrators had already indicated their eagerness for reform, and Poullain de Grandprey warned the Five Hundred that any further delay would "consummate the ruin of the forests."[206] Once again, however, legislators' woodland intentions were brutally out of step with the march of events. Before Poullain de Grandprey's committee recommendations could be enacted into law, Napoleon and his coconspirators overthrew the Directory on 18 Brumaire, year VIII (9 November 1799). As one of several dozen deputies who protested the coup, Poullain de Grandprey was unseated and barred from the new government. His committee's labors fell along with him.[207]

Building on the goal of controlling disorder in all its forms that had originated under the Committee of Public Safety and the Terror and expanded under the Directory, the new leadership installed by 18 Brumaire would consolidate France's shift from liberalization to centralization. The trend would be apparent in woodland reform, as well. Between 1789 and 1799, reform proposals – all of which remained mostly unrealized – came nearly full circle, from the suppression of the Eaux et Forêts and sale of *biens nationaux* to commentators like Rougier de la Bergerie's acknowledgment of the need to rein in proprietors' woodland freedoms.

The forest had suffered enormously from the license granted individuals in this period. By 1800, the nation's wooded domain was 100,000 hectares smaller than it had been in 1789.[208] Peasant pilfering, incursion,

[204] See Poullain de Grandprey's presentation of the project on 14 Ventôse, year VII (4 March 1799), in *Le moniteur universel*, no. 170, 20 Ventôse, year VII (10 March 1799), 696.

[205] Poullain de Grandprey presentation, continued in *Le moniteur universel*, no. 171, 21 Ventôse, year VII (11 March 1799), 700.

[206] Ibid., 703.

[207] Martyn Lyons, *Napoleon Bonaparte and the Legacy of the French Revolution* (New York: St. Martin's Press, 1994), 37–40.

[208] *Les Eaux et Forêts*, 303.

and abuse played a role in this decline, but more significant were the auctions of nationalized woodland properties to buyers – principally better-off landowners, manufacturers, and bourgeois – who quickly put them to the axe. Administrators like Quirot, the commissioner of the Doubs, made no bones about the havoc these sales wreaked on the nation's woodland health. "The citizens who have acquired [forests] from the Republic are stripping them bare; they leave absolutely nothing," he bristled in December 1797.[209]

Despite such criticisms, policymakers persisted in giving individual proprietors free rein and attributed abuses to communal and collective use rights instead. Though both Besson and Poullain de Grandprey had pointedly noted private forest owners' destructiveness in the preambles to their 1796 and 1799 reform proposals – models that were ultimately rejected for different reasons – they offered no real challenge to the status quo. By insisting on the liberty of the individual and the value of self-interest over communal accommodation and external controls, even when it came to resources best utilized and governed in common, France's revolutionary lawmakers created an insuperable obstacle to effective, enduring conservation. Their devotion to liberalism undermined the applicability and sincerity of the ideology of the *bien public* at the very moment that it was being elaborated. Far from strengthening and restoring customary rights already under siege during the Old Regime, and far from amending the inequities of woodland access and collective practices that, for all their faults, facilitated a multitiered model of forest use, the conservationist philosophy that took root during the Revolution intensified competition among stakeholders and accelerated the forest's deterioration. Collective woodland rights persisted, but their status remained conditional.

If, as Delisle de Moncel asserted in 1791, peasants had been "seduced by the word liberty," by 1799 an equivalent number were surely feeling jilted. The revolution they had embraced with such passion from the start had turned out to be a fickle stranger, a sower of discord and intensifier of trouble rather than a savior. Among Franche-Comté's *usagère* communities, disillusionment with the revolutionary romance and its revanchist drift would continue to mount over the first half of the nineteenth century as they found themselves facing new, more comprehensive constraints than ever on their woodland rights.

209 AN F^{1cIII} Doubs 6.

5

"Nothing is more respected ... than the right of property"*

The Creation of the 1827 Forest Code

In respecting forests [and] cultivating woods, men learn to cherish their duties and practice virtue.

Houry, ingénieur des mines, 1806[1]

We are not afraid to say that the most urgent law for the France of today, for the France of the future, is a Forest Code.

Isaac-Philibert Ardant, 1819[2]

The coup d'état of 18 Brumaire derailed the reorganization of the forest administration yet again, but advocates of change remained undaunted. Viewing the situation as a temporary setback, they pinned their hopes onto the comparatively unknown figure of Napoleon and hoped his ascent would finally clear the way for reform. Rougier de la Bergerie voiced the urgency felt by many when he observed,

For ten years, we have been awaiting a definitive organization of the forest administration, and, to France's misfortune, during all periods of the Revolution there has constantly been a fatal rotation at the center of the executive government that has put off or prevented from being heard the voices of wise men calling

* The quotation in the chapter title is from *Archives parlementaires de 1787 à 1860: Recueil complet des débats législatifs et politiques des chambres françaises, imprimé par ordre du corps législatif sous la direction de mm. J. Mavidal et E. Laurent. Deuxième série (1800 à 1860)*, 49:93 (29 December 1826) (hereafter *AP 2ème série*).

1 Houry, "Mémoire sur cette question: Quels sont les moyens d'augmenter la production de Bois, et de diminuer leur consommation sans nuire a l'agriculture?" *Mémoires de la Société d'agriculture, sciences, commerce et arts du département de la Haute-Saône* 1 (1806): 82.

2 Isaac-Philibert Ardant, *Projet de code rural et de code forestier* (Paris: Imprimerie Testu, 1819), pt. 2: 4.

for the conservation of the woods and the preservation of the fertility of the uplands.[3]

Daring to believe the wait might be at an end, he asked,

Have we at last arrived at the moment when it will be possible to stop the devastating hand of man, [which is] deteriorating the soil of the uplands everywhere by grievous and inconsiderate clearings and at the same time destroying the trees, woods and forests that nature has caused to grow with profusion, and which have made France the most fertile, salubrious, and happiest country in the world?[4]

Administrators knew all too well that new political upheavals might bring disruptions again. Nonetheless, they were cautiously optimistic. As the commissioner general of the Ourte remarked in a letter to the minister of the interior in January 1800, "if peace permits the government to apply all its attention to internal affairs, there is no doubt that the current state of the national forests will be changed in the most advantageous manner."[5]

For some observers, the advent of a powerful leader was itself cause for celebration. Confident that Napoleon could be as much a champion of the woods as he was of the battlefield, one enthusiastic admirer enjoined the First Consul to sponsor an immense, nationwide tree-planting festival to commemorate the "Peace of the Year IX" (the peace of Lunéville, signed with Austria in February 1801). "All that follows the 18th Brumaire should breathe heroism and be marked by great things," argued the proposal's author, a former ship's lieutenant named Besson, thus "why not mark the end of ten years of suffering, of sanctions, of the greatest anarchy … with a forest of more than four million trees aged 27 years old"?[6] Appealing to the state to assure the public good through woodland regeneration, Besson added, "the celebration of today's peace should resemble no other [and] its main goal should be a major service, making the moment convince the greatest skeptics that the Government only wants our happiness and to make France the first nation in the world."[7]

[3] Jean-Baptiste [Baron] Rougier de la Bergerie, *Mémoire et observations sur les abus des défrichemens et la destruction des bois et forêts; avec un projet d'organisation forestière* (Auxerre: Imprimerie de Laurent Fournier; Paris: Chez Madame Huzard, Imprimerie-Librairie, year IX [1800–1801]), 6.

[4] Ibid., 3.

[5] AN F^{10} 403, Le commissaire du Gouvernement près de l'Administration centrale du département de l'Ourte, au Ministre de l'Intérieur, 28 Nivôse, year VIII (18 January 1800). The Ourte is today the province of Liège in Belgium.

[6] AN F^{1cI} 103, Idée sur la fête de la Paix, 2, 3. Author is no relation to Alexander Besson.

[7] AN F^{1cI} 103, Idée sur la fête de la Paix, 2.

In the forest as in most other areas of French governance, the radicalism of the Revolution had come full circle. Having initially championed the decentralization of forest oversight, concerned legislators now sought to reinstate the 1669 Ordinance, or at least its powerful controls. Rougier de la Bergerie, for example, had strenuously pushed for woodland privatization in 1792, asserting that the national forests would be "incontestably better administered by landowners than by salaried personnel."[8] By 1801, however, he was wistfully recalling the policies of the Old Regime and urging present leaders to follow suit. The monarchy, he noted, "regarded the conservation of the forests as an essential duty … as a sacred stock to transmit to posterity the waters, the woods, the forests, the soil of the mountains." Articulating the increasingly dominant ideal of centralized, state conservation, Rougier de la Bergerie added, "The leaders of a republic should all the more look after this precious storehouse, repair the disasters, and work for the happiness of future generations."[9]

From the start of the nineteenth century, state authorities would seek to do just that. Reviving and reinforcing their woodland influence in pursuit of a more pliant and profitable natural resource, Napoleon and the Restoration governments that followed would justify their intensification of forest oversight on a broader notion of the *bien public* than the 1669 Ordinance's emphasis on naval timber alone. Yet even as woodland policymakers expanded their interests to agropastoral, industrial, and hydrological concerns, they became less tolerant than ever of customary and collective usage.[10] Redefined as an infringement of the national interest, communal use rights would bear the brunt of France's reinvigorated and restrictive woodland regime. Other forest elements, among them wolves and wild boar, brigands and bandits, were similarly conceived as

[8] Opinion, non prononcée, de Rougier-la-Bergerie sur l'administration des forêts, in *AP 1ère série* 39:313.

[9] Rougier de la Bergerie, *Mémoire et observations sur les abus des défrichemens*, 68–69. Translation of "*des générations*" here is based on the *Dictionnaire de l'Académie française*, 5th ed. (1798), 639.

[10] See discussion in Jérôme LaFargue, "Entre la ruse et l'impuissance: Les paysans face à la règle de droit au XIXe siècle," in *Sociabilité et politique en milieu rural: Actes du colloque organisé à l'Université Rennes 2, les 6, 7 et 8 juin 2005*, ed. Annie Antoine and Julian Mischi (Rennes: Presses Universitaires de Rennes, 2008); and Jacqueline Dumoulin, "Poursuites en réparation des délits et contraventions commis dans les bois communaux soumis au regime forestier au XIXe siècle," *Revue historique de droit français et étranger* 80, no. 2 (2002): 147–77.

nefarious to forest health and would become the targets of eradication campaigns in the nineteenth century.

This chapter examines the origins and emergence of the 1827 Forest Code, starting with a look at the deteriorated condition of France's forests at the start of the nineteenth century and an analysis of state efforts to clamp down on all manner of woodland disorder, and ending with an assessment of the debates that gave the law its final shape. A landmark of modern environmental legislation, the Code combined the rationalizing zeal of the Revolution and the emerging science of silviculture with the administrative consolidation and contempt for the poor characteristic of French governance in much of the nineteenth century. Though it brought much-needed clarity to the forest administration, it also presaged struggle for rural inhabitants on the margins. Ostensibly focused on the public good, the Code aggressively suppressed customary and communal forms of forest use, while making concessions to private and commercial interests that both undermined timber outputs and threatened woodland ecosystems. In privileging private property over common rights, even at the cost of the resources it purported to protect, the Code embodied the weakness and contradictions of modern conservationism.

"TODAY THE EVIL IS AT ITS PEAK"

After a decade of false starts and failed attempts by Revolutionary authorities to create a new forest administration, Napoleon did so by fiat only two months after he took power. Decreed on 16 Nivôse, year IX (6 January 1801), the new woodland hierarchy comprised five Paris-based superintendents, thirty *conservateurs* (equivalent to the former *grands maîtres*), three hundred subinspectors, and 8,500 guards.[11] The decree also freed the forest from the oversight of the Administration de l'Enregistrement et des Domaines, the division to which the Directory had consigned it, for financial reasons, in 1795.[12] Now under the supervision of the Ministry of Finances, the forest administration enjoyed an autonomy and authority that it would exercise with increasing confidence through the nineteenth century.

[11] *Loi relative à l'organisation d'une administration forestière*, art. III, 16 Nivôse, year IX (6 January 1801), in Baudrillart, *Recueil chronologique*, tome 1, 544.

[12] *Arrêt du comité des finances*, 4 Brumaire, year IV (26 October 1795). See discussion in *Les Eaux et Forêts*, 295–304.

As with other administrative innovations at the start of his rule – among them, the founding of a central bank, the rationalization of tax collection, and the organization of the prefectoral system – Napoleon proved pragmatically willing to adopt the proposals of his predecessors. Rather than stake out new terrain, the new forest administration incorporated both the thwarted reforms of September 1791 and the plans put forward in 1796 and 1799 by the committees of Alexandre Besson and Poullain de Grandprey. A product of expediency and necessity, the 1801 law was limited in scope. Beyond establishing the new administration – an achievement that for all their effort had eluded revolutionary lawmakers – its ten articles addressed only the rudiments of salaries and budget. On all other issues, including supervision, policing, and management, it invoked prior legislation.[13] As to the alarm expressed by many commentators over the need to rein in private landowners and launch reforestation schemes, the decree said nothing. Even so, administrators greeted it with enormous relief. "Already a wise law guarantees the conservation of the remainder of our forests that have escaped ten years of devastation," cheered the prefect of the Ourte in May 1801.[14]

The new woodland administration had its work cut out for it. Though the forest's deterioration was not as dramatic as the historian Jules Michelet asserted – peasants were so wanton in their abuse, he claimed, they "cut down two pines to make a pair of clogs" – the decade of the Revolution nonetheless had taken a heavy toll on the nation's forests.[15] By throwing woodland oversight into disarray, encouraging the division and clearing of the commons, auctioning off large portions of national forest, and augmenting warfare-related demand for industrial fuelwood, revolutionary legislators had unintentionally created perfect conditions for profiteering, despoliation, and abuse. Even those who dated France's forest troubles back to the physiocratic clearing statutes of the 1760s regarded the years after 1789 as cumulatively worse than those that had preceded it. The monarchy undoubtedly allowed "too fatal an increase in clearings," Rougier de la Bergerie remarked, "but this misfortune was much greater under the Revolution."[16] Other observers agreed. According to Houry, a mining engineer and agricultural society member

[13] *Loi relative à l'organisation d'une administration forestière*, art. X, 16 Nivôse, year IX (6 January 1801), in *Recueil chronologique*, tome 1, 544.

[14] AN F^7 3035 (3), 16 Floréal, year IX (6 May 1801).

[15] Jules Michelet, *Histoire de France*, 2nd ed., vol. 2 (Paris: Hachette, 1835), 53. Michelet is speaking specifically of the Pyrenees, not of France in general.

[16] Rougier de la Bergerie, *Mémoire et observations sur les abus des défrichemens*, 11–12.

from the Haute-Saône, "The lack of wood has been felt for a century, already our forbears were complaining of the scarcity of firewood," but, he concluded, "today the evil is at its peak."[17]

To a certain extent, comments like these served to deflect blame for the ills of the present onto previous, discredited regimes. They also functioned as expressions of hope. If the "evil" had reached its peak, surely things could only improve.

Either way, Houry's assessment was accurate. France's forests reached their nadir in the first quarter of the nineteenth century, both in extent and condition. At the time of Colbert's reforms, the country's wooded cover had been estimated at 13 million hectares. By the mid-eighteenth century, when Cassini was beginning his great topographical survey of France, this figure had fallen to 8 million hectares. The decline continued into the 1820s, when the nation's forests were estimated at 6.3 million hectares, their lowest extent before or since.[18]

The impact of this loss was especially noticeable in the south, as numerous reports attested. "Few departments would be as rich in wood as that of the Drôme, if they had been conserved," observed one Rhône region administrator in 1802. Instead "all the forests have been destroyed, and those that remain are due only to the laziness of the woodcutters, or to the lack of hands to destroy them."[19] The chief engineer for bridges and roads in the Ariège similarly testified, "in days gone by, a very great part of the department was covered" with woods, but "today many communities lack it, and these are the ones that had the most and that are located on the mountains."[20]

Even in the comparatively well-wooded departments of Franche-Comté, the forests' deterioration was noticeable. In the Haute-Saône, the overall wooded surface declined by 9 percent between 1789 and 1805, and the remaining trees were damaged, picked over, and reduced to the youngest and feeblest specimens. As the prefect, Jacques Paul de Vergnes, observed in 1801, "I cannot offer any precise evaluation of the wood harvest; a great variety exists in the per hectare yield per district, because of the devastations that have taken place for ten years."[21] In the

[17] Houry, "Mémoire sur cette question," 61–62.

[18] Louis Bourgenot, "L'histoire des forêts feuillues en France," in "Eléments d'histoire forestière," special issue, *Revue forestière française* (1977): 15.

[19] Citizen Collin, "Obsérvations sur la situation du département de la Drôme," in *Annales de statistique, ou Journal général d'économie politique, industrielle et commerciale; de géographie, d'histoire naturelle, d'agriculture, de physique, d'hygiene et de litterature* (Paris: Imprimerie de Valade, 1802 [year IX]), 2:391, 2:95.

[20] Citizen Mercadier, in ibid., 3:42.

[21] Vergnes, *Mémoire sur la statistique*, 11–12.

Jura, the statistician P. E. Herbin de Halle reported, "the forests of this department are beautiful and very extensive . . . but they have been greatly ravaged, and are in the highest need of being wisely managed."[22] As for the Doubs, the prefect Jean de Bry noted that although the "actual quantity of woods might still appear more than sufficient for the department's use . . . the knowledge of their quality and current state does not allow one to be unconcerned for the future; the situation is such that it demands the strictest supervision of their management and conservation."[23]

"LET US ... KEEP IN MIND THAT WE NEED TO SAVE OUR WOODS"

The forests' sorry state derived from a number of causes, for which commentators voiced a plethora of criticisms and, to a lesser extent, solutions. Among the most significant factors was clearing. Already on the rise in the eighteenth century, the auction of *biens nationaux* and division of the commons during the Revolution accelerated the conversion of forest and *friche* into arable. These initiatives, argued the agronomist abbé Tessier, compounded the harm of the already-excessive clearings of the 1760s. "That which is a good becomes an evil, when one exaggerates it, when one exceeds its aim," he warned in the 1796 *Encylopédie méthodique* volume on agriculture. "France has already begun to lack wood in many areas[;] every year one sees production decrease and consumption increase. We owe this shortage of a precious commodity in part to overly numerous and misguided clearings."[24]

When carried out on suitable terrain and assiduously maintained, clearings could and did increase the cultivable area available to a community and generate greater yields. Nonetheless, their influence on forests could not be denied. Joseph Marie Lequinio de Kerblay described the macabre effects of one such clearing in his *Voyage dans le Jura*, published in 1800.[25] As in so many other communities across the country, he noted, the inhabitants of Sellières divided their communal woods during the

[22] Herbin de Halle and Peuchet, *Statistique générale*, 5:536.

[23] Jean Antoine Joseph Baron De Bry, *Mémoire statistique du département du Doubs* (Paris: Imprimerie impériale, year XII [1804]), 16.

[24] L'abbé Alexandre-Henri Tessier et al., eds., *Encyclopédie méthodique: Agriculture*, vol. 4 (Paris: Agasse, 1796), 35.

[25] Joseph Marie Lequinio de Kerblay, *Voyage dans le Jura*, vol. 2 (Paris: Caillot, 1800 [year IX]).

Revolution with the goal of cultivating individual lots. Because removing "the vigorous oaks that had seen the passage of centuries" proved too onerous and costly, the enterprising new landowners ended up girdling the trees and leaving them in place to die.[26] The result, Lequinio de Kerblay observed, was a "bizarre and striking spectacle" of destruction and abundance: "hideous, colossal, and lifeless skeletons" surrounded by rich fields of alfalfa and sainfoin.[27]

The situation of woodland pasturage was no better. By reducing the area available to communities for grazing, clearing had accelerated animals' impact on the spaces that remained. Taking stock of the situation in 1802, De Bry, the prefect of the Doubs, observed that the forests of the department were "generally degraded because of pasturage."[28] Furthermore, he noted, private landowners were reluctant to replant scrubby or abandoned areas because of their "fear of seeing their work destroyed by pasturage."[29]

Population growth across France in the first half of the nineteenth century intensified these pressures. In Franche-Comté, the population of all three departments continued to rise until the start of the July Monarchy, fostered by agricultural innovation, the expansion of commerce, and the establishment of winter industries like pipe, clock, and eyeglass manufactures in the economically marginal uplands.[30] While the increase in prosperity improved inhabitants' standard of living, it also strained their already-diminished natural resources, damaging the forest further and precipitating discord.

Beyond clearing and pasturage, wood-powered industry posed perhaps the most rapidly mounting threat to France's forests in the first half of the nineteenth century. Iron manufacturing in particular had expanded during the Revolution, aided by a combination of regulatory confusion and military demand. As one observer in the Ariège remarked, echoing the opprobrium expressed in the *cahiers de doléances*, "instead of dividing the woods that they are exploiting into regulated lots, the majority of forge masters have cut nearly everything without thinking of the needs of the future; today the mountain forges have to import a great quantity of charcoal from the plains, where it is also beginning to run short."[31] In the Haute-Saône, the prefect noted that the department's ninety-eight

[26] Ibid., 18. [27] Ibid., 18–19. [28] AN F^{1cIII} Doubs 11. [29] Ibid.

[30] See report from the subprefect of Pontarlier on population growth and new manufactures, 14 Messidor, year IX (3 July 1801), in AN F^{1cIII} Doubs 6.

[31] *Annales de statistique, ou Journal général d'économie politique, industrielle et commerciale*, 3:45.

iron manufactures were buying all the firewood available from communal and national forests and driving up prices.[32]

The problem was not limited to ironworks. In nearby Lorraine, unscrupulous manufacturers set up unauthorized "salt refineries" to "recook" the leftovers from the national saltworks. These establishments produced a substance "of the worst quality" and prompted abuses by boosting the cost of wood, observed a memorandum from the region in May 1800.[33] "Because industries in the departments of the Meurthe and Vosges use more than half the needs of the people," the author lamented, "well-off Citizens can only buy firewood for an excessive price, and the poor have no other resources than crime."[34] "It is absolutely necessary to stop these infractions, otherwise in just a few years the forests will be in such a state of ruin that there will be nothing left to exploit, neither for the people nor for the factories," the document warned.[35]

Although prefect De Bry of the Doubs contended that wood-burning industry actually helped perpetuate forests that otherwise would have been cleared, many local notables cast a critical eye on manufacturing, viewing it not only as a culprit in woodland decline but also as a cause of public misery.[36] Justin Girod-Chantrans, the founder of the agricultural society of the Doubs, believed that when one took into consideration the harm that forges and other manufactures caused forests and communities, their negative effects outweighed the benefits. "In a civilized state," he opined, "one cannot consider advantageous a good that is acquired through the sufferings of the great majority."[37]

The conflict between industrial demands for fuelwood and the activities of the rural populace was the subject of much debate among Comtois agricultural societies.[38] In 1805, the agricultural society of the

32 See discussion in Vergnes, *Mémoire sur la statistique*, 11–12.

33 AN F[10]403, Mémoire sur les causes de dégradation des Forêts de la ci-devans Province de Lorraine, Lunéville, 26 Floréal, year VIII (16 May 1800).

34 AN F[10]403. 35 Ibid.

36 "Competition has increased, in recent years, the value of lands covered in wood, private interest is coming to the aid of the general interest, and the conversion of forests into pasturage or cultivable land is becoming more rare; it will perhaps be true to say that the department which owed its forges to its forests, will as a result owe its forests to its forges." De Bry, *Mémoire statistique du département du Doubs*, 99.

37 Justin Girod-Chantrans, "Quelques réflexions politiques sur l'accroissement des forges et des usines de l'ancienne province," in *Mémoires de la Société d'agriculture, sciences, commerce et arts de la Haute-Saône* 2 (1808): 6.

38 See discussion in *Rapport sur les travaux de la Société d'agriculture, sciences, commerce et arts ... depuis son établissement jusqu'au moi de floréal An XIII (1805)*, 5 Floréal, year XIII (25 April 1805).

Haute-Saône asked its members, "what are the means of increasing the production of wood and diminishing its consumption without harming agriculture?" The answer, respondents insisted, lay in finding alternative sources of energy like peat and coal. Speaking on the advantages of "carbonized peat" and its abundance in the Haute-Saône, a society member enthused, "One [can] derive tar for the caulking of buildings; lubricating oils for tanneries, paper mills, manufacturing in general; and a peaty area ten hectares in size is enough to sustain ten forges for twenty-five years!"[39]

Coal held even greater potential, if only it could turned to account. In Franche-Comté, local coal was said to be too "too sulfurous, too phosphorous" and produced an "acidic and brittle" metal.[40] Transport, too, remained a challenge. Unlike the gentle terrain and slow-moving rivers of England, which facilitated low-cost canal carriage, France's large size and varied topography hampered canal building and made road improvement expensive. Rather than haul cumbersome combustibles overland, French forge masters preferred to relocate when wood and ore ran short. Nonetheless, advocates of coal were insistent, particularly in the wake of Napoleon's embarrassment at Trafalgar, which saw more than half the French fleet destroyed. Summoning French patriotism as well as the nation's urgent naval situation, the mining engineer Houry observed, "Let us continue to keep in mind that we need to save our woods and focus their use on the needs of the navy, and that England owes its maritime strength only to the coal fires steadily lit in the palace of its kings."[41]

"EXTRAORDINARY AND FREQUENT FLOODING"

As woodland observers and administrators grappled with solutions to the nation's forest troubles, they also began to point to a new cause for concern: torrents of an intensity and violence heretofore unseen in France.

[39] J. A. Marc, "Quelques vues qui pourront servira à la solution de cette question: Quels sont les moyens d'augmenter la production de Bois, et de diminuer leur consommation sans nuire à l'Agriculture ni aux Arts?" *Mémoires de la Société d'agriculture, sciences, commerce et arts du département de la Haute-Saône* 1 (1806):, 56.

[40] Ibid., 54. On developments in the use of coal in the Doubs, see De Bry, *Mémoire statistique du département du Doubs*, 100–101. On resistance to the use of coal, see Belhoste, "La sylviculture du XVIe au XIXe siècle," 56.

[41] Houry and Descolins, "Rapport sur l'usage de la Houille et de son Escarbille, fait à la séance publique du 3 frimaire an 14," *Mémoires de la Société d'agriculture, sciences, commerce et arts du département de la Haute-Saône* 1 (1806): 139–42.

Floods had wreaked havoc in some regions for centuries, notably in the Dauphiné and Haute-Provence, where forge-related deforestation caused severe erosion at the end of the seventeenth century. The physiocratic clearing decrees of the 1760s amplified the problem. Nonetheless, it was not until the nineteenth century, when the nation suffered an astonishing string of flood-borne disasters, that the issue of erosion and watershed protection began to inform woodland policy in a substantive fashion. Beginning with recurrent overflows of the Seine between 1800 and 1810, and peaking in the catastrophic inundations of the Rhône, Garonne, and Loire in the 1840s and 1860s, flooding rose to be an object of near-universal distress in France.[42] In the process, it played a formative role in catalyzing state control over the forest.

Visiting the high upland village of Vescles in the Jura in 1799, the travel memoirist Lequinio de Kerblay described the traumatic effects of one such flood, albeit a relatively minor one compared with the disasters to follow. Having stopped to dine with friends, he was alarmed by the sudden outbreak of a thunderstorm, which pelted the mountains with rain and hail and forced the Ain from its banks. The next morning, the devastation was frightening: "eight houses were destroyed; animals, furnishings, and people, all had disappeared."[43]

Torrents of this sort were once a rarity in Franche-Comté, but now were becoming increasingly common. Prefect De Bry blamed revolutionary-era deforestation. The department's "long mountain chains" had previously been "covered with forests," he noted in a letter of 1804, yet "today no longer offer but arid rocks." De Bry cited in particular the "numerous clearings that have taken place since the Revolution, and principally since the law of 10 June 1793." These clearings, he concluded, were "not at all unrelated to the origins of the extraordinary and frequent flooding that for several years has caused so much damage in the department."[44]

Impelled by these calamities and empowered by Napoleon's organizational reforms, theorists and administrators from the end of the Revolution forward began addressing the forest problem in a new way. Emphasizing woodlands' broader ecological significance, these individuals voiced an

[42] André Lorion, "Les inondations de la Seine sous le Consulat et l'Empire," *Revue de l'Institut Napoléon*, no. 127 (1973): 59–62.

[43] Lequinio de Kerblay, *Voyage dans le Jura*, 2, 182.

[44] De Bry, *Mémoire statistique du département du Doubs*, 5. On flooding of the Doubs and Loue Rivers, see Herbin de Halle and Peuchet, *Statistique générale*, 355.

emerging consensus that restoring and protecting France's upland woods was vital not only for the production of fuelwood and timber but also for the nation's economy, climate, and long-term hydrologic health. The challenge lay in putting this vision into practice.

Among the first theorists to address the flooding problem was Jean-Antoine Fabre, an engineer for the division of roads and bridges in the Var whose pioneering work on floods systematically analyzed the forest's role in watershed protection. In his *Essai sur la théorie des torrens et des rivières* (1797), Fabre concluded that the "destruction of the woods that covered the mountains is the primary cause of the increase in flood development."[45] To inhibit future flood damage, including the destruction of farmland and blockage of river transport, Fabre advocated reseeding and reforesting slopes, building embankments, and instituting careful controls on upland clearing.[46] Above all, he called for goats to be banned from forests, asserting that "the tooth of this animal is murderous for budding trees."[47]

François Antoine Rauch reached similar conclusions in his 1802 *Harmonie hydro-végétale et météorologique*, a quasi-mystical study in which he claimed that excessive clearing and other outrages of the Old Regime had upset the natural balance created by God and brought about climate change, floods, drought, smaller harvests, and the decline of fish and game.[48] Like Fabre, Rauch pushed for government reforestation programs as well as controls on wood-burning industry. These measures, he argued, were essential to restoring humans' equilibrium with nature.[49]

"BETWEEN PENURY AND PREVARICATION"

In addition to flooding, a quieter, yet no less corrosive problem was undermining France's forest health: the perpetuation of the flawed management principles of the Ordinance, which had been kept largely in place throughout the Revolution. As a frustrated observer from Vosges noted in January 1799, the deterioration of the region's pine forests had many causes, but none of them "could compare to the multitude of [abuses]

45 Jean-Antoine Fabre, *Essai sur la théorie des torrens et des rivières* (Paris: Bidault, 1797 [year V]), 144–45.

46 Ibid., 64–65, 131–33. 47 Ibid., 131.

48 François Antoine Rauch, *Harmonie hydro-végétale et météorologique* (1802).

49 On Rauch's place among French ecologists, see Raphaël Larrère, "Rauch ou Rougier de la Bergerie: Utopie ou Réforme?," in *Révolution et espaces forestiers: Colloque des 3 et 4 juin 1987*, ed. Denis Woronoff, Collection Alternative[s] rurales (Paris: L'Harmattan, 1988), 247–56.

produced by their current method of management[;] it is that above all that is the seed of their near-complete devastation."[50]

France had once led Europe in the emerging science of silviculture. Building on the diligent autodidactism that had informed the careers of early *grands maîtres* including Maclot in the Jura and Louis de Froidour in the Pyrénées, theorists like Réaumur, Buffon, and Duhamel du Monceau had crucially shaped scientific understanding of forest dynamics and tree growth. Though their ecological interpretations occasionally missed the mark – Buffon, for example, insisted in his *Époques de la nature* that clearing the land was necessary to counteract the "slow loss of heat from the earth"– their contributions to early forestry were enormous.[51] The ascendant influence of physiocracy shifted the state's preoccupation to agriculture, however, and by the third quarter of the eighteenth century, plans for an overhaul of the Ordinance – widely acknowledged as necessary by experienced foresters like Telles d'Acosta – were languishing. This governmental indifference to woodland science would culminate in the misguided forest auctions of the Revolution and Restoration.

In everyday terms, insufficient silvicultural knowledge manifested itself in weak forest oversight and in the inadequate training of woodland inspectors, officers, and guards. Already problematic under the Old Regime, when venal office holding filled the top ranks of the Eaux et Forêts with wealthy opportunists, the situation was exacerbated by the decentralization of woodland authority during the Revolution and the resultant exodus of forest personnel. These jobs, when they were filled at all, were often taken by individuals ill suited to their task. "Most [guards] are cabaret owners, wood and livestock merchants, junk dealers, hunters or poachers, national guard officers and tax collectors," lamented a report on the forest administration in 1802. The report continued, "Amid all the matters and concerns that occupy them, one must expect the neglect of [their] duty, and no longer be astonished by the cries coming from all corners of the Republic concerning the devastation of the forests."[52] The reorganization of 1801 and its new hierarchy of

[50] AN F[10] 403, Le Citoyen Clemens à Epinal: Mémoire sur l'amélioration des forêts situées dans les Vosges, 29 Nivôse, year VII (18 January 1799).

[51] Buffon was more right than he realized; but at a time when fending off cold was an ongoing struggle, he considered global warming a positive goal. Buffon, 7th époque of *Époques de la nature*, 1788, reprinted as *Oeuvres d'histoire naturelle*, vol. 8, *Époques de la nature* (Berne: Chez la Nouvelle Société Typographique, 1792), 231–35.

[52] AN D[XVIII] 1, *Observations sur l'administration forestière*, 7 Brumaire, year XI (28 October 1802).

conservateur, subinspectors, and guards seemed to promise a new rigor, but, the same report observed, "with whom does the commission intend to fill [these positions]? … It takes twenty years to make a good forester."[53]

For want of guidelines more recent than the writings of Buffon and Duhamel du Monceau, a flurry of instructional manuals began appearing from 1801 forward, each of them aiming to educate officers, guards, and landowners alike in the art of woodland supervision.[54] Many of these tracts bemoaned the lack of systematic training for forest officers, such as that received by civil engineers at the *École nationale des ponts et chaussées*. "The ignorance of foresters is one of the biggest factors in the deterioration of our forests," despaired Jean-Baptiste Lorenz, author of an 1802 *Manuel du forestier*.[55] Spurred in part by Andreas van Recum's 1807 exposition of the issue in his *Obsérvations sur la nécessité d'établir en France des écoles forestières*, the push for a national forestry school would continue to gain ground over the next decade.[56]

In the meantime, French foresters looked to Germany for guidance. Jacques-Joseph Baudrillart, a career forest administrator who later became known for his directories of woodland law, was especially influential in promoting German management and techniques. In 1805 and 1808, he published two meticulously annotated translations of German forestry manuals, each of them singing the merits of clear-cutting, natural regeneration, and conifer plantations as practiced on the other side of the Rhine.[57] These ideas would be central to the curriculum when France at last opened a training school for forest officers in January 1825. Presided over by an Alsatian, Bernard Lorentz, the new *École royale forestière* based its curriculum largely on German models.[58]

53 Ibid.

54 See pamphlets, letters, and *mémoires* in AN F^{10} 403 dossier 4, years VII–VIII.

55 *Manuel du forestier*; cited in Badré, *Histoire de la forêt française*, 123.

56 Karl Georg Faber, *Andreas van Recum, 1765–1828; ein rheinischer Kosmopolit* (Bonn: Röhrscheid, 1969).

57 Jacques-Joseph Baudrillart, *Instruction sur la culture du bois, à l'usage des forestiers: Ouvrage traduit de l'allemand de G.I. Hartig, Maître des Forêts de la principauté de Solms, et Membre honoraire de la Société de physique de Berlin* (Paris: Levrault, 1805); and Jacques-Joseph Baudrillart, *Nouveau manuel forestier, à l'usage des agens forestiers de tous grades, des arpenteurs, des Gardes des Bois Impériaux et communaux, des Préposés de la Marine pour la recherche des bois propres aux constructions navales … Traduit sur la 4e édition de l'ouvrage allemande de M. de Burgsdorf, Grand maître des Forêts de la Prusse … et adapté à notre système d'administration d'après l'ordre du Gouvernement* (Paris: Arthus-Bertrand, 1808). See discussion of Baudrillart's career in *Les Eaux et Forêts*, 358–61, 461–63.

58 Lachaussée, "L'évolution de la sylviculture en France," 5–10.

Like the problem of insufficient training, inadequate salaries also posed a chronic obstacle to the effectiveness of the forest administration. During the Directory, national and communal forest guards had gone for as long as two years without pay. Although disbursements improved under the Consulate and Empire, salaries remained low and payment often came months late. As one inspector remarked in 1805, "forest agents, who have between their hands a trust precious to all society, are proportionally the worst paid of all Government employees." Under these circumstances, keeping guards and officers on the straight and narrow was well nigh impossible. "It is difficult and laborious," the official noted, "to supervise in the middle of the woods, a father of ten to twelve children who receives an annual salary of 300 to 400 francs, for which he is obliged to wait sometimes many months, and who every day has the opportunity to interact with delinquents!"[59]

Recognizing these challenges, state functionaries worked strenuously under Napoleon's leadership to build a more professional and effective forest administration. Between January 1801 and September 1804 alone, the government issued no less than 235 notices on all matters of management, regulation, and policing.[60] Many aimed at regularizing the woodland workforce. In keeping with a decision of April 1801, for example, forest personnel were required to wear matching green uniforms.[61] Two years later, the state consolidated national and communal forest guards into one unit.[62] To prevent communities from trying to circumvent the high cost of forest oversight through ad hoc arrangements such as had been common during the Revolution, the Consulate also issued an injunction against so-called honorary or assistant guards. Only individuals who had been formally recognized by the forest administration and who were on the state payroll could serve as guards. "If you have verbally or in writing given someone these qualifications," admonished a government notice of September 1804, "you should know to stop them immediately."[63]

[59] AN F[7] 3035, item 205. On salaries, see *Les Eaux et Forêts*, 308.

[60] *Les Eaux et Forêts*, 362.

[61] AN D[XVIII] 1, "Observations sur l'administration forestière" (n.s., n.d.), dossier 37.

[62] *Loi relative aux bois des particuliers, à la garde des bois communaux et des établissemens publics, et à l'organisation des gardes des bois domaniaux et de ceux des communes*, title 3, art. 17–18, 9 Floréal, year XI (29 April 1803); see also the instructions on this law, 7 Prairial, year XI (27 May 1803), both in Baudrillart, *Recueil chronologique*, tome 1, 637 and 640–42.

[63] Circulaire no. 236, 3 Vendémiaire, year XIII (15 September 1804), in Baudrillart, *Recueil chronologique*, tome 1, 637.

Together, these measures helped spur more rigorous woodland surveillance. In the Haute-Saône, registers of guards' reports show an enormous increase in both volume and thoroughness from the Directory forward, and judicial statistics bear out their resolve.[64] In year X (1801–2), the departmental judiciary issued no less than 2,191 decisions on forest offenses, with the highest number of them recorded in the heavily industrialized district of Gray.[65] "Rural and woodland policing is performed fairly well in this department as a result of the care that has been taken to establish salaried guards and because of the activity of the courts," remarked the prefect of the Haute-Saône, Vergnes, in an 1802 report to the minister of the interior.[66] The prefect of the Doubs indicated similar improvements: "The work of the forest administration has noticeably improved since taking action against agents convicted of dishonesty," he reported in 1806.[67] Nonetheless, salaries remained a weak spot. As the Doubs prefect explained, "One hopes to find more zeal among the guards, whose assiduity alone can guarantee the conservation of the woods; but they are poorly remunerated and too often are placed between penury and prevarication."[68]

Occasional intransigence on the part of local officials, who were inclined to be more sympathetic to the needs of their community than to rules that might impinge upon it, posed another ongoing challenge. Writing to the minister of the interior in November 1801, the leaders of the forest administration accused rural authorities of refusing to accompany guards on house searches, as the law required. "As a result," the letter noted, "the national forests are devastated and delinquents [are] nearly certain of the impunity that public officials' unresponsiveness supports."[69] Forest officers also criticized mayors for too freely awarding certificates of indigence, a status that exempted offenders from having to pay woodland fines.[70] Writing to the minister of police in 1806, the prefect of the Moselle observed,

> I cannot hide from you that the *Conservateur des forêts* complains of bias on the part of members of the [district] tribunal. He tells me they accuse the laws of

[64] ADHS 17 P 358 (1793–97) and 17 P 352 (1806–12).

[65] AN F^{1cIII} Haute-Saône 6. [66] Ibid.

[67] AN F^{1cIII} Doubs 12, Notes de tournée, September 1806. [68] Ibid.

[69] AN F^7 3035, Administration générale des Forêts au Ministre de l'Intérieur, Paris, 28 Brumaire, year X (19 November 1801). There are numerous other similar complaints in this carton.

[70] See the communication between the minister of the interior and prefects on this issue in 1801 in AN F^7 3035.

severity; that they accept official certificates, falsely obtained by delinquents through the weakness of certain mayors; that the least pretext suffices for them to absolve the accused; that they seek to make themselves popular by allowing in principle that cutting grass in coppices can not be regarded as an offense.[71]

In anticipation of such challenges, the state had established the forest service in January 1801 as an entity separate from the regular administration of civil affairs. The forest service's independence gave it greater sway in such situations. As the agency's power grew, it began holding departmental and municipal authorities accountable for transgressions that, for want of influence, it once might have left unpunished. This was the case in the autumn of 1802, when the mayor and lieutenant mayor of Villerschemin in the Haute-Saône were sentenced to forty days in prison and a fine of 500 francs "for having given the bad example of stealing the commune's wood."[72]

Prefects, for their part, found the expanded autonomy and assertiveness of the forest administration galling. Among other affronts, they noted that woodland officers were failing to report their presence when touring their departments.[73] A May 1801 directive issued by the Ministry of Finances and the Administration général des Forêts encouraged the rift by reducing prefects' jurisdiction over woodland offenses.[74] In cases where prefects' cooperation was needed, however, this approach backfired. Two prefects made their frustration plain in response to an April 1805 request by the *Conseiller d'état* to investigate abuses in their districts. "It is not up to me to pursue offenses of this type," wrote the prefect of the Eure-et-Loir, noting that "since its reorganization, the forest administration has always sought to isolate itself and make itself independent from the administrative authority."[75] The prefect of Mont-Tonnerre similarly observed, "Since the new organization of the forest administration[,] the materiel and personnel of this sector have, as it were, become strangers to the administrative authority of the department, which is no longer consulted, neither about the operations of the [forest] administration, nor for the selection of its agents."[76]

In a nation where authority was so strongly consolidated under one person, rivalries over the remaining morsels of influence were not

[71] AN F^7 3035. [72] AN F^{1cIII} Haute-Saône 6.

[73] See Circulaire No. 20, "Agens forestiers: Les égards qu'ils doivent aux préfets," 6 Thermidor, year IX (25 July 1801), in Baudrillart, *Recueil chronologique*, tome 1, 548.

[74] Decision of 7 Prairial, year IX (21 May 1801), discussed in C.A.R.A.N., *Histoire de forêts*, 102, and in AN F^7 3035, item 81.

[75] AN F^7 3035, item 81. [76] AN F^7 3035, item 105.

uncommon. Yet if the prefects felt as though their role in forest affairs had been marginalized, this sentiment was all the more true for inhabitants of woodland communities and even private landowners, who also saw their freedoms constricted under Napoleon.

REINING IN "EGOISM AND SELFISH MOTIVES"

In addition to strengthening the forest administration's powers of enforcement, the Consulate and Empire also sought to reassert state prerogatives over collective and individual woodland resources. Among the many mandates of this type issued during Napoleon's rule, two stood out: a March 1803 law requiring forest users to authenticate their access rights in national woods and an April 1803 law clamping down on clearing in privately owned woods. Prompted by military and strategic interests, both signaled the regime's seriousness about arresting forest decline, as well as its willingness to return to the controversial and draconian measures of the 1669 Ordinance to achieve it.

The law concerning clearing in private woodlands, issued on 9 Floréal, year XI (29 April 1803), was the bolder of the two.[77] Imposing a twenty-five-year moratorium on unauthorized clearing, it required landowners to inform forest officers of their intentions to clear six months in advance, during which time the Ministry of Finances could rule against it. Anyone found violating this statute would have to reforest an equivalent-sized parcel at their own expense and pay a fine proportional to the sale value of the uprooted timber.[78] In the same law, the state also reinstated its right to earmark and purchase any and all trees deemed appropriate for the navy on private land.[79]

Forest advocates had been strenuously demanding the revival of limits on private clearing for at least five years, arguing that the Revolution's removal of state oversight, land auctions, and *partage* decree had incited unbridled pillage in private woodlands by proprietors and trespassers alike.[80] One of the more insistent critics, the aptly named Antoine Dubois-Bellegarde, deplored the "absurd and destructive system of alienating the national woodlands" and the related "devastation of the woods

[77] *Loi relative aux bois des particuliers*, in Baudrillart, *Recueil chronologique*, tome 1, 636–37. See discussion in *Les Eaux et Forêts*, 324–26.

[78] *Loi relative aux bois des particuliers*, art. 1–3, in Baudrillart, *Recueil chronologique*, tome 1, 636.

[79] Art. 7–9, in Baudrillart, *Recueil chronologique*, tome 1, 637.

[80] See discussion in *Les Eaux et Forêts*, 323–25.

of individuals."[81] A former deputy from the Charente, Dubois-Bellegarde lamented that in the past one needed to obtain special permission before clearing, "yet since the Revolution, the quantity of woods destroyed to extract more revenues from them is incalculable."[82] A chorus of commentators seconded his observations. "The pillages that this license has brought about are no longer possible to reprimand by the laws that [currently] exist," exclaimed an engineer in the Ariège in 1802. An observer in the Drôme affirmed,

> The national woodlands were garnished with the most beautiful oaks, which the law prevented even [their] owners from cutting because it wanted to preserve them for the navy[;] these properties were sold, and the acquirers, seduced by the prices for which the cost of wood had risen, or in a rush to profit without future concern, cut down the majority of the trees.[83]

Some were even more pointed in their criticism. "Few woods of individuals are in a flourishing condition, authorized [as landowners are] by the Law to administer them by themselves," remarked the authors of an 1802 report on the forest administration.[84] These properties could be just as productive as the national and communal woodlands if they were subject to the same oversight, the report argued. "Why aren't they? Why are they privileged in a country where there is no longer privilege?" the report's authors fumed, adding, "One does not tolerate a farmer cutting his grain well before its maturity with the aim of depriving society of it, thus the flagrant abuse that has been perpetrated up to now in this important part of public resources places the government under the obligation of providing for their administration."[85] Anticipating resistance, the report's authors insisted that the state "can do it without violating the principle of justice and without affecting the rights of property, bearing in mind that it will produce order out of disorder and that in improving the property, it will enrich the proprietor."[86]

81 Antoine Dubois-Bellegarde [Dubois de Bellegarde], *Quelques idées sur les forêts et bois nationaux, par le Citoyen Antoine Dubois-Bellegarde, ci-devant Député du département de la Charente au Corps législatif* (Paris: Imprimerie de Lemaire, year VIII [1800]), 11.

82 Ibid.

83 Mercadier reporting on the Ariège and Citizen Collin reporting on the Drôme, in *Annales de statistique, ou Journal général d'économie politique, industrielle et commerciale* (Paris: Imprimerie de Valade, an IX [1802]), 3:42 (Mercadier) and 2:391, 2:395 (Collin).

84 AN D^{XVIII} 1, *Observations sur l'administration forestière*, 7 Brumaire, year XI (28 October 1802).

85 AN D^{XVIII} 1, 7 Brumaire, year XI (28 October 1802).

86 Ibid.

Among landowners' many critics, Rougier de la Bergerie stood out, given his prior advocacy of privatization. Calling for a temporary nationwide moratorium on clearing and strict prohibition of "all types of sales of woods belonging to the nation," Rougier de la Bergerie acknowledged the errors to which his earlier outlook had given rise.[87] "The revolution gave us the greatest of benefits, liberty; it inspired a universal enthusiasm that made the epoch the most memorable in the annals of all nations," he observed, "but let us have the courage to admit, for we are in need of strong truths; it also developed to a great degree egoism and selfish motives, so contrary and so fatal to the public good."[88] Like the anonymous authors of the 1802 report on the forest administration, Rougier de la Bergerie, too, recommended repealing the regulatory independence accorded forest owners in 1791.[89]

For commentators like these, the April 1803 announcement of limits on clearing in private woodlands was both a relief and a step in the right direction, for it appeared to serve notice to landowners, woodcutters, and clandestine clearers alike that they could no longer act with impunity and that forests were a resource whose social value exceeded individual profit. In practice, however, the state sent a mixed message. Within less than two years of the moratorium's promulgation, the Conseil d'état issued another, considerably better-known law: the Civil Code of January 1804. Carefully shepherded to completion by Napoleon, the Civil Code's 2,281 articles lucidly synthesized customary, Roman, and Revolutionary law on matters ranging from citizenship and divorce to contracts and mortgages. Above all, the Civil Code addressed issues of property and affirmed the Revolution's abolition of feudal privileges. By establishing partible inheritance and defining property ownership as individual, absolute, and exclusive, it effectively undercut the claims of collective and individual rights holders and challenged the idea of governmental authority over private woodlands.[90] To impede or prevent a proprietor from clearing or felling timber on his own land, the law's provisions seemed to suggest, the state would have to have a compelling reason, and rights holders would have to prove incontrovertibly the legitimacy of their claims.[91] Moreover, by guaranteeing proprietors' right to dispose of their

[87] Rougier de la Bergerie, *Mémoire et observations sur les abus des défrichemens*, 69.

[88] Ibid., 53. [89] Ibid., 69.

[90] *Code Napoleon; or, The French Civil Code*, trans. George Spence (London: William Benning, 1827), art. 544.

[91] See also article 552, "Ownership of the land carries with it ownership of what is upon and what is below it."

land as they wished, the Civil Code facilitated the breakup of private forests into smaller and smaller parcels. The result was ever-increasing fragmentation of the nation's forests and accelerated conflict between landowners, forest users, and the state.[92]

In elevating private property over collective possession, the Civil Code echoed a decree issued one year earlier limiting use rights in the national forests.[93] According to the law of 28 Ventôse, year XI (19 March 1803), groups and individuals claiming rights of pasturage, fuelwood, timber, and other usages in national forests had six months from the law's promulgation to demonstrate the validity of their claims. If they did not, their actions would be treated as unlawful trespass.[94] The statute's ultimate aim, as an April 1803 announcement noted, was to "achieve a complete accounting of all the use rights encumbering the national forests and submit them to a strict verification."[95] With this information in hand, the state hoped to "end the usurpations to which the forest domain has been subjected."[96]

This was not the first time the state had attempted such an accounting. The 1669 Ordinance contained a similar stipulation, as did a September 1797 decision issued under the Directory.[97] In each case, enforcement turned out to be almost insurmountably difficult, not only because woodland grazing and gathering were so vital to the rural economy, but also because the task of sorting out such claims, rooted as they were in distant transactions of kinship and commerce, munificence and need, expediency and ambition, was a deeply contentious and laborious process. The civil courts to which the assignment fell found the process overwhelming, and appeals dragged the process out further. That the six-month window was unworkable was demonstrated by the appearance of a follow-up law a year later extending the implementation timeline.[98] As it happened, the state did not clarify what it considered acceptable proof of access until 1807, by which time the momentum of the law had been lost.

[92] For examples of conflicts between customary rights holders and landowners, see Jean-Marie Augustin, "La propriété et les droits d'usage des habitants de la terre de Mouthe dans la forêt du Noirmont," in *Mélanges offerts à la mémoire de Roland Fiétier par ses collègues de Besançon*, ed. François Lassus, Annales litteraires de l'Université de Besançon (Paris: Les Belles Lettres, 1984), 56–57.

[93] *Loi relative aux droits de pâturage, pacage, et autres usages dans les forêts nationales*, 28 Ventôse, year XI (19 March 1803), described and announced 23 Germinal, year XI, in Baudrillart, *Recueil chronologique*, tome 1, 636.

[94] Ibid., art. 1. [95] Ibid. [96] Ibid.

[97] 1669 Ordinance, title XV; also *arrêté* of 5 Vendémiaire, year VI (16 September 1797). See discussion in *Les Eaux et Forêts*, 327.

[98] Law of 14 Ventôse, year XII (5 March 1804), discussed in *Les Eaux et Forêts*, 328.

Adding to the pressures on the nation's forests, Napoleon's growing taste for the trappings of power began to imbue the management of the nation's forests with shades of imperial caprice. Building on the law of 28 Ventôse, year XI, a wide-ranging ban issued in 1808 forbade fishing, swimming, washing linens, watering horses, collecting sticks and leaves, and gathering nuts and "all other fruits in the woods and forests of His Majesty."[99] Implicitly referencing the Emperor's recreational prerogatives, the law also expressly prohibited dogs off their leash and cutting and removing wood along hunting paths in the forest.

The growing clampdown on woodland access led to intense and acrimonious dispute among claimants competing for the same resources. This was the case in the upland Doubs village of Frasne, so called because of its abundant *frênes*, or ash trees. Spurred by the 28 Ventôse decree, Frasne's leaders sought and failed in 1804 to bar two former seigneurs from inclusion in the community's use rights.[100] On the losing side again in 1810, Frasne's mayor appealed a court ruling that stripped the commune of its forest rights and attributed them instead to the neighboring village of Bonnevaux.[101]

Farther to the west, the city of Dole found itself in a similar situation regarding its rights to firewood, timber, and grazing in the Forêt de Chaux. These rights, argued a member of the municipal council in 1812, had their origins "in the mists of time" and had been repeatedly affirmed by charters, rulings, and opinions dating back eight centuries.[102] That the forest administration was now challenging these rights at all, he added, was due only to the ignorance of the current mayor, who had not bothered to consult the city archives when making the city's case. Presented with all the facts, the council member contended, the state would surely restore Dole "the fullness and enjoyment of its rights."[103] Doing otherwise, he pleadingly concluded, would "violate the authority of the treaties, which is unimaginable under a government as enlightened as it is paternal and just."[104] Nonetheless, the appeal would drag on for another two decades.[105] Though such arguments were occasionally successful, the

99 AN BB^3 195.

100 See decision of the Cour de Cassation, reprinted in Baudrillart, *Recueil chronologique*, tome 1, 700.

101 AN F^{1cIII} Doubs 7, October–December 1810.

102 Vuillier, *Mémoire concernant les droits d'usage indéfini de la ville de Dole*, 2, 28.

103 Ibid., 24–27, 28. 104 Ibid., 28.

105 Georges Plaisance, "Les droits d'usage en Forêt de Chaux et leur cantonnement," in *Actes du symposium international d'histoire forestière, 24–28 septembre 1979* (Nancy: ENGREF, 1982).

fact remained that as long as maritime and industrial interests on the forest persisted, state and private landowners would continue their efforts to reduce or eliminate customary and communal usage.

"THE CLEAREST ENEMY OF THE TREE IS THE GOAT"

Silvopastoralism topped the list of customary and communal practices that the state sought to contain. The practice had long been denounced by forest advocates. "It is incontestable that nothing is more contrary to the growth of woods than bringing grazing animals into them," Duhamel du Monceau had asserted in his 1760 treatise, *Des semis et plantations des arbres*.[106]

In Duhamel's view, all animals, even wild ones, were harmful and should be kept out of the forest. "Horses and donkeys graze the buds ... pigs root up the earth with their snouts and eat the seeds ... red deer are like big livestock, roe deer like goats and sheep, wild boar like pigs, ... and rabbits, who dig up dirt and feed on tree bark also commit much disorder in young woods."[107] By the end of the Revolution, however, one species in particular – the goat – was being singled out for the havoc it wreaked upon trees. Considered the "poor man's cow," goats surged in number in the latter half of the eighteenth century, particularly in upland areas like the Ain, Isère, and Jura.[108] In the Doubs, the prefect, De Bry, reported astonishing growth, from a mere 53 goats in all the villages of the region in 1755, to a teeming 25,470 in 1804.[109] The phenomenon was also noticeable in less mountainous regions. Writing in 1794, an official from the Yonne observed that "the approximate number of goats ... is truly frightening."[110]

Extensive and repeated military requisitions contributed in part to goats' spread. As a report from Reims observed during the Revolution, communities throughout the district were complaining of the depopulation of their livestock due to the "requisition of goats, oxen, cows, and sheep for the use of the Armies," as well as demands for "an enormous

106 Duhamel du Monceau, *Des semis et plantations des arbres et de leur culture*, 330.

107 Ibid., 330–31.

108 Rougier de la Bergerie, *Traité d'agriculture pratique*, 190. On the rising goat population during the Revolution, see documents in Octave Festy, ed., *Les animaux ruraux en l'an III: Dossier de l'enquête de la commission de l'agriculture et des arts* (Paris: vol. 1, P. Hartmann; vol. 2, Tépac, 1941–46).

109 De Bry, *Mémoire statistique du département du Doubs*, 91.

110 AN F[10] 328, year III (1794–95).

quantity" of grain.[111] The grain requisitions were made more difficult by the division of the commons, which left the few animals that remained little place to graze. Cheaper to obtain and replace than cows and sheep, quick to multiply, and exceptionally adept at foraging, these cut-rate ruminants offered a welcome source of milk and cheese in a time of intense need.

Nonetheless, goats' benefits came at a cost. Almost as rapidly as they reproduced, these notoriously avid eaters began generating criticism and alarm. "These voracious and destructive animals are a public nuisance," cried one Committee of Agriculture correspondent in 1794. Though ceaseless warfare and the topsy-turvy revolutionary economy had contributed to pressure of all types on the forest, goats were assigned the highest blame for woodland dilapidation. As the agricultural committee correspondent argued, "Goats have contributed, perhaps more than any other cause, to the scarcity of wood that we are bemoaning; everywhere their number has increased, [and] the amount of firewood has diminished as a direct result."[112]

Critics described goats in terms of violence and even warfare. "The clearest enemy of the tree is the goat," contended one.[113] "The tooth of this animal is murderous for budding trees," averred another.[114] The image of homicidal incisors was particularly popular. As an agronomist from the Haute-Saône exclaimed in 1806, "This vagabond race errs everywhere, pitting itself against the young plants that it chomps, stripping bark with its murderous tooth."[115] Though the agronomist acknowledged that goats were a critical resource "to its indigent family," he considered them a scourge. "There is nothing more devastating, more harmful to the forests," the Haute-Saône observer remarked, adding that "a troop of goats scattered within them does one hundred times more damage than the axe."[116]

Tinged with moral indignation, these hyperbolic denunciations of goats appear exaggerated, even outlandish in comparison to the impact of auctions, clearing, and fragmentation. However, when viewed in light of the state's broader efforts to control unrest and its growing intolerance in the first half of the nineteenth century toward the needs and practices of the rural poor, the outcry makes sense. Stubborn, resourceful, and ornery, goats were closely associated with landless laborers, who constituted the

[111] AN F[10] 328. [112] AN F[10] 328. [113] AN F[10] 405.
[114] Fabre, *Essai sur la théorie des torrens et des rivières*, 131.
[115] Marc, "Quelques vues," 38. [116] Ibid., 37.

majority of goats' owners.[117] By framing goats as irremediably harmful to the forest and the public good, woodland advocates not only made a case for banning goats from the woods but also criticized by proxy the poorest sector of the rural economy and justified suppressing their collective and customary rights in general.[118]

TAMING THE WILD COUNTRYSIDE

The backlash against the rural poor and their collective rights was part of the broader state push for public order begun from the end of the Convention forward. One of the most important goals of revolutionary policymakers had been the expansion of rural production. This ambition drove myriad Committee of Agriculture campaigns to drain marshes and ponds, perfect breeding stocks, introduce fodder crops, and institute other projects. In the wake of the Terror, however, the idea of improvement increasingly became fused to a new all-consuming goal: the suppression of dissent and the safeguarding of state control.[119] Initially motivated by the threat of counterrevolution, the campaign to tame the wild countryside soon encompassed everything from four-footed predators to fugitive conscripts, slugs to grubs, mendicants to ruminants.

Brigands were the first target. Though not uncommon under the Old Regime, brigandage grew during the Revolution, driven by resistance to military conscription as well as the decline of ecclesiastical charity. Favoring forests as their hideouts, brigands infused these spaces with a sense of menace that led some to view trees themselves as adversaries.[120]

[117] On the association of the poor and goats, see Daniel Solakian, "De la multiplication des chèvres sous la Révolution," in *Révolution et espaces forestiers: Colloque des 3 et 4 juin 1987*, ed. Denis Woronoff, Collection Alternative[s] rurales (Paris: L'Harmattan, 1988), 53–62; Claude Journès, "L'utilisation d'une forêt dans l'Ain du début du dix-neuvième siècle: Enjeux et conflits," in *La coutume et la loi: Études d'un conflit*, ed. Claude Journès et al. (Lyon: Presses universitaires de Lyon, 1986), 44; and Denis Woronoff, "La 'dévastation révolutionnaire' des forêts," in *Révolution et espaces forestiers: Colloque des 3 et 4 juin 1987*, ed. Denis Woronoff, Collection Alternative[s] rurales (Paris: L'Harmattan, 1988), 50.

[118] See discussion in C. Kieko Matteson, "'Bad citizens' with 'murderous teeth': Goats into Frenchmen, 1789–1827," *Proceedings of the Annual Meeting of the Western Society for French History* 34 (2006): 148–62.

[119] McPhee, *French Revolution*, 166–67.

[120] On the presence of deserters and runaway conscripts in forests, see AN F^{1cIII} Doubs 8. On administrative attitudes toward forests as "sites of sedition," see Chauvaud, "Les représentations morbides de la forêt au XIX siècle," 367–74. On trees as mystical menaces in Comtois popular culture, see "Le chêne du Diable" and other tales in Gabriel Gravier, *Franche-Comté: Pays des légendes* (Lons-le-Saunier: Marque-Maillard, 1980), 1:185–86.

When faced with reports of royalist insurgents taking cover in the forests of Poitou and the Midi in 1798, for example, the army sought permission to raze the forest to flush the rebels out.[121]

Under the Directory, as anger over the interminable demands of war mounted and the government's abandonment of Jacobin social welfare legislation took effect, vagabonds, deserters, and conscripts on the run swelled the ranks of highwaymen and dreaded *chauffeurs de pieds* (foot burners).[122] In Languedoc, the Massif Central, Brittany, and the departments of the west, counterrevolutionary sentiments contributed to banditry's growth. There, royalists and refractory priests capitalized on antipathy toward the Revolution by recruiting gangs of *égorgeurs* – throat cutters – to exact bloody vengeance on Republican officials, constitutional priests, and purchasers of nationalized émigré properties.[123]

With its porous foreign border, history of religious obstreperousness, and residue of feudal influence, Franche-Comté proved similarly vulnerable to political brigandage. Reporting on the situation in 1797, Jean-Baptiste Quirot, commissioner of the Doubs *Directoire exécutif*, noted that *égorgeurs* had appeared in Besançon, intending to make it the "base of the [Prince de] Condé."[124] Quirot's fears were underscored by accounts like the one from the cantonal commissioner of Vuillafans, who testified, "The public spirit in this canton is very bad since the refractory priests returned, the aristocrats, the fanatics and the anarchists have reestablished themselves and form the majority of inhabitants of the canton[;] the theophilanthropes and those who follow the [constitutional Catholic church] have united and organized to support the government, [but] they are the minority."[125]

121 Permission was not granted. Buttoud, "Les projets forestiers," 12.

122 Gwynne Lewis, "Political Brigandage and Popular Disaffection in the South-east of France 1795–1804," 212, and Colin Jones, "The Politics and Personnel of Social Welfare," 56–72; both in *Beyond the Terror: Essays in French Regional and Social History, 1794–1815*, ed. Gwynne Lewis and Colin Lucas (Cambridge: Cambridge University Press, 1983); also Howard G. Brown, *Ending the French Revolution: Violence, Justice, and Repression from the Terror to Napoleon* (Charlottesville: University of Virginia Press, 2006), 213–21.

123 Lewis, "Political Brigandage and Popular Disaffection," 201–10.

124 AN F^{1cIII} Doubs 6, Le Commissaire du Directoire exécutif près l'Administration du département du Doubs, au Ministre de l'Intérieur, 11 Vendémiaire, year VI (2 October 1797).

125 AN F^{1cIII} Doubs 11, Copie de l'extrait du Régistre des arretés et déliberations de l'administration municipale du canton de Vuillafans, 20 Vendémiaire, year VI (11 October 1797). Theophilanthropes were members of a deistic civic sect established in 1796 and suppressed in 1801.

Concerted government repression, particularly under Napoleon, eventually brought these movements to heel, but the fear that they or other grass-roots rebellions might reignite remained a central concern of administrators. At the same time, the state's pursuit of order also carried over to other, nonhuman fauna. In a regime devoted to powerful authority and upholding the "inviolable rights" of property, no threat, be it badgers, caterpillars, parasites, or wolves, could be dismissed. From the Directory forward, lawmakers initiated broad extermination campaigns against "all detrimental and dangerous animals, from quadrupeds to insects."[126]

Like the domestic goat – the pariah of livestock – wolves were considered the foremost menace among wild animals.[127] Because of their double peril to life and property, wolves had been the subject of government eradication efforts since the time of Charlemagne. Yet these programs, led by the hierarchical and exclusive *lieutenants de la louveterie*, were widely regarded as ineffective and costly.[128] Together with laws that prevented the peasantry from carrying guns and granted seigneurs monopolies on dovecotes, rabbit warrens, and hunting, the *louveterie* received intense criticism in the *cahiers de doléances* of 1789. As the residents of Rioz in the Haute-Saône complained, "there is a great quantity of wolves that destroy livestock, [and we are] unable to defend ourselves from them, not having any firearms and not even daring to hunt a hare eating from a garden, without being watched by the seigneur's guards and incurring a fine."[129]

In response to demands by scores of communities for the "right to shoot game, devastator of grain," as well as "wild boar, wolves, and other savage beasts that damage the countryside," the National Assembly

[126] The campaign was launched at the very end of the Convention but gained force under the Directory. See instructions to the Committee of Agriculture and the Arts to "immediately [draft] simple and precise guidelines to the most reliable methods of destroying wolves and all detrimental and dangerous animals, from quadrupeds to insects," in "Projet de décret relatif à la destruction des loups," 13 Frimaire, year III (3 December 1798), in Gerbaux and Schmidt, *Procès-verbaux des Comités d'agriculture*, 3:338–40. On caterpillars, see law of 26 Ventôse, year IV (16 March 1796). On the pursuit of *nuisibles* under the Old Regime and in the early nineteenth century, see also AN F[10] 459–489 and AN F[7] 3035.

[127] Jura folklore depicted wolves and goats as wily, resourceful, and devilish. See "La louve et la chèvre sorcière, " in Gravier, *Franche-Comté: Pays des légendes*, 2:30–31.

[128] Jean Richard, "Les Loups et la communauté villageoise: Quelques documents," *Annales de Bourgogne* 21 (1949): 284–96.

[129] *Cahier de doléances*, Rioz (Haute-Saône), 21 March 1789, in Godard and Abensour, *Cahiers de doléances du bailliage d'Amont*, 2:325.

democratized the hunt at the start of the Revolution.[130] Yet instead of turning their weapons on the fanged foes threatening their herds, many rural inhabitants seized the opportunity to kill more gustatorily appealing animals. From a predator-management standpoint, it was a serious setback. Within a decade, seigneurs' cosseted pigeons, partridges, pheasants, and hares plunged in number, while wolves and wild boar enjoyed a fleeting renaissance.[131] The social disruptions caused by conscription, emigration, and civil war, moreover, meant that *battues* (organized hunts) also occurred less frequently.[132] As the deputy from the Somme, Antoine-Joseph Lemarchant de Gomicourt, observed in June 1797, "While we were acquiring our liberty, wolves also acquired [a liberty] very detrimental to flock owners, sheep, and sometimes men, of which many in the past and present years have died of the effects of rabies."[133]

Invoking the liberal logic that an appeal to self-interest was the best way to uphold the public good, Lemarchant de Gomicourt – who would himself soon be pursued in the coup of 18 Fructidor, year V (25 August 1797) – advised the Council of Five Hundred to revive the bounty system introduced by the Convention two years earlier.[134] The bounties marked the start of a wave of state-sponsored wolf killing. Between 1797 and 1799 alone, nearly 12,000 wolves were killed and presented for compensation.[135]

Under Napoleon, expanded policing and increased bureaucratic efficiency helped intensify the state's pursuit of brigands and beasts.[136] As with its revival of central elements of the 1669 Ordinance, the Napoleonic administration brought back Old Regime approaches to animal control

130 *Cahier de doléances*, Charquemont, March 1789, in Jouvenot, *Le bailliage de Baume-les-Dames*, 247; *Cahier de doléances*, Gonvillars, 17 March 1789, in Godard and Abensour, *Cahiers de doléances du bailliage d'Amont*, 26–27.

131 McPhee, *Revolution and Environment in Southern France*, 125–26.

132 On the *battue* and bounty system in the Doubs, see Clade, *Mutations, permanences, ruptures*, 4, chap. 13.

133 A.-J. Lemarchant-Gomicourt [Lemarchant de Gomicourt], *Rapport fait ... sur la destruction des loups* (Paris: Conseil des Cinq-Cents, year V [1797]), 2. Lemarchant de Gomicourt was spokesperson for the Directory's committee on wolf destruction.

134 Lemarchant-Gomicourt [Lemarchant de Gomicourt], *Rapport fait ... sur la destruction des loups*, 1–7.

135 On numbers of wolves killed, see Gilles Ragache, *Le retour des loups* (Paris: Éditions Ramsay, 1990), 188–89.

136 On policing under Napoleon, see Jean Tulard, *Joseph Fouché* (Paris: Fayard, 1998).

like the *louveterie* and the office of *grand veneur*.[137] At the same time, it embarked on a related, though far more complex project of rural pacification and organization: the survey of land distribution and value known as the *cadastre*, aimed at analyzing and clarifying issues of taxation, ownership, and access.[138]

Overall, these campaigns to impose order had mixed results. The incidence of brigandage declined, but mounting national prosperity, good harvests, and rising wages between 1802 and 1810 contributed as significantly to this shift as state repression. Wolves declined in number, as did wild boar. Customary rights were more closely scrutinized and forest crimes better reported. In many ways, however, the advent of a muscular state presence also served to provoke resistance among rural administrators and inhabitants. The preferential hiring of army veterans for guard and officer spots – a policy upheld throughout the nineteenth century – further inflamed conflict. Armed with sabers, rifles, and muskets, these recruits brought to the job a battlefield mentality that reflected the forest administration's intolerance for delinquency and belligerent attitude toward woodland rights holders, whom they increasingly regarded as enemies.[139]

Incorporating absolutist models of power and the radical Revolution's faith in totalizing solutions, the ascendant model of conservation from the Directory forward espoused hierarchy, hegemony, and heedlessness of local practice as its guiding principles. As long as the war and economy were going strong, discontent with this approach remained muted, but after France's military fortunes began to deteriorate and the British blockade led commerce to falter, open political opposition as well as more subaltern forms of dissent escalated.

Between July and September 1812, for example, some 657 verdicts on forest infractions were pronounced by the courts of Besançon and

[137] The high cost of bounties together with the ongoing consolidation of state rule led to the *louveterie*'s reestablishment. See decree of 8 Fructidor, year XII (26 August 1804); also Organisation de la louveterie, 1 Germinal, year XIII (22 March 1805), discussed in Vion-Delphin et al., *Les hommes et la forêt en Franche-Comté*, 109; Durand-Vaugaron, "Le Loup en Bretagne pendant cent ans, 1773–1872, d'après les documents inédits," *Annales de Bretagne* 70, no. 3 (1963): 304–5; and *Les Eaux et Forêts*, 345.

[138] On the implementation and uses of the cadastre, see Hugh D. Clout, *The Land of France, 1815–1914*, London Research Series in Geography 1 (London: Allen & Unwin, 1983), 8, 127–33.

[139] On being armed, see C.A.R.A.N., *Histoire de forêts*, 106. On the preference given at all ranks of the forest administration to army veterans, see *Les Eaux et Forêts*, 373.

Baume-les-Dames alone.[140] Forest-related decisions typically outnumbered civil cases by two to one in these districts.[141] The onset of harsh winters and bad harvests in 1811 and 1812 added to the trouble. Reports from prefect De Bry – now holding the titles of "Baron de l'Empire, Chevalier-Commandant de la Légion d'honneur" – illustrate the region's shifting mood. "Policing and enforcement have been particularly occupied this trimester with provisions, the progressively increasing cost of which has served as a pretext for [those of] ill will to excite and alarm people and cause disorders," he wrote in March 1812.[142]

Deserters and draft evaders also kept authorities busy. In one such case, a group of fugitives was caught while posing as woodcutters. In another, the search for a group of seventeen who had fled their detachment at Strasbourg led to the death of a local boy, who was mistaken for a deserter and shot. None of these incidents improved the dimming popular regard for the state, nor did they help De Bry to, in his words, "contain those who seek to profit from circumstances by fomenting trouble."[143]

Coming amid a string of military losses, Napoleon's abdication in April 1814 threw Franche-Comté into turmoil. Pierre Georges de Scey-Montbéliard, De Bry's replacement as prefect of the Doubs, remarked in his annual report for 1814,

> To better appreciate the effects of the Restoration in the department overall, it suffices to consider the situation at the time of the 1st of May: an invasion, of which, by its location on the border, it was the first and last witness, the losses that followed, rising to more than 11,000,000 [francs]; the blockade of its capital that necessitated the destruction of all the surrounding houses; an epidemic, the common fruit of war; a parasitic disease that ruined communes where animals comprised their entire wealth; multiple fires that wiped out villages and nearly an entire large town; such are the ills that afflicted this department in the short space of a year.[144]

De Scey-Montbéliard went on to assure his superiors at the Ministry of the Interior that that the material damage was being repaired and the public mood was improving. The recent passage of His Royal Highness "gave inhabitants a happy opportunity to show their devotion to the King, and nothing troubled their unanimous joy," he asserted.[145]

140 AN F1cIII Doubs 8, 1812–19. 141 AN F1cIII Doubs 8.
142 AN F1cIII Doubs 8. See also ADHS 17 P 352, January 1811–8 March 1812.
143 AN F1cIII Doubs 8, July 1813.
144 AN F1cIII Doubs 8, dossier 1, Rapport annuel 1814. 145 Ibid.

Nonetheless, the Emperor's short-lived return two months later would see a brief surge of Bonapartism in the region.[146]

While the Francs-Comtois, like the rest of France, accepted Napoleon's second abdication and the installation of Louis XVIII as necessary to the stability they so desired, they remained fiercely committed to the freedoms they had won through the Revolution. "Perhaps the ancient memory of mortmain and feudal rights, under whose crushing weight this region previously suffered, reinforces the constitutional system here more than elsewhere," observed the prefect of the Doubs in 1819.[147] Nonetheless, those who relied on woodland rights – rights not protected in the constitution – would see their livelihoods and liberties increasingly constricted in the years to come.

"INCREASED THE OBSTACLES RATHER THAN REMEDIED THE DEFECTS"

Marked by political polarization, aristocratic nostalgia, intensifying repression, and widening socioeconomic disparity, the Restoration in many ways combined the worst facets of the Old Regime and the Revolution. In upholding the abolition of privilege and its replacement by wealth and property as the sole legitimate avenues to power, Louis XVIII's constitutional Charter and associated land settlement cemented state opposition to collective use and customary rights.[148] At the same time, the restored monarchy expressed mounting anxiety about the poor, whose unpredictability – though now largely stifled – continued to terrify the propertied elite. Typical was the remark of one Comtois official, who noted that poverty and homelessness, "the incurable leprosy of the social order," were best treated with a combination of relief and repression to minimize their interference with the productive sectors of society.[149] This outlook carried over to the administration of the forests as well. Though the crown celebrated the ascension of Louis XVIII, Charles X, and other

[146] On Bonapartism in Franche-Comté (and the general lack thereof), see Vincent Petit, "Étude sur l'identité comtoise au XIXe siècle, telle qu'elle est conçue par les milieux politiques et littéraires de l'époque . . .," *Lettres comtoises, revue de l'Association du livre et des auteurs comtois*, no. 9 (2004): 17–35; and Mayaud, *Les secondes républiques du Doubs*, 17–20.

[147] AN F[1cIII] Doubs 8, dossier 1, Compte annuel de situation administrative, January 1819.

[148] Maurice Agulhon, *Apogée et crise de la civilisation paysanne, 1789–1914* (Paris: Éditions du Seuil, 1976), 87–89.

[149] AN F[1cIII] Doubs 8, January 1819.

occasions with amnesties for forest infractions, these demonstrations of mercy were exceptional.[150] The general trend under the Restoration would be severity toward woodland users and rigidity in woodland management.

Following the advice of his cost-conscious minister of finances, Joseph Dominique, the baron Louis, one of Louis XVIII's first moves upon assuming power was to authorize massive sales of state and communal woodlands as well as auctions of timber reserves. With millions in reparations owed to Europe and damages promised to former émigré nobles, the restored monarchy was anxious to find an easy source of revenue. Like every government before it, it looked to the forests for assistance, putting up 300,000 hectares for auction as security for royal bonds.[151] Less than half this amount actually sold, however, and the proceeds – driven down in part by oversupply – yielded less than 12 percent of the funds needed to cover the nation's debts.[152]

In addition to alienating woodlands and selling timber, the monarchy also attempted to cut administrative costs by slashing the number of forest conservation districts from thirty to six.[153] This reduction, announced in May 1817, was partly a function of the return of the nation to its former borders and the reinstatement of some 480,000 arpents of forest to returning royals and nobles.[154] Even so, its severity went beyond prudence. At the same time, the crown incorporated the forests into the Administration de l'Enregistrement et des Domaines, a tactic the Directory had tried two decades earlier and found wanting.[155]

Almost immediately, officials began complaining about the changes' impact. "The reintegration ... of the administration of the forests with that of the Enregistrement et des Domaines seems to have increased the obstacles rather than remedied the defects of the forest service," observed the Doubs prefect, Paul Étienne de Villiers du Terrage in 1819.[156] Not only was woodland oversight divided between a *conservateur* in Colmar

[150] AN BB^{18} 945, July 1816, "Amnistie: Instructions sur l'application de l'amnistie aux délits forestiers." See also "Étendue de l'amnistie prononcée en faveur des prévenus de délits forestiers, 1814," in *Bulletin des lois*, 1814, 1:303–4.

[151] Under Charles X, the prime minister, the comte de Villèle pushed through the indemnities for émigrés' losses. *Les Eaux et Forêts*, 412–23.

[152] Only 121,957 hectares were purchased. *Les Eaux et Forêts*, 422, 415. See criticism in AN F1cIII Doubs 8 (dossier 1).

[153] Decree of 22 May 1817, discussed in C.A.R.A.N., *Histoire de forêts*, 102 and 108.

[154] Ardant, *Projet de code rural et de code forestier*, pt. 2, 10.

[155] *Les Eaux et Forêts*, 430–33.

[156] AN F^{1cIII} Doubs 8.

and *directeur de l'enregistrement* in Besançon, causing delays in communication as well as coordination, but the latter only intervened "in matters of conflict or bookkeeping," while the *conservateur* was laden with everything from carrying out fellings and reforestation to choosing and supervising guards.[157] The arrangement encouraged misconduct, de Villiers du Terrage contended, and permitted municipal officers, "above all the mayors and assistant mayors," to carry out illicit sales of communal timber and coppice "little by little, without public bidding."[158]

As always, the debate over how to improve the situation fluctuated between those who advocated greater state control, and those who felt that power should be vested with the community. Prefect de Villiers du Terrage fell into the former camp. "Never has the need for the intervention of a watchful authority over the communal administrations been so keenly felt, nor more obvious," he lamented.[159] Others argued that the state should attribute more authority to the municipalities, a cost-saving measure in itself. "At a time when the communes, stripped of all their lands and revenues, find themselves more than ever in the position of having to apply the greatest economy to their expenses," asked an anonymous report from the Doubs, "will one continue to force on them the costs of guarding their woods, which often surpasses the annual revenue of these woods, above all when experience shows that this guarding is far from being as beneficial as it was before the new forest organization?"[160] "The needs of the communes are imperative," the report continued emphatically, arguing that "these needs demand that their lands are restored to them, it is the aim of the king's will, and it would be misinterpreted if they were forced to spend money on guards and administration."[161]

The lack of controls on private forests also continued to be a source of concern. Arguing that the absence of oversight spawned impunity in private forests, one essayist pushed for the laws protecting national and communal woods to be extended to individually owned properties. "Landowners can only succeed in [repairing losses and preventing new disturbances] through very thorough and strict surveillance," the essayist Vautrin observed, "and the existing laws do not permit them to set up this

157 Ibid. 158 Ibid. 159 Ibid.

160 AN F^{icIII} Doubs 12 (dossier 1), Réflections sur l'organisation des autorités administratives et sur les Moyens de fixer dans chaque localité des administrations speciales qui puissent être des intermédiaires utiles entre l'administration générale et les administrés (n.d., 1818?).

161 AN F^{icIII} Doubs 12 (dossier 1).

surveillance."[162] By contrast, the prefect of the Jura advocated putting brakes on landowners themselves. In light of recent flooding in Lons-le-Saunier, Dole, Poligny, and elsewhere, he asserted in 1821, "It seems natural that a landowner could take part of his uncultivated land and clear it[,] but he also needs to be guided in this undertaking."[163] Noting that erosion from newly cleared terrain was taking off the topsoil and destroying land downstream, the official added, "furthermore, it is sometimes necessary to forbid [clearing]. The disasters that accompany storms would be less frequent."[164]

"WE HAVE BECOME POOR": THE PUSH FOR THE FOREST CODE

Amid these complaints, demands for a comprehensive, coherent, and effective forest code grew louder. "We have become poor," lamented Isaac-Philibert Ardant, a senior official in the Conseil d'état, noting that France's forest cover was lower now than it had ever been. "The *Ordonnance des eaux et forêts*, this monument to the wisdom and the foresight of our fathers, regarded as a masterpiece of reason, justice, and policy, has fallen nearly into tatters amid our new institutions; it exists anymore only in fragments," he intoned in 1819.[165] Hoping to catalyze the creation of legislation that would "save what we have and return to us what we have lost," Ardant proposed a version of his own, one that, like the 1669 Ordinance, included vigorous limits on private landowners as well as collective use rights. With less than half of France's total forest cover – a combined 3,000,000 hectares as opposed to 3,500,000 hectares of private woods – state and communal forests were unable to supply all the fuelwood and timber the nation needed to flourish, Ardant noted.[166] For this reason, he insisted, the obligation of a *quart en réserve* in state and communal woods should be extended to private woods as well. The experience of the 1791 forest law had already shown that, left to their

162 M. Vautrin, Chevalier de Saint-Louis, *Mémoire sur la nécessité d'étendre aux bois des particuliers, notamment des grands propriéetaires, les lois établies pour l'administration des forêts royales, des communes et des établissemens publics, et pour la répression des délits qui s'y commettent*, (Paris, 1816), 1.

163 Compte annuel du Préfet, 25 June 1821. On earlier flooding, see Rapport trimestriel, 12 August 1812, both in AN F1cIII Jura 8.

164 AN F1cIII Jura 8. 165 Ardant, *Projet de code rural et de code forestier*, 11.

166 Ibid., pt. 2, 10–12. Ardant gave the figures in arpents; the conversion corresponds to the figures given by Martignac in his address to the Chamber of Deputies. *AP 2ème série* 51:155 (11 April 1827).

own devices, private landowners were quick to profit from timber sales and imprudent clearing. "The denuded state of individuals' woods," Ardant asserted, "shows better than all reasoning the necessity of a law which teaches them what is in their interest to want and to do, and that property and liberty are only precious possessions in so far as one uses them as a good citizen for the well-being of oneself, one's family, and society."[167]

Confronting the problem at last, the crown initiated a series of ambitious forest improvement projects from 1820 forward. It began by disentangling the woodland administration from the clutches of the Enregistrement and restoring its independence under the direct authority of the ministry of finances.[168] Building on the work of writers like Ardant as well as the guidance of specialists like Etienne-François Dralet, author of the seminal *Traité du regime forestier*,[169] the crown next turned to devising a new forest code. An initial version, drafted in large part by the woodland jurist Baudrillart and backed by the prime minister, the comte de Villèle, was put out for comments in 1823. Running to some twenty sections and 372 articles, this first effort retained substantial elements of the 1669 Ordinance, including technical principles and clauses on fishing and hunting mingled with sections on administration, policing, and punishment.[170]

Especially striking was the text's strictness on private property. In keeping with the constraints proposed by Ardant, the Code's 1823 draft required landowners to obtain authorization before clearing woods larger than two hectares, or face fines of up to 1,500 francs per hectare and the obligation of replanting the entire area.[171] It also included procedural requirements like the number of seed-bearing timber trees proprietors must retain per hectare.[172] These elements constituted an about-face from

[167] Ardant, *Projet de code rural et de code forestier*, pt. 2, 10.

[168] Ordinance of 11 October 1820 created three *administrateurs* and one secretary general under the direct authority of the Ministry of Finances. *Les Eaux et Forêts*, 469–71.

[169] Étienne-François Dralet, *Traité du régime forestier, ou Analyse méthodologique et raisonnée des arrêts, règlements, décisions, instructions et circulaires concernant l'organisation des officiers et employés forestiers, et la partie administrative de leurs fonctions* (1812).

[170] AN BB^{18} 1270, Projet de code forestier soumis par le minister des finances, January 1823.

[171] AN BB^{18} 1270, Projet de code forestier, art. 228, 229, 231. For a forest to be cleared, it could not be enclosed, over two hectares in extent, contiguous with other woods, or on the slopes or summit of a mountain.

[172] Ardant, *Projet de code rural et de code forestier*, 71; AN BB^{18} 1270, Projet de code forestier.

the National Assembly's abandonment of controls on forest proprietors in 1791.

While the content of the Code was still being worked out, the crown, now held by Charles X, announced plans for the creation of a forestry school to train woodland officers in silviculture, forest law, surveying, and other subjects.[173] Opened in January 1825 in the Lorraine city of Nancy, close to the German forests where many of the instructors had gotten their start, the forestry school rapidly became the cornerstone of a movement to professionalize the forest service and affix its work to a sound scientific foundation.[174]

A second draft of the Forest Code appeared later that spring and was swiftly distributed for comment to the nation's prefects, conseils généraux, forest administrators, and tribunals. Replying in October 1825, the Conseil général of the Doubs took particular issue with measures that interfered with local traditions and customary rights. "Our mountains, which constitute the majority [of the department], are primarily covered with parcels locally called *près-bois* [edgewoods], in which grow bushes and shrubs whose principal function is to shelter pasturing animals against the heat," noted the Conseil général. "Will the forest agents become their masters, planting them with trees to transform them into true woods? The communes, owners of these *près-bois*, will they be unable to prune and clear bushes whose excess growth impedes pasturage?"[175] Citing a multitude of prior rulings, the council argued that the prefect and the Conseil général, not forest officers, should decide when to permit woodland pasturing and how exploit the communal forests.[176] Council members also objected to the forest agency's high fees, a perennial source of complaint that predated even the 1669 Ordinance. "The government has no more right over communal properties than over those of individuals," asserted the Conseil général, denouncing a 10 percent surcharge on timber sales to pay for forest administrators' assistance.

[173] The Ordinance of 26 April 1824 created a director general of forests to lead the forest administration and announced that a forestry school would also be be created (article 8). See discussion in *Les Eaux et Forêts*, 474–76.

[174] An ordinance of 1 December 1824 laid out the organization of the school, which opened a month later on 1 January 1825. On the German influence on silvicultural training at the school, see J. Pardé, "Des temps gallo-romains aux temps contemporains: Premiers pas et progrès des aménagements," In "L'aménagement forestier hier, aujourd'hui, demain," spec. issue, *Revue forestière française* (1999): 8.

[175] AD Doubs P 420, Extrait du Régistre des deliberations du Conseil général du department du Doubs, 26 October 1825, Examen du projet du Code forestier, 6.

[176] AD Doubs P 420, Examen du projet du Code forestier, 7, 16.

"If indeed certain properties were to merit more respect than others," the report added, "it would be [the property] of communes and public institutions, these entities being considered minors, to which the law and judges accord a special protection."[177]

THE BATTLE OVER AFFOUAGE

Of the many debates that shaped the Forest Code's final form, the struggle over the proposed law's stance on *affouage* was among the most contentious. The issue revealed more clearly than any other the difficulty of developing a one-size-fits-all policy for France's forests as well as the ongoing tension between the ideology of individuation and the actualities of woodland usage.

During the Revolution, demands by the landless poor and others on the margins of the rural economy for a more equitable distribution of communal firewood and timber achieved little of substance. A 1794 decree authorizing allocation *par tête* proved problematic, and the events of Thermidor that soon followed rendered the law's radical intentions obsolete. Unable to attain meaningful reform, inhabitants without *affouage* rights continued to meet their needs as best they could on their own, picking forest edges clean and pilfering wood wherever possible. "If someone does not stop them soon," railed one observer in 1800, "in a century there will be nothing but one tree left in the service of the French empire."[178] The imperial Conseil d'état took a different tack, declaring in two opinions of July 1807 and April 1808 that communal fuel and timber would henceforth be evenly distributed *par feu* (per hearth, or head of household) rather than per person. As with the abortive 1794 decree, however, these rulings gained the force of law but did not gain a following.[179] This was the case in Franche-Comté, where local notables and powerbrokers in a majority of villages were able to maintain their *affouage* benefits in accordance with ancient custom or even petition for their reinstatement.[180] Disputes, which continued unabated, increasingly ended up in court.[181]

177 AD Doubs P 420, Examen du projet du Code forestier, 16.

178 AN F^{10} 403, M. Cabanez, dept Moselle, au Ministre de l'Interieur, 19 Messidor, year VIII.

179 Dalloz et al., *Répertoire méthodique*, 25, 546.

180 Vion-Delphin et al., *Les hommes et la forêt en Franche-Comté*, 113–15, 23; also R. Weipert, *Du mode de jouissance des bois communaux et de la distribution de l'affouage aux habitants des communes: Necessité de la révision de l'art. 105 du Code forestier* (Lure: Imprimerie de Bettend, 1852), 11–12 and 20.

181 See cases in AN F^{1cIII} Jura 8, Comptes analytique des arrêtés, décisions et travaux de la Préfecture du département du Jura, Trimestre d'Avril 1813.

Seeking to settle the *affouage* issue once and for all, the authors of the draft Forest Code proposed yet another solution: allocate firewood per household and make timber available for purchase by those who need it, at a price to be determined by "experts." Not only would the change assign greater authority to the forest administration, it would also limit the tendency of better-off recipients to sell their unused allocations at a profit. Proceeds from the sale of any excess timber would go to the commune instead.[182]

The Doubs Conseil général was incensed. The draft Code's proposal deviated significantly from the Comtois tradition of granting landowners an annual share of standing timber, proportional to the extent of their buildings. Composed of the very citizens who benefited most from the status quo, the Conseil général fumed that landowners who purchased timber would be "no more rewarded on the price than a stranger" and that workers "without wife nor children who spend a great deal of the year outside their home" would receive a surfeit of firewood while households with extensive family and servants went without. "This equality of allocation is one of the most shocking equalities that could be imagined," declared council members. "The spirit of society," they insisted, "is not a society per person nor per head of household but a society whose principal aim is agriculture and industry." Just as communal taxes were "not divided per household nor per person but with regard to each person's property," the council believed that the benefits of the commons "must also be matched with property."[183]

The Conseil général's insistence on upholding inequitable *affouage* and other property-based privileges demonstrates the importance of use rights across the socioeconomic spectrum. Across Franche-Comté, private forests constituted less than a quarter of total forest cover.[184] In upland areas, especially, inclusion in communal usage and access to the wooded commons were as vital to prosperity as individuated landownership.

182 AD Doubs P 420 and *Code forestier, Sanctionné le 21 mai 1827, et promulgé le 31 juillet suivant*, art. 105, in Jacques-Joseph Baudrillart, *Traité général des eaux et forêts, chasses et pêches*, part 1, tome 3, *Recueil chronologique des réglemens sur les forêts, chasses et pêches, contenant les lois, ordonnances royales, arrêts de la Cour de Cassation, décisions ministérielles, et les circulaires et instructions administratives* (Paris: Imprimerie de Madame Huzard, 1824; hereafter *Recueil chronologique*, tome 3), 538.

183 AD Doubs P 420.

184 The figures were 21, 22, and 18 percent in the Doubs, Haute-Saône, and Jura, respectively. By contrast, private forests comprised more than three-quarters of total forest elsewhere. *AP 2ème série* 51:651–54 (8 May 1827).

Yet the council's stubborn defense of woodland custom also reflected the powerful particularist sentiment underpinning local politics. Hawking a fictive, though no less formative, history of a supposedly more harmonious feudal past, free of political struggles and clashes over resources, this vision of Franche-Comté rejected industrialization in favor of pastoralism, bourgeois land acquisition in favor of noble pedigree, and state centralization in favor of communal control. As evoked in essays published by the departments' learned societies celebrating the region's chivalric past and Sequani civilization, as well as later writings like Charles Nodier's *Voyages pittoresques et romantiques dans l'ancienne France*, Comtois particularism emerged after 1801 when noble influence was resurgent at all levels of office holding and flourished under the Restoration.[185] Like the broader project of rural pacification that accelerated with such force from Napoleon onward, this ideology masquerading as impartial outlook sought to refigure inhabitants as tractable and tranquil, eliding internal clashes over forest rights and emphasizing instead Franche-Comté's lost autonomy. Espoused predominantly by the ultra and legitimist elite, this blend of chauvinism and nostalgia served not only as a rebuke to the ideals of the Revolution but also as a brake on state interference in local affairs.[186]

In the case of the Forest Code, although officials from the Doubs shared the ultra royalist sensibilities of the comte de Villèle, the Code's ministerial backer, they opposed key elements of the proposed legislation, starting with the value of having a general law at all. Expressing the same insistence on self-determination that had given rise in 1793 to the "petite Vendée" on the Doubs plateau and the Federalist movement in the Jura, the council members noted, "In our department, the allocation of firewood, half by household and half in proportion with taxation, and the allocation of timber according to the measurements of houses, is in keeping with the ability of the forest and the needs of the inhabitants, but this correspondence may not exist in other departments where the

[185] Claude-Isabelle Brélot, "Le Sentiment provincial en Franche-Comté pendant la première moitié du XIXe siècle: Persistances et sociologie." In *Provinces et états dans la France de l'Est: Le rattachement de la Franche-Comté à la France*, ed. Maurice Gresset, Cahier d'études comtoises (Besançon: Annales litteraires de l'université de Besançon, 1977), 109–26. Charles Nodier et al., *Voyages pittoresques et romantiques dans l'ancienne France, la Franche-Comté* (Paris: Imprimerie de J. Didot l'Ainé, 1825; repr., Éditions Jeanne Laffitte, Marseille, 1977).

[186] See discussion in Alan B. Spitzer, "The Elections of 1824 and 1827 in the Department of the Doubs," *French History* 3, no. 2 (1989): 153–76.

woods are more abundant or more rare."[187] Such cases, the council concluded, proved the weakness of a "universal and equally applicable rule for all of France."[188]

Xavier Chifflet, deputy from the Doubs and president of the department's Conseil général as well as a member of the Chamber of Deputies' committee on the Code, kept up this line of argument during the legislative debates of March and April 1827.[189] Largely because of his haranguing, the deputies eventually settled on an amendment laying out a more equivocal version of *affouage* than was proposed in the original draft.[190] As voted into law, article 105 stipulated that communities could distribute wood *par feu* if there was "no title or custom to the contrary." Otherwise, existing methods held sway.[191]

A victory from the standpoint of upholding regional custom and communal usage, this confirmation of *affouage*'s status quo was but one way in which legislators rebuffed the egalitarian inclinations of the Code's authors in favor of preferential treatment for the most prosperous members of society. Proposed limitations on property rights would prove even more contentious.

RECONCILING "THE NEEDS OF ALL WITH THE RIGHTS OF EACH"?

Presenting the draft of the Forest Code to the Chamber of Deputies in December 1826, the vicomte de Martignac, director general of the Enregistrement, noted that conservation of the forests was "one of the primary interests of societies and thus one of the primary duties of governments."[192] "All the needs of life are linked to this conservation[;] agriculture, architecture, nearly all the industries seek in it essentials and resources that nothing could replace," he added.[193] In addition to assuring the ongoing availability of firewood and timber, the Code was motivated by the growing scientific awareness of the forest's vital ameliorative and stabilizing powers. As Martignac explained, "It is not only through the riches that are offered by wisely planned forest exploitation that one should judge their utility: their very existence is a benefit protecting and

[187] AD Doubs P 420 (p. 14). [188] AD Doubs P 420.
[189] *AP 2ème série* 50:221–22 (12 March 1827).
[190] On Chifflet's role, see Marlin, "Un problème d'affouage," 171–72, also Vion-Delphin et al., *Les hommes et la forêt en Franche-Comté*, 116.
[191] Code forestier, art. 105, in Baudrillart, *Recueil chronologique*, tome 3, 538.
[192] *AP 2ème série* 49:87 (29 December 1826). [193] Ibid.

nourishing springs and rivers, whether they are undergirding and supporting the soil of the mountains or exerting a felicitous and healthy influence on the atmosphere."[194]

Both a revision and synthesis of the 1669 Ordinance and the disparate decrees issued on the forest since then, the Forest Code stated its mission as "reconcil[ing] the needs of all with the rights of each," while conserving France's "forest riches, the foremost object of its concern."[195] This mission would require some sacrifices, Martignac noted, but, he assured the deputies, the "independence of private property" would be subject "only to the restrictions required by obvious general interest."[196] Thus, while the Code's conservationist measures were portrayed as "indispensable" to the public good, their restrictions would be felt most by rural communities, whose communal usage and customary rights would come under unremitting scrutiny and whose missteps would be subject to harsh punishment under the new law.

Martignac made the state's position on customary rights clear. "These rights represent, for the property of the state as well as for private property, the most redoubtable of dangers, and the most fecund source of damage and abuse," he averred in his presentation to the deputies. "Numerous and powerful efforts have been made to suppress or reduce them, but these have only produced very weak results," he noted.[197] The new Code, Martignac argued, would have greater success, for its regulatory clarity would be combined with the power of an increasingly professionalized forest administration.

Blaming customary usage for the degraded conditions of the state's woodlands, the Code's authors devised a long-term strategy for eliminating or at least reducing the scope of use rights through rigorous substantiation, supervision, and repression. Articles 61 through 85 of the completed Code formed the keystones of this effort. Article 62 forbade the granting of any more use rights "of whatever sort and under whatever pretext" in the forests of the state.[198] As for rights currently in practice, article 61 announced the termination of all that could not be demonstrated as valid within two years from the law's announcement.[199]

[194] Ibid.

[195] *AP 2ème série* 49:88 (29 December 1826). See also Martignac's introduction of the Code to the Chamber of Deputies in Duvergier, *Collection complète des lois*, 27:142.

[196] *AP 2ème série* 49:88 (29 December 1826).

[197] *AP 2ème série* 49:91 (29 December 1826).

[198] *AP 2ème série* 51:165 (11 April 1827).

[199] *Code forestier*, art. 61, in Duvergier, *Collection complète des lois*, 27:200.

This provision, which echoed Napoleon's largely ineffectual 1803 decree, set off a frantic search by villagers for titles, rulings, letters – anything that could certify their long-standing claims to the forests. As draconian as it seemed, however, article 61's final form was a considerable improvement from the initial draft, which proposed terminating all use rights that had not already been authenticated or were not in the process of being confirmed. In one of several amendments aimed at lessening the law's severity, the committee of deputies charged with examining the Code noted that the measure was particularly unfair to rights holders whose claims had never been contested and who, therefore, had never been required to prove their authenticity. A grace period, they argued, would allow these claimants time to plead their case.[200]

Even so, gaining official recognition was not a guarantee of lasting access. Under the provisions of article 63, the state reserved the right to liberate a forest of use rights on a case-by-case basis, through the method of *cantonnement*, the limited attribution of ownership and confinement of rights onto a smaller parcel of woods, generally one-third of the whole. After 1827, and particularly from the July Monarchy forward, *cantonnement* would come to be state and private landowners' most effective – albeit litigious – method of marginalizing use rights.[201]

Pasturage rights, by their nature, were unsuitable for *cantonnement*, but the Code allowed them to be eliminated through financial compensation, the amount of which would be determined, like *cantonnement*, on a case-by-case basis. This proposal raised alarm among the administrators of upland departments, including deputy Chifflet of the Doubs. In his capacity as a member of the Deputies' committee on the Code, he pushed for a softer stance on woodland grazing. "As you know, gentlemen, there are areas where grazing is so indispensable to inhabitants that they have no other income or resource beyond the yields of the animals they raise," noted the committee to fellow deputies in March 1827. "If you strip them of this sole means of existence, you will force them to abandon the soil that has seen them grow, where they have led a peaceful and laborious life, where they exercised a form of industry useful not only to themselves

200 *AP 2ème série* 50:216–17, 50:238 (12 March 1827).

201 See discussion in Andrée Corvol, "La privatisation des forêts nationales aux XVIIIe et XIXe siècles," *Histoire économique et financière de la France: Études et documents* 2 (1990): 219–20; Goujon, "Legislation et droits d'usage," 15–25; and Whited, *Forests and Peasant Politics*, 36–37. On the experience of Franche-Comté, see Plaisance, "Les droits d'usage en Forêt de Chaux et leur cantonnement."

but also to commerce."[202] Calling on "His Majesty's paternal governance," the committee pushed through an amendment that would allow woodland pasturage to persist in state forests where it was an "absolute necessity" for inhabitants' existence. Either way, pasturage would be limited to beneficiaries' personal needs and would not be permitted for commercial aims. Moreover, the forest administration would determine when, where, and how many animals were permitted to graze, and it would be authorized to levy harsh fines – from two to fifty francs, plus prison for repeat offenses – for violations like grazing in unauthorized areas or omitting to bell one's beasts.[203]

Communal woodlands were subject to comparable rigor. Perpetuating many of the terms of the 1669 Ordinance, the Forest Code upheld prohibitions against clearing and division, required communities to maintain a *quart en réserve*, and stipulated that forest officials oversee all marking, felling, and auctions.[204] Here, too, the deputies amended measures they considered too restrictive. In particular, they demanded that issues like the hiring and firing of guards and whether to preserve the *près-bois* for shelter and grazing be decided at the prefectoral and municipal level.[205] It was at this stage that the deputies of the Doubs successfully resisted the Code's controversial democratization of *affouage*. They also worked to reject a costly surcharge on timber and firewood revenues to pay for the cost of the forest administration.[206]

In this way, the Chamber of Deputies tempered aspects of the state's stance on communal usage and customary rights. Yet despite their modifications, few lawmakers disputed the premise that use rights were chiefly to blame for forest deterioration. "These insatiable burdens, as they are justly called, have continued to exist, and never has the danger been greater and more widely acknowledged than in the current era, when one is rightly alarmed by the ever-growing destruction of the woods of the realm," declared Favard de Langlade, the secretary of the Forest Code committee.[207] This outlook carried over into the Chamber of Peers, as well. Arguing in support of the Code before that body, the spokesperson for the Peers' committee on the Code, the comte Antoine Roy, observed that while "everyone can see that the breeds of animals which normally

[202] *AP 2ème série* 50:218, 50:239 (12 March 1827).
[203] *Code forestier*, art. 66–79, in Baudrillart, *Recueil chronologique*, tome 3, 535–36.
[204] *Code forestier*, art. 91–93, 100, in Baudrillart, *Recueil chronologique*, tome 3, 537.
[205] *AP 2ème série* 50:220–21, 50:242–43 (12 March 1827).
[206] *AP 2ème série* 50:221–22, 50:243 (12 March 1827).
[207] *AP 2ème série* 50:216 (12 March 1827).

live in the woods present only degenerated specimens," woodland pasturage was still in "very rare" cases "an indispensable necessity" and thus could be tolerated under the strict supervision of the forest administration.[208]

"THE INTEREST WE MUST PRINCIPALLY PROTECT IS ... THAT OF THE LANDOWNER"

Even as the new law reined in rights holders, offering only the barest concessions in cases of critical need – and this only at the urging of upland legislators like Chifflet – it ended up being remarkably lenient toward private landowners. As Martignac stated before the Chamber of Deputies in 1826, earlier statutes, especially the 1669 Ordinance, had infringed too much on the rights of individual proprietors.[209] By contrast, the 1827 Code promised to exempt private landowners from the forest administration's control. The law "does not dictate to them, nor does it bar any mode of management; conversely, it assures them the most complete protection," observed Martignac.[210] Landowners were free to choose their own guards, confine use rights through *cantonnement*, terminate other customs as they felt necessary, and apply the same penalties to delinquents as in the woods of the state.[211]

Nonetheless, the Code proposed two important constraints. The first authorized the navy to continue to inspect, mark, and exercise preferential purchasing power over timber trees on private woodlands. This prerogative, known as *martelage*, had been eliminated from private woodlands by the law of 1791 and reinstated in 1803. The other constraint imposed a twenty-year temporary limit on clearing in private woods – essentially an extension of the twenty-five-year moratorium decreed by the law of 9 Floréal, year XI (29 April 1803), which was poised to expire in April 1828.

Attempting to justify these requirements, Martignac noted that while privately held forests accounted for slightly more than half the nation's total, state and communal forests were insufficient to meet current needs. Furthermore, he added, the law of 1791's emancipation of private forests had caused landowners to "abuse this unaccustomed liberty," "to the point that in many areas the crumbling of cleared land and the

[208] *AP 2ème série* 51:639 (8 May 1827).
[209] *AP 2ème série* 49:87, 49:93 (29 December 1826).
[210] *AP 2ème série* 49:93 (29 December 1826). [211] Ibid.

deforestation of the mountains caused the vegetated earth to disappear and left bare rocks."[212] These arguments echoed the conclusions of scores of earlier writers on the subject, but they would not be accepted easily among legislators. Hoping to deflect their inevitable criticism, Martignac avowed, "Nothing is more respected, sirs, than the right of property; by its nature this right allows but few limits; it includes, as we know, the authority to use and abuse."[213] "All the same," he concluded, "this important concept must itself yield, as you know, before the still-greater consideration of social necessity and collective upkeep."[214]

Legislators were not convinced. Despite the Code's overall indulgence of private woodlands, the proposed constraints provoked adamant and indignant argument in both chambers. Among the deputies, the comte de Charency protested, "Eh! Why then can I not enjoy my property without authorization? Why is the administration always going to the length of directing us in our agricultural and industrial activities?"[215] The deputy from Corsica, General Sébastiani, similarly bristled, "Do you believe that wood owners are imprudent, ignorant, greedy to the point of losing sight of their true interests, those of their family? I don't think so."[216] In the Chamber of Peers, the comte Roy observed, "The right that is demanded for the naval service ... is not only an inconvenience but a veritable attack on the right of property."[217]

The geography of woodland proprietorship went a long way in shaping deputies' responses to the proposed limitations on landowners. Deputies who raised objections came from departments where forests were relatively few and where private ownership accounted for a high proportion. Whereas only 20 percent of the woods of Franche-Comté were privately held, the comte de Charency's home department, the Orne, ranked in the bottom half of wooded departments, had an almost equal extent of privately owned and state forests, and contained only 288 hectares of communal woodlands. The situation in Corsica was similar.[218] In his home region, the Corsican legislator Sébastiani reasoned, private ownership perpetuated rather than depleted the forests. "Never forget that the wealth of the State is only the result of the fortune of individuals," he admonished, noting that woods were protected in the hands of large

[212] *AP 2ème série* 49:90, 49:93 (29 December 1826).
[213] *AP 2ème série* 49:93 (29 December 1826). [214] Ibid.
[215] *AP 2ème série* 51:126 (9 April 1827). [216] *AP 2ème série* 51:127 (9 April 1827).
[217] *AP 2ème série* 51:647 (8 May 1827).
[218] Figures cited in *AP 2ème série* 51:651–53 (8 May 1827).

landowners, "where the spirit of family is powerful and enlightened."[219] Charency similarly scoffed, "Your committee assures us that there would be a danger in revoking the clearing prohibition. ... All the same, allow me to suggest, that for very strong reasons and extremely powerful motives, one would not clear." In the "very canton where I live," he continued, "the woods bring in a revenue [that is], relatively speaking, reasonably profitable. And indeed, gentlemen, although poor, we do not clear, we plant."[220]

On the basis of this reasoning, Charency proposed applying the clearing restriction only to upland and mountain regions where, he noted, citing the state's own words, forests "nourished and protected springs and rivers."[221] His suggestion was quickly rebuffed, but the deputies nonetheless approved a lesser change: they doubled from two to four hectares the maximum extent of woods that could be cleared without the state's consent.[222] *Martelage* faced even more resistance. Noting that the navy's lumber needs amounted to no more than 36,000 cubic meters, or 48,000 trees per annum – still an astonishing amount in a time of timber crisis – the Chamber of Deputies' committee on the Code argued that state and communal forests should be able to supply this amount, provided their management improved as the new law prescribed.[223] Thus, committee members urged, the obligation of *martelage* should be limited to ten years.[224] In addition to this amendment, the deputies approved several others, including one that reduced the fine for felling timber without authorization from the "exorbitant rate" of forty-five francs to the much lower charge of eighteen francs per meter around.[225]

Overall, the deputies' challenges to the Code illustrate how much more insistently they defended the interests of private property compared with customary and communal usage. Whereas *martelage*, *quarts en réserve*, and the preservation of seed-bearing *baliveaux* remained permanent requirements of communal woodlands, the amended Code limited

219 *AP 2ème série* 51:127 (9 April 1827).

220 *AP 2ème série* 51:126 (9 April 1827).

221 *AP 2ème série* 51:125 (9 April 1827).

222 Discussion in *AP 2ème série* 50:230, 50:257 (12 March 1827); final version in *Code forestier*, art. 223, in Baudrillart, *Recueil chronologique*, tome 3, 546.

223 *AP 2ème série* 50:224 (12 March 1827).

224 Discussion in *AP 2ème série* 50:224 (12 March 1827); final version in *Code forestier*, art. 124, in Baudrillart, *Recueil chronologique*, tome 3, 539.

225 *Code forestier*, art. 125, in Baudrillart, *Recueil chronologique*, tome 3, 539; see also the amendments to art. 126, 128–29, and 131–33, discussed in *AP 2ème série* 50:224–25, 50:246–47 (12 March 1827).

landowners' power to fell and clear for only ten to twenty years – a particularly short period in the lifespan of a tree. In nearly all other ways, private landowners had complete freedom over their forests. Moreover, should they need greater intervention, they were guaranteed the same protections by governmental force as state and communal woodlands.[226]

Of the use rights and easements that the Chamber of Deputies did defend, the two they were most insistent about protecting – *affouage* and *affectations* – involved disproportionate benefits to commercial interests or individuals who were already well-off. As noted earlier in this chapter, the Code that was initially presented to the deputies sought to democratize the distribution of firewood and timber to village inhabitants, but this plan was crushed in favor of preserving modes of *affouage* that benefited the largest taxpayers. As for *affectations* – the term for long-standing easements allowing industries to obtain firewood in certain areas of state forest – these had become, in Martignac's words, "veritable alienations" and a "privilege irreconcilable with the free competition that enriched the country."[227] Accordingly, the draft Code announced that all *affectations* would be terminated as of 1 September 1837.[228] Hoping to thwart rights holders from claiming that their *affectation* had been granted in perpetuity, the draft further noted that in case of appeal, claimants would lose the ten-year grace period.[229]

This proposal did not sit well with the Deputies' committee on the Code, which mounted a vigorous defense. "These *affectations*," explained the committee, "were all accorded with the double goal of encouraging the development of industry and creating new means of consumption for forests that lacked them."[230] Most common in Alsace, Lorraine, and Franche-Comté, only seventeen such concessions were left, "as opposed to a multitude of use rights." Arguing that because *affectations*, like regular use rights, consisted only of the concession of firewood and not land, the committee insisted that they should be negotiated in the

226 *Code forestier*, art. 164 and 189, in Baudrillart, *Recueil chronologique*, tome 3, 542, 544. See Chamber of Peers discussion on this point in *AP 2ème série* 52:159 (19 May 1827).
227 *AP 2ème série* 49:91 (29 December 1826).
228 Martignac explained that the *affectations* had been granted in clear violation of title 20, art. 11, of the 1669 Ordinance, and were therefore invalid. *AP 2ème série* 49:91 (29 December 1826); presented in art. 58 of the draft *Code forestier*, in *AP 2ème série* 49:100 (29 December 1826).
229 Art. 58 of the draft *Code forestier*, in *AP 2ème série* 49:100 (29 December 1826).
230 *AP 2ème série* 50:215–16 (12 March 1827).

same way: before the courts, on a case-by-case basis.[231] The argument carried the day, as did an amendment to award the ten-year grace period even to *affectations* under appeal, regardless of outcome.[232]

These legislative maneuvers notwithstanding, it was in the punishments and prosecution of woodland violations that the Code's biases became most apparent.[233] Spelled out in titles X through XII, the rules prescribed remarkably high fines for activities that were not only common practice in most forests but also critical to the survival of rural communities. Some of the most sweeping prohibitions were contained in article 144, which warned that

> all unauthorized extraction or removal of rocks, sand, minerals, soil or turf, peat, heather, gorse, grass, live or dead leaves, fertilizer on the forest floor, acorns, beechnuts and other fruits or seeds of the woods and forests will give rise to fines set as follows: loads taken by wagon or cart, 10 to 30 francs per draft animal; loads taken by pack animals, 5 to 15 francs per load; loads carried per person, 2 to 6 francs.[234]

Far from moderating this or other draconian clauses, the Deputies' committee on the Code heartily seconded it, noting, "For the conservation of the forest, establishing rules and principles and setting up an active and methodical surveillance is not enough. . . . The prompt and severe repression of infractions is in truth the most efficient means of preventing the devastation of the woods."[235] In the Chamber of Peers, the only uncertainty that legislators raised about the measure was why the removal of organic material should ever be allowed at all. "It would be desirable," remarked the comte de Mailly, "if the extraction of leaves and heather were forbidden in the most absolute manner."[236] As with the other use rights that benefited the poorest sectors of society, this concession was grudgingly defended on the argument that in some regions, refusing it "would render farming impossible and compromise the very existence of the population."[237]

Though deputies also protested the Code's harsh fines for grazing violations, which could reach as high as ten francs per animal in repeat

[231] Ibid.
[232] *Code forestier*, art. 58–60, in Baudrillart, *Recueil chronologique*, tome 3, 534.
[233] *Code forestier*, title XI, art. 159–87, in Baudrillart, *Recueil chronologique*, tome 3, 542–44.
[234] *Code forestier*, art. 144, in Baudrillart, *Recueil chronologique*, tome 3, 541–42.
[235] *AP 2ème série* 50:227 (12 March 1827).
[236] *AP 2ème série* 52:147 (19 May 1827).
[237] Le Marquis de Bouthillier, *Directeur général* des forêts, in *AP 2ème série* 52:147 (19 May 1827).

incidents, their demands for a reduction were rejected.[238] "There are years when a sheep, after it has been shorn, is worth no more than two or three francs," lamented M. de Montbel of the Indre during the Chamber's final debate on the Code.[239] As Montbel's meek protest suggests, for the most part, deputies' levied their most effective objections against the penalties that impinged on landownership. Chifflet of the Doubs underscored this point during the discussion of article 192, regarding fines for cutting or removing trees. "The interest we must principally protect," he contended, "is not that of the Treasury in establishing strong fines, but that of the landowner, to whom harm has been done and who should be compensated." "The majority of forest delinquents are not rich," Chifflet added, calling for the per-tree penalties to be reduced. "If you keep such high penalties," he argued, "they will absorb their worth [in costs and fines], and there will be nothing left to pay the damages."[240] His amendment passed.

In practice, the Code's punitive measures tended to affect the poorest of the poor the most, not only because their lack of property and inability to pay high wood prices made them more likely to rely on customary rights, but also because they rarely contested their charges.[241] Larger violations carried out by or on behalf of manufacturers or landowners were harder to prove and more likely to be challenged – a situation that led to these offenders' being let off more frequently. In addition, unlike minor offenders caught in the act of gathering firewood without permission, landowners received a warning before being reported.[242] The Code stipulated that convicted offenders be held until they had paid all fines, damages, and other penalties. However, if deemed insolvent, violators could be incarcerated for two months and twice that in cases of repeat offense.[243] Sentences handed down by the courts could extend the imprisonment further. Legislators accepted these proposals with few objections. In the Chamber of Peers, the sole protest raised was by the duc de Praslin, who complained about landowners' having to foot the bill for feeding

238 *Code forestier*, art. 199–202, in Baudrillart, *Recueil chronologique*, tome 3, 544–45.

239 *AP 2ème série* 51:123 (9 April 1827). 240 *AP 2ème série* 51:103 (7 April 1827).

241 *Code forestier*, art. 179: The accused who wanted to contest a forest guard's report had to do it in writing and in person by a notarized proxy, or court clerk before the audience indicated in the citation. See discussion in Dumoulin, "Poursuites en réparation des délits," 176.

242 Dumoulin, "Poursuites en réparation des délits," 169.

243 *Code forestier*, art. 213–14, in Baudrillart, *Recueil chronologique*, tome 3, 545.

offenders while under arrest, if they wished to recover the damages to which they were entitled.[244]

"TIMBER [IS] THE PRINCIPAL AIM OF CONSERVATION"

Their disagreements over aspects of the Code aside, legislators on the whole concurred with the state – here represented by Villèle, Martignac, and the ultra-royalist director general of the forests, Constantin-Marie-Louis Léon de Bouthillier-Chavigny – concerning the primary importance and urgency of protecting France's woodlands. On the minds of many was the imminent expiration of the moratorium on clearing in private forests that had been imposed by Napoleon in 1803. As the comte Roy warned his fellow Peers in May 1827, beseeching them not to challenge the Chamber of Deputies' amendments, "We must not lose sight of the fact that the delay prescribed by law which, for twenty-five years, has prevented clearings without prior authorization, expires next April, and if the deliberations of the other Chamber becomes useless in this session, the law could very well not be proclaimed before great disorders [erupt]."[245] The Peers heeded his warning and voted to approve the Code on 21 May 1827.

Roy's exhortation to his fellow lawmakers reflected the deep misgivings that many of them felt about granting landowners unfettered control of their forests. Martignac made no secret of his concern, noting, "in an unlimited liberty there is danger for the health of many departments, for the consumption of woods necessary for naval and civil construction."[246] Despite this admission, the Code's drafters proposed extending clearing restrictions for a mere two decades. For all its administrative rigor and crusading rhetoric regarding woodland preservation and the protection of the general interest, when it came to politically unpopular measures, the Code employed the same evasive maneuvers and timorous tactics as every other attempt at forest reform from the Revolution forward.

Scrupulously careful to avoid impinging on the rights of private property, the Code was all the more exacting of common rights holders. By inserting the forest administration at every level of woodland use, from firewood distribution to mushroom picking, the Code rendered rural communities dependent on a growing state bureaucracy and prevented

[244] Charles-Raynard-Laure-Félix Choiseul, Duc de Praslin, in *AP 2ème série* 52:152–53 (19 May 1827).

[245] *AP 2ème série* 51:650 (8 May 1827).

[246] *AP 2ème série* 51:128–29 (9 April 1827).

them from upholding and adapting their own modes of allocation to be more equitable. Individuals and groups whose customary practices could not be validated by documentation or decree were recast as delinquents. For those who were able to authenticate their rights, the Code encouraged landowners and the state to carry out *cantonnement*. Squeezed onto a fraction of their former territory, forest users were also subject to increased scrutiny by an ever more efficient guard force. In this way, the Code engendered greater conflict between rural inhabitants and the forests' enforcers.

What could explain this double standard in a law ostensibly informed by science and experience, and devoted to the conservation of forests "for present and future generations"?[247] There are a number of possible answers. One lies in the theory of civil equality ushered in by the Revolution, which rejected the imprecise currency of hierarchy and human relations in favor of the conviction that in absolute private property lay individual liberty. In this new social and political order, which was confirmed by the Civil Code, there could be no room for the old notion of property as a bundle of unevenly distributed rights. Rather, the vehemence of bourgeois legislators' denunciations of communal tenure must be understood as a necessary reinforcement of their theoretical right to power.[248]

Though some condemned the growing effort to curb use rights as a "serious attack on the public interest" and a violation of "the most sacred of pacts," the persistence of even the mildest forms of customary and communal usage was a thorn in theorists' side – a reminder than a historically viable alternative to atomization and exclusion existed.[249] For this reason among others, political theorists as much as agronomists encouraged curbing rights holders' claims through *cantonnement*. Already commonplace prior to the Forest Code, *cantonnement*

[247] *AP 2ème série* 49:87–88 (26 December 1826).

[248] See discussion in Paolo Grossi, *An Alternative to Private Property: Collective Property in the Juridical Consciousness of the Nineteenth Century*, trans. Lydia Cochrane (Chicago: University of Chicago Press, 1981), 9, 16, 23. These ideas are discussed in Pierre-Joseph Proudhon, *What Is Property?*, ed. and trans. Donald R. Kelley and Bonnie G. Smith (Cambridge: Cambridge University Press, 1994); and in Peter Ørebech, *The Role of Customary Law in Sustainable Development* (Cambridge: Cambridge University Press, 2005), 12–24.

[249] Latruffe-Montmeylian, *Des droits des communes sur les biens communaux, ou examen historique et critique des démembremens des Usages communaux, opérés autrefois sous les noms de Réserves et de Triages, et aujourd'hui sous celui de Cantonnement* (Paris: Libraire de la Forest, 1825), 1:xvi.

proceedings would rise to a flood after 1827. As one author presciently observed in 1825, "The condition of our commons is much more precarious than one thinks; ... really, in my opinion, I would dare say that given the jurisprudence and doctrines currently in force, not a single one of our villages can hope to escape a demand for *cantonnement* directed against its inhabitants."[250]

In addition to ideology, economic expediency drove the Code's clampdown on customary and communal usage. While the conservation of timber for maritime use remained a primary concern – "timber trees," Roy reminded his fellow legislators, "are the principal aim of conservation"[251] – raising the state's forest earnings was also paramount. In times of fiscal need, the state continued to regard its forests as a backup source of revenue, yet the land auctions of the Restoration, less lucrative than expected, had shown that use rights were an impediment to sales. In this respect, the goal of making woodland parcels more marketable helped shape the Code's sections on divesting use rights through *cantonnement*.[252]

The Code's indulgence of private landowners and related timbering and manufacturing similarly derived from economic concerns. The law provided few specifics on wood-burning industry and commerce, apart from its broad prohibitions on unauthorized cutting and its clause limiting the proximity of forges and factories to national forests.[253] The dearth of directives concerning manufactures may have reflected the fact that such establishments were more likely to be located on and powered by private woods, in which the forest administration had no lasting regulatory authority. Nonetheless, the Code's relative silence on such a significant cause of woodland deterioration, as well as its guarantee of a ten-year grace period to *affectations* – five times longer than the grace period granted to common-rights holders – tacitly signaled the state's support of industrial and commercial pursuits. By not holding forest owners to the same rules of management as applied in the state and communal forests, the Code effectively encouraged them to profit from their forests as they saw fit, while designating the lion's share of responsibility for timber conservation to national forests, "naturally assigned to the needs of the state," and communal woodlands, "more particularly required to provide for the needs of society."[254]

[250] Ibid., 1:x. [251] *AP 2ème série* 51:639 (8 May 1827).
[252] Corvol, "La privatisation des forêts," 218–19.
[253] Code forestier, art. 151–52 and 157, in Baudrillart, *Recueil chronologique*, tome 3, 541.
[254] The comte Antoine Roy discussing *martelage*, in *AP 2ème série* 51:647 (8 May 1827).

Recognition of the forests' ecological significance and the dangers of deforestation was another area in which the Code's severity toward customary and collective usage contrasted sharply with its modest stance on landowners' rights. From the 1860s forward, the forest administration would begin a massive reforestation campaign of the sort that theorists like Jean-Antoine Fabre and François Antoine Rauch had urged for decades, but these projects, like the Forest Code, would impinge on communal autonomy and practice more than on individual proprietors.

A fourth contributing element was the state's goal of suppressing dissent – human and otherwise – and creating an acquiescent and industrious citizenry. As the Haute-Saône engineer Houry argued, "in respecting forests and cultivating woods, men [would] learn to cherish their duties and practice virtue."[255] Founded on contempt for activities and animals that could not be easily contained within prescribed boundaries, like woodland pasturage, goats, and human foragers, the Code's multitudinous restrictions and the forest administration's accompanying adherence to rational exploitation techniques functioned as a ruthless form of social control in areas where, from Louis XIV forward, the state had previously had little success in enforcing order.

The Code's mission of forest regeneration and woodland protection was not in itself unsympathetic, nor was it unwarranted. Nonetheless, the vision of conservation that it ultimately imposed was fundamentally flawed by its failure to acknowledge all the forests' stakeholders and meaningfully address their needs. Emphasizing a statist conception of the public good and undermined by its concessions to bourgeois liberalism, the Forest Code fell short of its mission of impartial oversight and prudent exploitation. The new law was just a weak copy of the 1669 Ordinance, and the minister of finances' forest plans were guided only by "men who are only financiers, and *savants* who do not know or who look down upon rural issues!" as the disappointed and disillusioned Rougier de la Bergerie fumed in an essay of 1831.[256] These missteps on the part of France's lawmakers would lead rapidly to opposition and tumult among woodland inhabitants.

[255] Houry, "Mémoire sur cette question," 82.

[256] J. B. Rougier de la Bergerie, *Mémoire au Roi et aux Chambres législatives sur la déstruction des Bois et sur les graves conséquences qui peuvent en résulter* (Paris: G.-A. Dentu, 1831), 53.

6

"Not even a branch of wood has been granted to us"*

> Here in sum are the principles that will be inscribed in the *Démocrate Franc-Comtois* as the basis, the rule, the torch of its mission, if democracy will lend its initiative, its strength, its faith[:] ... the revision of the forest laws and customs, above all those concerning the allocation of timber trees and the overly absolute ban on woodland grazing and pasturage.
>
> Description of a proposed new "journal of popular interests," 6 December 1849[1]

The product of years of proposals, deliberations, drafts, and debates, the Forest Code was approved at last on 21 May 1827.[2] An accompanying *ordonnance réglementaire* containing transitional procedures and details of implementation appeared nine weeks later, on 1 August.[3] Together, these measures armed the forest administration with a level of authority and focus that it had lacked since the time of Colbert a century and a half before.

Though it would take years before the Code was fully executed, a hardening administrative attitude was apparent almost immediately.[4]

* The quotation in the chapter title is from AN BB[18] 1460 (5248A), Petition of the residents of St. Laurent de Cerdans to the Commissaire de gouvernement, Perpignan, 1848.

1 The proposal appeared in *Le Franc-Comtois, Journal de Besançon et des trois départements*, 6 December 1849, p. 1.

2 *Bulletin des lois*, 21 May 1827, no. 176.

3 The Code was promulgated on 31 July 1827. For full text of the *Code* and the accompanying *ordonnance réglementaire* that appeared the next day, see Baudrillart, *Recueil chronologique*, tome 3, 530–47 and 548–64.

4 On the law's implementation, see AD Doubs P 425, *État général des bois communaux et d'établissements publics de département du Doubs, soumis au Régime forestier*, May 1828, and other related correspondence in the same carton, 1828–29.

In early July, for example, Villèle sent a letter rejecting the mayor of Besançon's request to allow inhabitants to cut coppice stems to bind their sheaves of grain – a practice he had consented to in the past. Though the harvest was only a month away, the minister was unbending. "Toleration of this sort has always been considered contrary to the conservation and growth of woods," he asserted, "and if, in 1825, I authorized it under certain conditions, I was persuaded less by the preventive measures that were taken to monitor it, than by the fear of suddenly depriving the communes of a resource on which they had been able to rely." Cultivators should take this as an opportunity to develop other means of binding their harvest, Villèle added.[5]

The state's response to a minor uprising the following winter in the southwestern town of Auch showed similar resolve. As the *procureur du Roi* explained in a letter of March 1828, the recent sale of the town's communal woods, "which provided resources to the indigent for their heating during the winter," had "excited the discontent of the common people," who were now being cited daily for trespass by the new owner's guards.[6] "Since the sale," the *procureur* noted, "the poor have not ceased to go there, in keeping with custom, to make their faggots."[7] Thirty were convicted in six weeks alone. When the new landowner's guards arrested eight women one morning for gathering firewood, popular outrage ignited, and some one hundred fifty people, "the majority belonging to the worker class," descended upon the town hall to demand their freedom. "Around seven o'clock the mayoral plaza was jammed," reported the beleaguered *procureur*. "Down with the mayor, long live the woods of Auch, let the women go, were the cries that could be heard all around."[8] Local authorities ordered soldiers to "cross bayonets" and "suppress the curious and ill-intentioned." The courts weighed in next, sentencing the uprising's alleged instigators (four men and four women) to sentences of six months in prison for the men and a year each for three of the women. The fourth woman was exiled from the community for fifteen months, followed by ten years of surveillance.[9] "I think that justice has been sufficiently done," remarked the state's prosecutor approvingly in a letter of May 1828.[10]

Examining conflicts like the Bisontins' bid to gather sheaf fastenings and the Auscitains' insistence on pursuing their former firewood rights, this chapter shifts back from the political and ideological context of

[5] AD Doubs P 420. [6] AN BB^{18} 1159 (332). [7] Ibid. [8] Ibid. [9] Ibid.
[10] Ibid.

woodland reform explored in Chapters 4 and 5 to an analysis of the way forest policy was implemented, experienced, opposed, and altered at the local level. In the years after the Forest Code's promulgation, state and woodland administrators, emboldened by the new law, ruled against customary and collective practices with increasing rigor and conviction. Rural inhabitants, in turn, developed dogged and sometimes destructive ways of upholding their forest access. The 1848 revolution would give the peasantry a second chance to obtain the forest freedoms they had demanded in their *cahiers de doléances* a half century before. Armed with a clear political agenda based on specifically rural concerns, the Francs-Comtois would seek, once more, to achieve the woodland autonomy they had long ago enjoyed and ever since sought to restore.

CLAIMS, CONTESTATION, AND *CANTONNEMENT*: THE FOREST CODE'S RECEPTION ACROSS FRANCE

As the disturbance in Auch suggests, the elimination of access to forest resources disproportionately affected inhabitants on the margins. Many had no choice but to carry on with their now-illegal activities and face the consequences of getting caught. The trend is evident in departments' annual "Inventory of Offenders against whom arrest for debt has been carried out, in accordance with articles 211, 212 and 213 of the Code."[11] Based on figures submitted by the criminal courts of each district, the tally for the Jura in 1828 shows that while a handful of delinquents were able to pay their fines and quickly obtain release, the vast majority – 104 of a total of 127 convicted – were classified as insolvent and remained in prison for an average of one month.[12]

Following the lead set by the Forest Code, an increasing number of private landowners also began revoking or restricting communal access to their forests after 1827. As in Auch, a great number of cases concerned trespass charges against villagers who pursued their long-standing customs without proof of title. In the Haute-Saône, for example, a member of the former seigneurial family of de Beauffremont sued the tiny community of Neuvelle-lès-Scey for leading their flock through his woods. Although the villagers held a legitimate right to water their animals at a spring on the other side of the forest, they had no documentation substantiating their

[11] AN BB^3 195.

[12] AN BB^3 195, Cour royale de Besançon, *Etats des délinquans forestiers contre lesquels la contrainte par corps a été exercée en vertu des articles 211–3 du Code forestier.*

rights to cross through it. As a result, the court upheld de Beauffremont's complaint and ruled that inhabitants must henceforth find another route to the spring.[13]

To avoid difficulties of this sort in state woodlands, *usagère* communities hastened to substantiate their use rights within the two-year grace period the Forest Code allowed. Between 1827 and 1829, scores of villages across France scrambled to procure the evidence needed. In the Doubs district of Pontarlier alone, twenty-three communes and one individual – the widow Jobez, of the prominent forge master clan – filed to uphold their claims. Of these, nearly half registered with the Doubs prefecture only a few months or in some cases only a few days before the July 1829 deadline.[14]

Overwhelmed by the deluge of paperwork, the state put up little resistance in the region. Rather, the director of Enregistrement et des Domaines, who was responsible for coordinating deliberations among his own agency and the forest administration, urged the mayors and lawyers of the Doubs to accept a six-month delay of their hearings, "and not to insist that the Court rule immediately on their claim."[15] Perhaps anticipating the clashes that would ensue when the state started divesting its forests of use rights through *cantonnement*, the office of Enregistrement opted initially for conciliation. As the director noted, "the Government will have no difficulty accepting those [usages] that are founded on titles and in any case the communes are guaranteed recourse to the courts."[16]

Nonetheless, competing priorities kept easy resolution at bay. While the director of Enregistrement feared "creating unnecessary costs" to fight users' claims in court, the minister of finances, under whose purview the forest administration fell, emphasized the Code's enforcement.[17] Dismayed by the leniency shown toward woodland practices, the minister issued a bulletin to all prefects demanding to know if "the rights of the State had been seriously examined and suitably defended," and instructing them to appeal every decision made in claimants' favor.[18] Such was

[13] *De Beauffremont c. Habitans de la Commune de Neuvelle-les-Scey*, in Cour de Cassation, chambre criminelle, 20 March 1830, *Journal du Palais* 23 (1830–31), 289.

[14] AD Doubs P 579, *Droits d'usage dans les forêts royales: Communes qui réclament des droits d'usage devant les tribunaux conformément à l'article 61 du Code forestier*, November–December 1829.

[15] AD Doubs P 579 (30), Direction générale de l'enregistrement et des domaines, Doubs, Direction de Besançon, à M. le Préfet du Doubs, 8 December 1829.

[16] AD Doubs P 579 (30), 8 December 1829. [17] Ibid.

[18] AD Doubs P 579, Le Roy, Minister of Finances, to the Prefects of France, 20 January 1829.

the situation in Aix, where the Cour de Cassation overturned a lower court's decision to permit a certain "sieur Pecoul" to pasture goats along with his sheep in the communal woods of Selonnet, in the Alpes-de-Haute-Provence. Pecoul had argued that a certain number of goats were needed to lead his otherwise doltish sheep through the forest, but this contention was decisively rejected by the appeals court. "The courts cannot introduce an exception to the general and absolute prohibition against introducing goats into the forests," concluded the judges.[19]

Even when claimants succeeded in legitimating their use rights, the forest administration found other ways to thwart them. In the Doubs hamlet of Hauterive-la-Fresse, a tiny assemblage of farms just north of Pontarlier, municipal authorities petitioned in 1828 to guarantee the commune's long-standing right to dead wood in the nearby crown forest of Ban.[20] Although the state strenuously contested it, the tribunal of Pontarlier twice upheld the commune's claim.[21] Nonetheless, the forest administration refused to concede failure, and without its permission, villagers could not obtain the wood to which they were entitled. After prevailing in court for the second time, in 1832, Hauterive-la-Fresse's municipal council beseeched the prefect of the Doubs to pressure the regional forest inspector to comply with the court's decision. Emphasizing the urgency of the situation, the mayor noted that "the hamlet of La Fresse, located on the highest mountain in the Doubs, is covered with snow and frost nearly eight months of the year."[22] The wood had to be delivered by September, or else the community would freeze.

Ten months later, after what must have been a very cold winter, the mayor was still awaiting a response. In desperation, he turned to the subprefect of Pontarlier. "I very humbly inform you," he wrote in April 1833, "that I continue to be tormented by the inhabitants of La Fresse about their right to the dead wood in the Forest of Ban, a right which the forest administration has long denied them, but which the court judged in their favor a year ago January."[23] Confronted with the threat of yet a

[19] *Forêts c. Pecoul*, in Cour de Cassation, chambre criminelle, 7 May 1830, *Journal du Palais* 23 (1830–31), 452.
[20] AD Doubs P 579 (30).
[21] See ruling of 9 January 1832, Tribunal de Pontarlier, discussed in AD Doubs P 579.
[22] AD Doubs P 579 (19), Letondal, Mayor of Hauterive-la-Fresse, to Derville-Maléchard, Prefect of the Doubs, 8 June 1832.
[23] AD Doubs P 579 (19), Mayor of Hauterive-la-Fresse to Demesmay, subprefect, 1 April 1833.

third round of litigation as well as ongoing depredations in the woods, forest officials finally relented.

The experience of Hauterive-la-Fresse was typical of villages in Franche-Comté in the years after the Forest Code. While in this instance the conflict was resolved in the inhabitants' favor, outcomes just as often went the other way.[24] In either case, the forest administration's stringency won it few friends. By challenging rural inhabitants' needs and norms, the state sowed resentment where it hoped to achieve respect. Chafing against the Code's dictates as well as against guards and officers' increased surveillance, individuals and groups rebelled in a variety of ways.

The most flamboyant and well-known of these responses unfolded in the Pyrenean department of the Ariège, where forge masters' purchase of large tracts of woods had reduced the fuelwood and pasture available for inhabitants. Conflicts over the situation predated the Forest Code, but by empowering and encouraging landowners to divest their properties of use rights, the new Code ignited inhabitants' discontent. Starting in the late winter of 1828, a time of year when access to fuelwood was particularly vital, groups of peasants retaliated against the charcoal burners, and forest guards who had obstructed their traditional woodland practices. The rebels – in this case all male – were given the name "demoiselles" because their untucked tunics made them appear vaguely feminine. Disguised in feathered headdresses and blackened faces, they assailed guards who tried to impound their flocks, burned down the huts of charcoal burners, and drove off any remaining usurpers with gunfire. The rebels also sent intimidating letters to authorities. After three years of effort, the state finally suppressed the revolt by the end of 1831. Nonetheless, sporadic violence continued. By the 1840s, arrests and convictions for forest crimes in the Ariège would reach into the thousands.[25]

The outbreak of revolution in July 1830 provided an impetus for more widespread protests against the Forest Code. Opposition was especially fierce in the eastern and southern areas of the country. While the spring had been marked by isolated incidents of woodland violence – in the Loiret, for example, inhabitants "exasperated by the severity of some

[24] See for example, Brélot, "Pour une histoire des forêts comtoises dans la première moitié du XIXe siècle: Le procès de la Haute-Joux." *Société d'émulation du Jura: Travaux* (1979): 181–225.

[25] Whited, *Forests and Peasant Politics*, 40. On the uprising, see Goujon, "Legislation et droits d'usage," 20; François Baby, *La guerre des demoiselles en Ariège, 1829–1872* (Montbel: Éditions Cairn, 1972); Merriman, "The Demoiselles of the Ariège, 1829–1831," 89–90; and Sahlins, *Forest Rites*.

forest officers" set the Forest of Orléans ablaze – the momentary lapse in authority brought about by the ouster of Charles X and his replacement by Louis-Philippe provided the opportunity for broader resistance.[26] In the Puy-de-Dôme, villagers angry at the subjection of their communal woods to the state administration assaulted a forest guard, while in Belfort, inhabitants pursued their customary rights in the state forest in direct challenge to woodland authorities.[27] In the district of Sarrebourg, in the Moselle, the *procureur général* reported in mid-August that "most serious abuses" had been committed in the royal forests and that forest guards and *gendarmes* had been "met with a rebellion of more than one hundred fifty people armed with rocks and sticks."[28] Though these incidents were on the whole short-lived, they demonstrated widespread hostility among rural inhabitants toward the new Forest Code.

In Franche-Comté, reactions to the 1830 Revolution centered primarily on the taxes levied on commercially sold wine. Besançon and Arbois – the latter was long a hotbed of republican and Bonapartist sentiment – both saw the outbreak of winemakers' riots in September 1830. In Arbois, the *procureur général* reported, "a considerable gathering of individuals of all sexes and all ages" converged at the tax office, smashed in the door, heaved the furniture and tax registers out the window, and burned the whole lot, "to the acclaim of the multitude."[29] National guardsmen assigned to disperse the crowd either joined in or slunk away.[30] The next day, the tax comptroller who had come to investigate was "recognized by the throng, which threatened to maltreat him," and had to flee into the countryside, where he was chased by "a dozen individuals" and escaped only by hiding in a cellar for hours.[31] Nearly six months passed before the purported ringleaders were brought

[26] AN F^7 9336, Lettre du secrétaire général de la préfecture du Loiret au ministre de l'Intérieur, 5 April 1830. For incidents of woodland unrest after the 1830 revolution, see AN BB^{18} 1187–88; also AN BB^{18} 1195 (5218).

[27] AN BB^{18} 1186, August 1830; AN BB^{20} 53. See Roger Price, "Popular Disturbances in the French Provinces after the July Revolution of 1830," *European Studies Review* 1, no. 4 (1971): 338, 341.

[28] AN BB^{18} 1186 (3976), Délits commis dans les forêts de Sarrebourg (Meurthe), August 1830.

[29] AN BB^{18} 1188 (4269).

[30] AN BB^{18} 1188 (4269), Troubles au sujet des contributions indirectes, Arbois, October 1830–March 1831. The winemakers of the Battant quarter of Besançon were known as "*bousbots*." See discussion in Fiétier, *Histoire de la Franche-Comté*, 384–85; also Pamela Pilbeam, "Popular Violence in Provincial France after the 1830 Revolution," *English Historical Review* 91, no. 359 (1976): 287 and 293.

[31] AN BB^{18} 1188 (4269).

into custody, a lag that attested either to local authorities' intimidation or to their support.[32]

The change in royal regime seemed to barely register with Franche-Comté's woodland citizenry, who were focused on the increasingly distressing ramifications of the Forest Code. The Code's preservation of inequitable *affouage* customs was particularly contentious. Denied a share of their community's timber resources, two hundred people from the village of Petit-Noir, near Dole, rioted in the spring of 1831. *Affouage* protests also occurred in the Haute-Saône in January 1832.[33]

In addition to *affouage*, inhabitants also found themselves struggling against *cantonnement*, the method by which landowners and the state divested their forests of customary rights. Founded on legislators' certainty of communal usage's ills, and authorized and encouraged by the Forest Code, *cantonnement* was virtually impossible to combat except through the rearguard action of revolt. Although the procedure increased the size of communal holdings, the final acreage available to inhabitants was generally smaller than the forest they had previously been able to access through their use rights. *Cantonnement* further excluded rural communities by clearing the way for the forest administration to enact strict management regimes on the newly unencumbered national woodlands.

Deeply unpopular, *cantonnements* proceeded apace in state forests from 1830 forward. They were particularly widespread in the Jura, where the state's forest holdings were exceptionally large.[34] Jurassien resistance, in turn, was fierce. In March and April 1831, the *cantonnement* of the large state-owned Forêt de la Fresse, near Champagnole, gave rise to open rebellion. Thiessé, prefect of the Jura, blamed local leaders as well as the forest administration for the uprising, declaring, "One has seen this magnificent forest of pines overrun by entire populaces, the *curés* leading the way, and the guards themselves encouraging it or letting it happen."[35] "In general," he added, "the forests of the Jura call for an active and enlightened management ... they are very much behind those of the Doubs, [though] they form one of the greatest advantages of the department."[36]

32 Ibid. 33 Fiétier, *Histoire de la Franche-Comté*, 385.

34 Vion-Delphin et al., *Les hommes et la forêt en Franche-Comté*, 123.

35 AN F^{IcIII} Jura 8, M. Thiessé, Préfet du Jura à M. le Ministre des Travaux publics, Rapport général sur la situation physique & morale du Jura, 5 September 1831.

36 AN F^{IcIII} Jura 8.

"FAR FROM REESTABLISHING PUBLIC TRANQUILITY, [IT] HAS ONLY MADE THE PROBLEM WORSE"

Critics of the Forest Code across the political and professional spectrum cited incidents like these to emphasize what they regarded as the law's obdurate and unworkable emphasis on state control. The ever-vigilant agronomist Rougier de la Bergerie once again advocated local oversight, arguing in an 1831 epistle to the government,

> Must [the evidence] penetrate all the sanctuaries and recesses of the top administration for one to be convinced that a mayor and municipal officials, elected and appointed, can better judge the importance and urgency of an undertaking useful to their commune than a head clerk in Paris, who knows nothing about it and who will make them wait two or three years for the ministry's decision?[37]

"The government and the Chambers must not deceive themselves about the centralization that an imprudent minister finds admirable," Rougier de la Bergerie warned, "for, during civil unrest, it can result in a grave local revolution."[38]

The editors of the legitimist mouthpiece the *Gazette de Franche-Comté*, Jacques Curasson and Albert de Circourt, also railed against the Forest Code's failings. Contending that the Code, like other projects extending the state's power over the departments, had "divided the people into two large classes: the people of the cities and the people of the countryside," de Circourt and Curasson played on local outrage over forest restrictions to advocate their vision of a revitalized feudalism and a return to Franche-Comté's ancient autonomy. "Liberalism has forgotten [the people of the countryside] in its plans for regeneration," fumed de Circourt, condemning the government's coercive tactics.[39]

Professional foresters like E. Tourney, the former inspector of operations in the Bois de Boulogne, similarly contended that the Forest Code had ham-fistedly inflamed tensions at a time of exceptionally high wood prices. Citing the demoiselles of the Ariège, revolts in the Meuse, and recent incidents of woodland arson in the Aube and Yonne, Tourney asserted that "far from reestablishing public tranquility, [the Forest Code]

[37] Rougier de la Bergerie, "Mémoire au Roi et aux Chambres législatives sur la déstruction des Bois," 71.

[38] Ibid.

[39] Albert de Circourt, in the *Gazette de Franche-Comté*, 18 January 1832, cited in Brélot, "Le sentiment provincial en Franche-Comté," 119, 117.

has only made the problem worse."[40] With no recourse but to "procure [wood] by fraud," he noted, trials for forest delinquency were increasingly rapidly. Cases initiated by the forest administration alone had risen 21 percent – from 57,002 cases to 68,964 – between 1825 and 1828.[41] "If private landowners pursued as many allegations as the government, proportionate to the quantity of wood that they possess," Tourney observed, "there would be three to four thousand convictions in France every year; and soon the entire population surrounding wooded areas would be prosecuted before the courts."[42]

Rather than changing the law, Tourney suggested a silvicultural solution. The means to reducing the cost of wood and averting social conflagration, he argued in his 1832 *Nouvelle méthode d'aménagement et d'exploitation des forêts*, lay in implementing periodic clear-cutting and renouncing selective felling, in planting imported tree species proven to be hardy and high producing, and in greatly expanding the technical training available to guards and junior officers through the establishment of *écoles normales des forêts* across France.[43]

In addition to criticizing the Forest Code's adverse social consequences and inefficient techniques, woodland observers also denounced the state's abiding view of its forests as little more than a cash reserve. Incensed by a March 1831 proposal by Louis-Philippe's minister of finances, Lafitte, to sell 300,000 hectares of state woods, Rougier de la Bergerie declared that "a nation that will not perish must preserve [its forests] in large extent, and defend against the designs of individuals."[44] "Be the enemy at the gates of its capital, it must never cede to private hands that which represents its needs and those of its descendants; it must always fear the passions of the gambler and the spendthrift," Rougier de la Bergerie insisted."[45]

The Forest Code's weak and temporary restrictions on private forest owners were especially controversial. Long after they passed into law, they continued to attract criticism. Private landowners seeking to clear their woods easily bypassed the twenty-year moratorium on clearing, provided they could secure the necessary authorization.[46] Even when they

[40] E. Tourney, *Nouvelle méthode d'aménagement et d'exploitation des forêts; suivie de la 3e édition de la Nouvelle méthode de semis, plantation et d'aménagement* (Paris: M. E. Tournay, 1832), 8.

[41] Tourney, *Nouvelle méthode*, 11. [42] Ibid. [43] Ibid., 15, 31, and 36.

[44] Rougier de la Bergerie, "Mémoire au Roi et aux Chambres législatives sur la déstruction des Bois," 57.

[45] Ibid., 57–58. [46] *AP 2ème série* 99:756 (27 January 1836).

did not obtain the required authorization, they were seldom prosecuted because their offenses were hard to prove.[47] By contrast, communities seeking to clear a portion of their woods confronted a near-ludicrous bureaucratic labyrinth.[48] As the moratorium's scheduled expiration date of July 1847 crept closer, critics of the provision grew more strident. The nation's forests had not improved in the past two decades, they pointed out. More enduring, behavior-changing approaches were needed.[49]

Liberal theorists believed the solution lay in tax exemptions. "To stop the clearings and destruction, at least provisionally," Rougier de la Bergerie argued, "the government has nothing more urgent and wise to do than to ... completely suspend the property tax on woodlots."[50] Likewise, he contended, the government could protect big timber by reducing the tax on woods more than thirty years old.[51] The philosopher Charles Comte agreed. Unless new measures were implemented, the expiration of the prohibition on landowner clearing would prove that "no revolution will have transpired, neither in the nature of our world, nor in human nature." Describing the threat of unchecked clearing, Comte noted in his 1834 *Traité de la propriété*,

> The water that falls on the steep mountains, where the forests will have been destroyed and the soil plowed up, will form into floods and will drag rocks and grit into the valleys, the same as in our time [and] the inhabitants of the mountains will be no more interested then than now in leaving the soil in a form that is more useful for the inhabitants of the valleys than for themselves.[52]

For government measures to be effective, he contended in classic liberal form, the people must be shown their true interests and they will willingly obey. Building on this belief, Comte proposed that the government grant forest owners a "permanent exemption [from the land tax] proportionate to the sacrifices demanded of them."[53]

[47] Unlike the landless poor, landowners were given a warning before being written up. Dumoulin, "Poursuites en réparation des délits," 167–69.

[48] See "Heureux effet de la centralisation [de l'administration forestière]," *Sentinelle du Jura*, 1 February 1832, p. 4.

[49] See the deliberations of the Vosges Conseil général on this issue, in AN AD^{XIXi} 1 Vosges 1844, Rapport de la Commission du Conseil général, 30 August 1844; also the debates on "Modes d'exploitation des bois" and "Deboisements" of the Société d'Agriculture de la Haute-Saône, 8 March 1846, published in *Recueil agronomique, industriel et scientifique*, vol. 5 (Vesoul: Imprimerie de L. Suchaux, September 1849), 209–12.

[50] Rougier de la Bergerie, "Mémoire au Roi et aux Chambres législatives sur la déstruction des Bois," 69.

[51] Ibid., 69–70.

[52] Charles Comte, *Traité de la propriété* (Paris: Chamerot / Ducollet, 1834), 1:247.

[53] Ibid., 256, 248.

For all these writers' suggestions, however, force, not reform, remained the state's primary approach to forest conservation. Under Louis-Philippe's leadership, the forest bureaucracy greatly expanded – in Franche-Comté alone, conservation and inspection districts increased in number by nearly 70 percent between 1833 and 1845.[54] A conscientious manager of his own woodlands, Louis-Philippe saw to it that enforcement of the Code intensified and that the sequestration of use rights via *cantonnement* multiplied. He also supported the formation of a commission to study the feasibility of state-sponsored reforestation. An improvement in the eyes of forest administrators, the July Monarchy's more active stewardship and the concomitant rise in personnel and policing did not fail to irritate the rural populace.[55]

TUMULT, MURDER, AND MAYHEM: THE FOREST CODE IN THE JURA

Reaction was especially fierce in the department of the Jura, where a combination of extensive state holdings, entrenched customary usage, and administrative intransigence sparked a firestorm of resentment. Jurassiens' struggles against the Forest Code began in 1833 with the designation of the department as its own forest district, the thirteenth, and the appointment soon thereafter of an officer named Cotheret as its woodland commissioner.[56] A devotee of German silviculture, including regular clear-cutting and monocrop conifer plantations, Cotheret regarded use rights as "bloody wounds that the forest administration has not yet succeeded in healing."[57] Science, he emphasized, not social convention, should be the guiding force in woodland management.[58] Despite or perhaps precisely because of his calculation that ninety-eight communities relied upon the Jura's sixteen state forests for "all the woods necessary for their consumption," Cotheret launched a concentrated effort to end all

[54] Vion-Delphin et al., *Les hommes et la forêt en Franche-Comté*, 125.

[55] See John M. Merriman, *Police Stories: Building the French State, 1815–1851* (Oxford: Oxford University Press, 2006), chap. 2, chap. 7.

[56] Cotheret was commissioner from at least 1836 forward. M. Perrod, *Table générale récapitulative des travaux et mémoires de la Société d'émulation du Jura depuis sa fondation, 1818–1917* (Lons-le-Saunier, 1918), 26; and M. Cotheret, "Sylviculture: Discours et considérations générales sur les forêts" in *Travaux de la Société d'Émulation du département du Jura* (Lons-le-Saunier, 1837), 26–35.

[57] M. Cotheret, "Considérations générales sur les forêts," in *L'annuaire du département du Jura* (1842): 299–300.

[58] Ibid., 295; also 299–301.

behaviors barred by the Code, among them the gathering of deadwood, a practice vital to the poor that had been largely tolerated by his predecessors. Under his watch, guards and officers also upheld the Code's onerous restrictions on woodland pasture and carried out controversial court rulings concerning *cantonnements* – among them denying the entire city of Dole firewood rights in the Forêt de Chaux.[59]

As it had been for more than a century following the introduction of the 1669 Ordinance, the Forêt de Chaux proved a focal point of Comtois resistance to the new Forest Code. Though the forest was enormous by the standards of France, virtually every twig within its twenty thousand hectares was spoken for by the multiple forges, glassworks, saltworks, sawmills, potteries, brickworks, and other industries that lined its perimeter. More than thirty communities also held customary rights in the Forêt de Chaux, rights that they had secured through their own protracted demoiselles' uprising in 1765.[60] Nonetheless, when the communities petitioned to confirm their rights in 1828, as required by the Forest Code, their claims were ignored. On the eve of the deadline, thirty-two Chaux-area villages filed suit against the state to demand that their rights be upheld.[61] Ten years of legal wrangling ensued. When, at last, the *tribunal de première instance* at Dole ruled in favor of the villages, the state quickly appealed the decision, thus intensifying inhabitants' ire. Taking stock of the situation in October 1839, the prefect of the Jura warned that the case was "producing a lively unrest that could perhaps translate into disorders if the state obtains victory."[62]

The appeals court would eventually side with the villagers, issuing a decision in 1841 that maintained the use rights of the nearly all the communities surrounding Chaux.[63] By then, however, the state had

[59] Jean Rousseau, "Les droits de la ville de Dole dans la forêt de Chaux: Procès contre l'État (1835–1843)," *Société d'émulation du Jura Travaux*, no. 1980 (1981): 357–67; Vion-Delphin et al., *Les hommes et la forêt en Franche-Comté*, 123.

[60] AD Doubs P 579, Le contingent annuel des communes riveraines de la forêt de chaux après le nouvel aménagement du 19 août 1766.

[61] See discussion in *Arrêt de la Cour royale de Besançon, pour les communes riveraines de la forêt de Chaux, contre l'État: 23 décembre 1841*, 5–10; and Rousseau, "Les droits de la ville de Dole dans la forêt de Chaux," 359.

[62] ADJ M 25, M. le vicomte Déjean, Directeur, Ministère de l'Intérieur, Direction de la Police générale, à M. Thomas, Préfet du Jura, 22 October 1839.

[63] *Jugement du Tribune de première instance de Dole (Jura) pour les communes riveraines de la forêt de Chaux [les maintenant dans l'exercice de leurs droits de panage, pacage et affouage dont elles jouissaient avant et lors de la promulgation du Code forestier de 1827] contre l'Etat (Du 28 août 1839)*; and *Arrêt de la Cour royale de Besançon, pour les communes riveraines de la forêt de Chaux, contre l'État: 23 décembre 1841*, 1–4, 10.

begun taking a different line of attack: isolating the claims through *cantonnement.*

Choosing not to wait around for the outcome of the appeal, inhabitants, too, went on the offensive. The first sign of the violence to come occurred in the Forêt de Chaux in February 1840, when a *machine infernale* – a booby-trapped gun tied to a snared thrush – exploded in the face of a hapless wood gatherer. The forest officer in charge of reporting the incident was later fired on by a hidden shooter. Together, these incidents led the prefect to speculate that the original explosion had been intended for a forest guard.[64]

During the same period, a string of suspicious blazes ignited the Forêt de la Serre, a large state-owned woodland just to the north of the Forêt de Chaux.[65] Although some of the flare-ups could be attributed to the dry spring weather, the prefect of the Jura, Napoléon Thomas, inferred malice in every unexplained incident. "These circumstances are all the more serious in that they provide indication of an organized system of vengeance against the forest agents," Thomas warned Cotheret in March 1840.[66] Forest officers had a slightly simpler interpretation of the blazes: reporting on a suspicious fire in a municipal coppice in May, one noted that it was "a ruse known to attract guards to a given point."[67]

Forests were not the only subject of Jurassiens' ire that spring. In early April, a particularly difficult time of year for the food supply because winter stocks were exhausted and new crops had yet to bear forth, a riot broke out in and around Lons-le-Saunier over the sale of potatoes. The uprising began when a representative of the nearby Château de Courlans, the site of several water-powered mills, showed up at the market to buy a half-dozen sacks of potatoes. A group of women who spotted the agent were incensed, for they blamed his master for having driven up the price of wheat during the winter, and now they feared the price of potatoes would go up as well. According to prefect Thomas's sensational and ultimately self-aggrandizing report, the agent eventually fled, but not

64 ADJ M 25, M. Thomas, Préfet du Jura à M. le vicomte Déjean, Directeur, Ministère de l'Intérieur, Direction de la Police générale, 17 March 1840.

65 ADJ M 25, M. Armand Marquiset, sous-préfet à Dole, à M. Thomas, Préfet du Jura, 3 March 1840; and M. Cotheret, Conservateur des Eaux et Forêts, à Thomas, Préfet du Jura, 12 March 1840.

66 ADJ M 25, M. Thomas, Préfet du Jura à M. Cotheret, Conservateur des Eaux et Forêts, 13e division (Jura), 16 March 1840.

67 ADJ M 25, incendies dans les forêts, 1840: Cernans, 18 April; Vaudioux, 29 April; Quintigney, 2 May; Champagnole, 3 May; and Champvans, 4 May.

before the gathering crowd had turned its anger on a group of potato-laden wagons, overturning two and threatening to upend four more.[68] Thomas responded by impounding the wagons in the town's garrison, intending to have them depart in the dead of night, but at the appointed hour their drivers were "in a state of complete inebriation" and could not take the reins. The next day, the wagons left under heavy guard, pelted by rocks "thrown by men, women, and children."[69] Upon entering neighboring Montmorot, the convoy was surrounded by a throng of "two to three thousand" peasants, some of whom had "hastened down from the mountain tops" to join the fray.[70] "The crowd rushed at the wagons," explained the prefect, "the wagons were overturned, the sacks ripped open and the potatoes scattered or carried off."[71] Thus emboldened, the group continued on to the Château de Courlans, six kilometers away, where they upending furnishings, terrified the owner and his family, and engaged in a two-day battle with the local national guard. Thanks only to the prefect's heroics (as he described it in his report), order was finally restored.[72]

Dramas like this were unusual, but the sentiments that provoked them were not. The humblest sector of rural society was being squeezed on all sides: their woodlands rights were being denied or marginalized to ever smaller sections of the forests, the price of foodstuffs was increasing, and newcomer landowners were even less tolerant of customary rights than former seigneurs had been. In August 1840, for example, the municipal councils of Dole and Poligny called for the suppression of *vaine pâture*, the right to pasture animals on privately owned fallow land.[73] Invoking the oft-used argument that *vaine pâture* posed a burden to private property and discouraged agricultural improvement, the councils contended that pasturage in municipal forests could make up the difference – an unlikely scenario given the forest administration's implementation of ever-stricter limits on woodland grazing. The proposal galled the peasantry, who relied on *vaine pâture* to sustain their flocks.[74]

Recourse to the courts was one option, but as the experience of the villages surrounding Chaux demonstrated, even a favorable ruling was no guarantee of access. In this context, many began to resort to aggression.

68 ADJ M 25 (8), Prefect Thomas to the Ministry of Interior, 4 April 1840.

69 ADJ M 25 (8). 70 Ibid. 71 Ibid.

72 See also Gustave Duhem, "La révolution des pommes de terre à Lons-le-Saunier en 1840," *Société d'émulation du Jura: Travaux* (1948): 17–32.

73 AN AD^{XIXi} 1 Jura 1840.

74 Agulhon, *Apogée et crise de la civilisation paysanne, 1789–1914*, 126–28.

The violence took a variety of forms. In December 1840, a private forest guard was murdered in the central Jurassien hamlet of Le Fied.[75] The following June, officers reported that scores of trees had been systematically mutilated in the Forêt de Chaux.[76] The year after that, in July 1842, Chaux's guards discovered that fifty-four more trees had been destroyed by girdling, an incident they attributed to the ex-mayor of Chatelay, a hamlet on the southern edge of the forest. The official was also suspected of having set a guard's barracks ablaze.[77]

The violence of these acts expressed both the helplessness and the outrage of the inhabitants surrounding Chaux, who were trapped in a struggle at once maddeningly bureaucratic and brutally consequential. The villagers of La Vieille-Loye, for example, were accused of having cut down several trees to fill boundary ditches that had been dug by the forest administration. Yet, noted the mayor, forest officials were really to blame for having dug the ditches in the first place, for they cut straight across the village's principal arteries and interrupted commerce with neighboring towns. "The Mayor of Bellemont and I complained many times already about the suppression of the road between our villages," wrote the mayor to the subprefect, "and the forest administration rejected these complaints, contending that the road was useless."[78] In response, villagers had taken justice into their own hands.

Cotheret, the woodland commissioner, accused the subprefect of Dole, Marquiset, of fomenting inhabitants' defiance through his open criticism of the Forest Code. Among other things, Marquiset had encouraged communities to sell their *affouage* wood to pay off their communal debts, in direct contravention of article 83 of the Code. "If the writings of the subprefect each time contain such opinions," Cotheret demanded in a letter of January 1842 to prefect Thomas, "what must pass between him and the mayors in their own meetings?" Cotheret was particularly irate at Marquiset's characterization of a recent peasant incursion in the Forêt de Chaux as "natural" in light of "the excessive harshness of the cold [weather] that we have had for a few days." If "these devastations, these acts that the Penal Code ... has classed under the heading of *destruction* [and] *devastation*; these acts which have satisfying vengeance or hatred as their goal and which arise solely from spite; these threats made against

[75] ADJ M 25, 11–12 December 1840. [76] ADJ M 26, June 1841.
[77] ADJ M 27, Forest Inspector Bizot to Thomas, Prefect of the Jura, 13 July 1842.
[78] ADJ M 27, Vernois, Mayor of Vieille-Loye to Armand Marquiset, subprefect of Dole, 27 January 1842.

guards, hatchets raised above them, are not *disorders* in the eyes of the subprefect, what more must one expect?" the forest commissioner railed.[79]

Cotheret's imputation of insubordination on the part of the district's mayors only served to infuriate them. In a barrage of incensed rebuttals informed by the forest administration's long record of aggravation and affront, the mayors and municipal councils of the villages surrounding Chaux denounced the commissioner's accusations as "nothing but calumny" and "exaggeration." The mayor of Falletans was especially unrepentant. Decrying forest agents' actions as "inhuman, vexatious and tyrannical," he noted that "guards may well experience some difficulty in their duties, but this is their fault, in light of the manner in which the forest authority exercises its powers."[80] "I believe there would not be damage committed by [inhabitants]," the mayor added, "if they were not aggravated in an appalling manner."[81]

Saddled with enforcing unpopular regulations, the forest administration's lowest-ranking personnel unsurprisingly bore the brunt of residents' frustration. Though not all may have been as unsympathetic as the mayor of Falletans described, the majority of guards and officers hailed from a military background, which may have rendered them more willing to carry out brutal commands and to view inhabitants as adversaries.[82] When threatened, they did not hesitate to use deadly force, as was the case in January 1842 when a guard encountered three men stealing coppice shoots in the communal woods of Mont-sous-Vaudrey. The guard demanded that the men stop, he reported, but when they instead slashed at him with their billhooks, he shot and killed one of them.[83]

Cotheret accused Marquiset of only promoting guards who indulged rights holders' concerns. For the most part, however, the relationship between guards and the community appears to have been tense.[84]

[79] ADJ M 27, Plainte du Conservateur des forêts contre le sous-préfet de Dole, 28 January 1842, 11, 6, and 3–4 (Cotheret's emphases). The conflict appears to have precipitated Marquiset's removal from office. Max Roche and Michel Vernus, *Dictionnaire biographique du département du Jura* (Lons-le-Saunier: Arts et Littérature S.A.R.I., 1996), 346.

[80] ADJ M 27, Gauthier, Mayor of Falletans, to Marquiset, subprefect at Dole, 8 February 1842.

[81] Ibid.

[82] From 1811 forward, veterans were given preference for guard and officer positions. *Les Eaux et Forêts*, 373.

[83] ADJ M 27, Le Lieutenant de Gendarmerie à M. le Préfet du Jura, 25 January 1842.

[84] ADJ M 27, Plainte du Conservateur des forêts, 12.

The year 1843, in particular, saw assaults on guards increase dramatically. In May, the guard squad of the municipal woods of Doucier, in the Jura, fended off fifteen masked attackers. In early June, the field and forest guard of St. Julien was ambushed and nearly lost his life. On 17 June, four assailants chased, disarmed, and wounded the guard of the *cantonnement* of Dole.[85] Finally, on 23 June, Sergeant Ruty, an officer in the Forêt de Chaux who had been threatened previously, was found lifeless in the woods, dead of a blow to the head.[86] A day laborer from La Grande Loye was arrested for the crime.[87]

The Jura's forest agents were not alone in being targeted for attack. Nationwide, assassination had become an occupational hazard for woodland guards by the early 1840s. In addition to the bloodshed in the Forêt de Chaux, murders also took place in 1843 in the Ariège, Ardèche, Meuse, Lorraine, and the forest of Versailles.[88] The following year, at least six other guards were killed from Aquitaine to Alsace. This trend would continue through nearly the end of the century, with every year from 1840 forward marked by one or more assassinations of forest guards on the job.[89]

Fire remained a second weapon of choice, though the danger it posed was to the forest itself.[90] While most blazes were quickly controlled, they occasionally caused more extensive damage. In the Forêt de Chaux, a forty-hectare fire caused 500 francs' worth of damage in 1844. Two years later, a much larger conflagration destroyed coppice and timber valued at 18,000 francs.[91] In both cases, authorities were unable to identify the culprits.

As the unrest grew, some municipal and departmental leaders appealed to the state to ease the measures introduced by the Code, hoping in this way to stave off further turmoil. The Conseil général of the Vosges was especially vocal. In light of the forest administration's apparent plan to push through the *cantonnement* "of all use rights," the council exhorted

[85] Possibly the same guard whose barracks had been set on fire a year earlier by the ex-mayor of Chatelay.

[86] All incidents mentioned in this paragraph are described in ADJ M 28, St. Ouen, Conservateur des forêts, à Thomas, Préfet, 30 June 1843. Ruty and his colleagues had been threatened in January 1842 by inhabitants of La Vieille Loye. See ADJ M 27, Report of threats and injuries filed by guards Ruty, Thibert, Darq, 5 January 1842.

[87] ADJ M 28, Emenjaud, Sous-préfet de Dole, à Thomas, Préfet, 29 July 1843.

[88] *Les Eaux et Forêts*, 494. [89] Ibid.

[90] On the prevalence of intentionally set fires across France, see ADJ M 25 (3), Le sous-secrétaire d'État de l'Intérieur à M. Thomas, Préfet du Jura, 23 March 1840.

[91] ADJ M 28 (2–3), Incendies dans la forêt de Chaux, 9 April 1844; 24 February 1846.

the government in August 1845 to scale back its schemes, lest the land division lead to "the ruin of a great number of *usagère* communities." "Too great a reduction in allocations could create a serious danger for the state's forests, which will be exposed to the depredations of a people squeezed by misery and need," observed the council, adding that the *cantonnement* rulings issued by the courts were inconsistent and flawed, because they were based on a faulty calculation of use rights' value and a serious underestimation of communities' needs.[92] For "twenty-five to thirty years," noted the Vosges representatives, the forest administration had reduced the annual delivery of wood to communities from the state forests, to the point that it was now "less than a third of what it had been in the past."[93] Underscoring that forest rights "still today comprise [communes'] sole resources for resisting the long and bitter winters, which often last for eight months," the council implored the government to seek "means of reconciling the interests of the State with the interests and the extremely ancient and respectable rights of poor inhabitants."[94] Instead of heeding the representatives' plea, however, the Conseil général placed even greater limits on woodland grazing and gathering through its revisions to the Forest Code in 1846.[95]

At the very same time that the Conseil général was tightening its stance on woodland customs, it was also debating what to do about the impending expiration of the Forest Code's twenty-year prohibition on clearing in private forests without prior approval. In 1844–45, lawmakers drafted a bill and submitted it for comment to the departmental conseils généraux, nearly all of which favored extending it. Other commentators agreed, arguing that the temporary restrictions were insufficient, that private forest owners were too indulged, and that the nation's overall forest cover had suffered as a result. Nonetheless, entangled with the self-interest of well-propertied council members, the bill quickly became bogged down in the debate over whether to ban upland clearing, modify felling techniques, or initiate reforestation. With time running out and the risk of seeing the Forest Code's limited constraints on clearing on private land lapse, the council elected merely

92 AN AD^{XIXi} 1 Vosges 1845, Rapport du Préfet au Conseil général, session du 1845. The council repeated the resolution in its proceedings of 1846 and 1847.

93 AN AD^{XIXi} 1 Vosges 1845.

94 Ibid.

95 On the 1846 revisions, see Peter McPhee, *A Social History of France 1780–1880* (London: Routledge, 1992), 172.

to extend the restrictions for another three years, leaving the text and scope of the law unchanged.[96]

The persistence of this regulatory double standard, which expanded harsh restrictions upon forest users on one hand and perpetuated porous controls on private landowners on the other, generated escalating conflict. In 1847, for example, multiple villages surrounding Chaux petitioned the forest administration for an extension of the period in which they could pay for and remove their annual wood allocation. The mayors had been informed of their townships' allocations only on 10 June, and the deadline for delivery and fees was only a month later. This time frame was too brief, argued the municipal council of Champagne, given that

> nearly two-thirds of inhabitants of the commune find it impossible to pay the excise tax on the *affouage*[;] that they can only provide for their most basic needs with great difficulty; that there are even some who go into debt to buy bread at the exorbitant price that it costs nowadays; that there are poor unfortunates responsible for families who know not where to borrow to pay for a loaf of bread; [nor] where they can borrow the 9 francs 30 [centimes] price of the wood tax, plus nearly ten francs for carriage costs.[97]

The council requested that the delivery period be extended until after the harvest, when "each family (with a few exceptions) would have wheat and would no longer have to purchase [it]."[98] Despite the soundness of the argument, the forest inspector of the Jura refused. Writing to the prefect in July, he noted that inhabitants had a "morally sufficient time to fulfill their obligation" and that the "landowner" – the state – was uninterested in petitioners' reasons for wanting "to remove, at their convenience as it were, products that they are obtaining dirt cheap."[99] Increasingly, the nation's forests were being managed in an exclusionary fashion: focusing on a view of the *bien public* that did not include the overlapping interests of rural communities, forest officials condemned customary and collective usage as impediments to the production and protection of timber.

[96] On the debate and failure of the private land clearing reform effort, see *Les Eaux et Forêts*, 496. On reactions during the 1848 revolution against the 1846 tightening of the Code, see McPhee, *Social History of France*, 176.

[97] ADJ 11Qp215, 17 June 1847. After 1906 the village became known as Champagne-sur-Loue.

[98] ADJ 11Qp215, 17 June 1847.

[99] ADJ 11Qp215 M. Picard, Inspecteur à Lons-le-Saunier, à Thomas, Préfet du Jura, 10 July 1847. Picard extended the deadline by two weeks but refused to push it past the harvest as the municipal council of Champagne and others requested.

The forest administration's muscle flexing was all the more brutal in that it came at a time when customary and collective rights were more vital than ever. Nationwide, France's rural population was greater in the 1840s than it had ever been. In Franche-Comté, a stretch of favorable weather, respite from warfare and disease, faster transport and communication, and small improvements in industry and agriculture contributed to higher birth rates and delayed mortality.[100] For the most part, however, the Francs-Comtois achieved this increase through the intensification of traditional techniques, like woodland pasturage, the gathering of dead and fallen wood, and myriad other forms of scrimp and salvage, with all their attendant potential for environmental deterioration.

Given the region's long winters and the marginal capacity of much of its land for production, this expanded population may have been unsustainable from the outset. Nonetheless, the forest administration's zealousness and the stringency of the Forest Code significantly accelerated the process of attrition and immiserated the populace at the same time. So ruthless were forest guards in the Haute-Saône and the Jura that the value of fines for infractions exceeded the amount of the departments' state taxes.[101] Deprived of woodland access, many inhabitants soon found themselves seeking their livelihood elsewhere.

Because Franche-Comté was so overwhelmingly rural and its inhabitants' reliance on forest rights so vital, the outward flow of residents began earlier and was more pronounced than in many other parts of the country, where the "*grand départ*" to towns and cities did not gain full force until the 1860s and 1870s. More than 86 percent of Francs-Comtois lived in villages of less than 2,000 people in 1836, and emigration took its greatest toll among the poorest of these communities, where the balance of natural resources and human need was most precarious.[102] The Jurassien hamlet of Champagne, near Chaux, was typical. Throughout the first two decades of the century, Champagne's population grew from 322 people in 1800 to a high of 380 in 1820. By 1851, however, the village population had declined to 250 – a loss of more than a third.[103] Internal migration stemmed some of Franche-Comté's decline – in the Doubs, relocation to Besançon and other towns deferred overall

[100] Fiétier, *Histoire de la Franche-Comté*, 353.

[101] Philippe Vigier, "Les troubles forestiers du première XIXe siècle français," in "Société et Forêts," spec. issue, Revue forestière française (1980): 132.

[102] Fiétier, *Histoire de la Franche-Comté*, 356.

[103] Figures from Rousset, *Dictionnaire géographique*, 1:407; and "Des villages de Cassini aux communes d'aujourd'hui" http://cassini.ehess.fr/cassini/fr/html/index.htm.

departmental population loss until the 1860s – but the Haute-Saône and the Jura ceased growing completely after 1841.

"THE MASTERS OF THEIR WOODS"

In light of the mounting demographic pressure, scarcity of resources, and the embittering restrictions of the Forest Code, the outbreak of forest protests following the February overthrow of Louis-Philippe in 1848 was almost inevitable. Momentous and full of promise, the revolution of February ushered out the last of France's ruling royals and made way for a return of republican government, this time more radical than the first. The sense of possibility and popular excitement was palpable throughout France. Among those who had fought for more equitable distribution of woodland rights, enthusiasm at the overthrow quickly translated into expectations that their demands would now be vindicated. Other woodland stakeholders who had suffered under the Forest Code believed that liberation from the state's repressive measures was at hand – and if it didn't happen, they would make it so.

For woodland officers who had come to expect violence and conflict in their work, the ferocity of the resulting upheaval was astonishing. Throughout the country, people acting alone and in groups unleashed two decades of accumulated resentment against the Forest Code, compounded by a century and a half of dislike of the 1669 Ordinance.[104] With state, departmental, and municipal administrations in disarray and the sympathies of local national guard forces divided, forest guards had to retreat. By the end of March, the forest administration had largely withdrawn from the turbulent uplands of the Vosges, Jura, Pyrénées, and Alps.[105]

To be sure, self-interest played a role in the tumult, particularly in the first few days following the revolution. In the Haute-Saône, for example, the first two weeks of March were marked by numerous incidents of oak trafficking and tree thefts from municipal woods.[106] In one such case, the acting mayor of Delain, twenty-five kilometers from the river-port town

104 AN BB^{18} 1460 (5248A); AN BB^{18} 1461 (5337 and 5423); and AN BB^{30} 358–66.

105 See discussion in Suzanne Coquerelle, "L'armée et la répression dans les campagnes (1848)," *Société d'histoire de la Révolution de 1848: Études* 18 (1955): 121–59.

106 On the trafficking of branches from naval oaks, see ADHS 17 P 23, Registre des jugements forestiers, Tribunal de Gray, *M. le Directeur de l'administration des Eaux et Forêts (et al.) contre Louis Jacquemard père (et al.), Brotte, Haute-Saône*, 15 March–17 July 1848.

of Gray, authorized the removal and sale of an oak to the son of a local landowner. Though the mayor may have intended the proceeds for the commune, he and the purchaser were sentenced as individuals.[107] Oak, the most lucrative form of timber, was the target of another case reported on the fourth of March. In this instance, thirty-seven people, predominantly landowners and their children, were convicted of having "dug up and removed with picks, hatches, and hoes an oak of 2 meters 30 centimeters around." They were also charged for damaging the surrounding coppice in their zeal to carry off the tree, whose existence was all the more vital because of its service as a *baliveau*, or seed-bearing standard.[108]

The records of these incidents suggest no particular political intent among their perpetrators. Rather, they appear to have been simply taking advantage of the moment to take wood and timber that would otherwise be off-limits or subject to lengthy bureaucratic approval.[109] Yet while the forest commissioner of the Jura, M. de Saint-Ouën, hastened to blame episodes like these on "BAD CITIZENS, most of them professional delinquents ... running amok in the forests of the Republic, communes, and public establishments," rural responses to the revolution soon evolved into something much greater.[110] By the middle of March, woodland incursions began taking the form of coordinated and purposeful attempts to reclaim the lost promise of 1789 and throw off the yoke that, under liberal, imperial, and restoration regimes, had settled ever more oppressively upon the land. Far from being limited to a handful of habitual criminals and social deviants, the turmoil that engulfed the forests of Franche-Comté in 1848 developed into a full-fledged peasant revolt, involving at times communities' entire populations and coordinated by their highest-ranking officials.[111]

The uprising in Rouhe is a case in point. One of several dozen communes in the canton of Quingey, just to the east of the Forêt de Chaux,

[107] ADHS 17 P 23, Jugements forestiers, Tribunal de Gray, 1848.

[108] ADHS 17 P 23, Jugements forestiers, Tribunal de Gray, 1848, *M. le Directeur de l'administration des Eaux et Forêts (et al.) contre Robert François dit Marcot; Lenoir, Claude (et al.); Vaudry, H-S*, March–July 1848.

[109] See incident described in ADJ M 30, Dévastations dans la forêt de Poligny.

[110] ADJ M 30, M. de Saint-Ouën, Conservateur des forêts, to Jules Grévy, Commissaire du gouvernement, 6 March 1848.

[111] See discussion in John M. Merriman, *The Agony of the Republic: The Repression of the Left in Revolutionary France, 1848–1851* (New Haven, CT: Yale University Press, 1978), 91–95.

Rouhe possessed a sizeable communal forest over which it had tussled with both the state and the seigneur for decades. Despite the village's small population – only 166 in 1846 – its woods barely sufficed for its needs. Instead, as Rouhe's *cahier de doléances* attested in 1789, much of the forest was allocated to the saltworks at Salins, some thirty kilometers to the south. Their demand at the time, that "His Majesty to return [the woods] to their control . . . either for their individual use, or to contribute to communal expenses and upkeep," apparently remained unmet.[112] Rouhe's residents again tried to fend off forest competitors in 1809, this time by suing their longtime seigneur, Mouret de Châtillon, for "undue exploitation" of their communal woods.[113] The introduction of the Forest Code in the region in 1828 amplified the community's annoyance. Fed up with outside interference and filled with hope by the February 1848 revolution, inhabitants once more seized the opportunity to drive out the forest administration.

The confrontation began on 21 March, when the guard-general of the canton and a communal guard named Chalon were passing through Rouhe's communal woods and happened upon some two dozen inhabitants, including the village's mayor, in the midst of felling oaks that had not been marked for removal. When the officers rebuked them, the villagers replied "with a resolute air, that they would cut as much as they wanted, that they would chop down the entire stand, and indeed the whole Forest; that they were the masters of their woods."[114] Heavily outnumbered and feeling threatened by the crowd, the guards retreated, pursued by shouts of "down with the guards!" They immediately went to warn a local manufacturer not to purchase any of the ill-gotten wood. "This action aroused the most energetic agitation in Rouhe," noted the state prosecutor in his report on the incident. "Immediately the drum sounded in the fields, and a group comprising all the inhabitants of the area, led by the Mayor and his deputy, set out to find the guards."[115] When they tracked down the lower-ranking officer, Chalon, in the neighboring community of Courcelles, the growing crowd surrounded him and demanded his resignation, all the while repeating that "they were free;

112 *Cahier de doléances*, Rouhe, 18 March 1789, in ibid., 108, II-5 art. 4.

113 AN F^{lcIII} Doubs 7, 16 April 1809. Mouret de Châtillon may have been expropriating wood to use for the forges at Châtillon, less than a kilometer away on the banks of the Loue River.

114 AN BB^{30} 359, Rapport du procureur général près la Cour d'appel de Besançon, April 1848.

115 AN BB^{30} 359.

that they no longer wanted guards, that they would administer their woods according to their own will."[116] Held captive in the home of the mayor of nearby Palantine, threatened by "men armed with guns and appearing ready to use them," and intimidated by shouts of "resign, or you will end up at la Motte" (the local cemetery), the unfortunate Chalon at last ceded to the crowd's demand. Encouraged by their success, the inhabitants of Rouhe seized another forest officer the next day and paraded him among their supporters in Courcelles and Palantine before accepting his terrified resignation.[117]

Other substantial outbreaks of woodland unrest in Franche-Comté followed, including days of forays into the communal forest of Poligny by winemakers seeking to obtain stakes for their vines. While offenses of this sort were common in this winemaking district of the Jura, the scale of the incident, in which four to five hundred inhabitants of both Poligny and neighboring Arbois were said to have carted away some 900 loads of wood, suggests that the events of 24 February had given them confidence that their actions, which they regarded as vital to their livelihood, would no longer be considered a violation.[118]

Residents along the far eastern edge of the Forêt de Chaux were motivated by similar faith. Soon after the February revolution, the inhabitants of Fourg learned that "nearly all the bordering communes of Chaux [had gone] to gather standing dead wood, already half rotten, and that the foresters authorized similar actions." "Wanting to profit from this tolerance," attested a petition signed by more than fifty residents, they asked the forest guards at their end of the woods if they could do the same. "These officers responded that they would not stop them from going; that on the contrary they preferred that the inhabitants of Fourg profit from this dead wood like the others, but that they should take care to cut it properly," the residents contended. Acting on these assurances, the village's entire population, along with that of neighboring Villars-Saint-Georges, went into the Forêt de Chaux in early April and loaded up their wagons in the presence of at least four forest guards. Within days, however, the villagers learned that the guards had written them up for theft and vandalism. "The said inhabitants of Fourg have never been taken for delinquents," they protested indignantly in their petition to Jules Grévy, who as government commissioner of the Jura had taken on the responsibilities previously held by the prefect. Indeed, they argued,

[116] Ibid. [117] Ibid. [118] ADJ M 30 Saint-Ouën, 6 March 1848.

they had "always respected the Forêt de Chaux." "Only recently," they added, "when a fire erupted [there], they left their very urgent work in the countryside to extinguish [it]."[119] Grévy was not persuaded. Unwilling to ignore the woodland incursions, he called for and received 250 infantrymen, who he stationed across the Forêt de Chaux so as to swiftly reach "the places of the forest where their presence is necessary."[120]

Driven by a combination of righteousness and wishful thinking, Comtois communities like those of Rouhe, Poligny, and Fourg took the 1848 revolution as a signal to seize what they believed was their due by dint of both custom and proximity. That they did so openly and collectively, articulating their actions in terms of justice and redress, indicates that these were acts of principle and protest rather than simple opportunism.[121]

Similar protests were repeated across France in the first months after the revolution.[122] Summing up the frustration that the enforcement of the Forest Code had aroused and the misery that incited them to challenge it in 1848, petitioners in the Pyrénées-Orientales village of Saint-Laurent-de-Cerdans declared,

> Know how odious this forest administration was to us under the deposed government. ... A poor miserable father went during the hardships of winter to gather some scrub for his fire, and immediately he was convicted. ... During two whole years, the *affouage* that the law accords us has been absolutely refused. Not even a branch of wood has been granted to us.[123]

Despite these pleas, the fledgling Republic did little to change the conditions that led to the upheaval. Rather, the new government overrode departmental commissioners and other authorities' scattered efforts at conciliation with the use of overwhelming force. Lawmakers who had helped usher in the Republic tended to view the military as a last resort, but as they lost their hold on power, they increasingly deployed the army from April forward to halt unrest before it could gain ground. "Under the current circumstances," remarked the attorney general at the court of appeals in Besançon, ruminating on the best way to handle rebellion in Rouhe, "a moderate punishment has happier results than too great a

119 ADJ M 30, Inhabitants of the commune of Fourg to Grévy, commissioner of the Jura, April 1848.

120 ADJ M 30, correspondence of 5 April and 7 April 1848.

121 See discussion in Pierre Merlin, "Poligny et l'idée démocratique sous la Seconde République, février 1848–2 décembre 1851," *Patrimoine polinois* 15 (2000): 91–101.

122 AN BB[18] 1460 (5248A); also BB[30] 358–66. 123 AN BB[18] 1460 (5248A).

severity; above all when remorse almost immediately follows the misdeed."[124] Overall, the state deployed 18,000 troops to quell forest troubles and another 15,000 to subdue tax riots during the spring and summer of 1848.[125] As the summer turned to fall and the rural elite gradually reestablished its hold on power, the forest's lowliest stakeholders would once again find themselves without a voice in the decisions governing their most basic survival.

SEEKING AN END TO "INIQUITOUS CUSTOM"

It is in the challenges to the old, inequitable forms of *affouage* that one can see most clearly the reestablishment of the traditional power elite and their extinguishment of the briefly ignited hopes of 1848. As noted previously, *affouage* had been a growing source of discontent since the first third of the eighteenth century, when a combination of Eaux et Forêts scrutiny, demographic growth, and seigneurial encroachment led to the exclusion of whole segments of rural communities from access to communal wood and timber. Repeated efforts to address the problem ended in failure. This was the case in 1790, 1794, and 1808, when lawmakers in the National Assembly, Convention, and Empire one after another opted to use slapdash solutions rather than take on the colossal and politically treacherous task of overhauling *affouage* systematically. The issue arose again in 1827 during the drafting of the Forest Code. Persuaded by the deputy from the Doubs, Chifflet, who emphatically denounced the proposal to distribute wood and timber in equal portions *par feu*, legislators upheld custom, granting communities the right either to allocate the *affouage* as they always had or to institute a more equitable mode of allotment, including auctioning off their timber for the benefit of the community as a whole.[126] Governed by the very people who gained most from ancient custom, most Franche-Comté municipalities chose the former. Twice thwarted, *affouage*'s critics saw the revolution of February as their chance to overturn "iniquitous custom" at last.[127]

124 AN BB^{30} 359, Proc. gen. près la Cour d'appel de Besançon, à M. le Ministre de la Justice, à Paris, Besançon, 10 April 1848.

125 Figures cited in Ted W. Margadant, *French Peasants in Revolt: The Insurrection of 1851* (Princeton: Princeton University Press, 1979), 42. On the tax riots in Salins Dole, Arbois, Champagnole, and Poligny, see ADJ M 30, Rapport mensuel, March 1848.

126 On modes of *affouage* in the Doubs, see Mayaud, *Les secondes républiques du Doubs*, 93–96.

127 AN BB^{24} 348, Reg. S3 No. 8131.

The struggle unfolded in every corner of Franche-Comté. Lowland districts were especially turbulent, among them the Doubs cantons of Marchaux and Rougemont along the Ognon River; the communities abutting the Forêt de la Serre in the Jura; and the western region of the Haute-Saône, where communal forests tended to be more expansive, more productive, and better maintained than the heavily grazed woods of the Doubs and Jura plateaus. With more at stake, demands in these areas for timber to be divided equally, regardless of the size of one's house, tax payments, or proprietary status, were fierce. As the editors of the conservative *Journal de la Haute-Saône* observed in June 1848, "The distribution of *affouage* wood in the department this year has raised challenges that, in many places, have degenerated into deplorable scenes of disorder."[128]

Among the protest sites was the canton of Jussey, in the northwestern corner of the Haute-Saône, where struggles over *affouage* broke out in at least four villages between April and June of 1848.[129] In one of them, Magny-lès-Jussey, five men were imprisoned for a month for defying "Monsieur the Mayor and his anti-Republican and reactionary sect" over the timber allotment.[130] As with most other Comtois communities, Magny-lès-Jussey's woodland resources were stretched thin at midcentury. The village's population had increased by more than 45 percent between 1794 and 1841, and as demands on the village's forest mounted, the custom of apportioning timber to landowners alone appeared intolerably unjust. For those least favored by *affouage* – artisans, day laborers, and other marginally landed or landless inhabitants – the revolution of February offered an impetus and language with which to challenge the system and demand a more equitable form of sharing communal benefits. "In this we ask only the equality proclaimed by the government," noted the protestors.[131] Rebuffed by the "local administration, accustomed to despotically governing the unfortunate class," Magny-lès-Jussey's would-be *affouage* reformers turned to the newly installed *commissaire du gouvernement* of the Haute-Saône for help.[132] The commissioner, a staunch Republican known as Boileau, was ideologically receptive to their demands, yet even he was unwilling to abandon custom completely.

[128] *Journal de la Haute-Saône* (hereafter *JHS*), 23 June 1848.

[129] *JHS*, 29 April, 26 May, 23 June, and 26 July 1848.

[130] AN BB^{24} 348, Reg. S3 No. 8131, Demande en Grace, Cour d'appel de Besançon, dépt de la Haute-Saône, Tribunal de Vesoul. Benez (Joseph) et al. à Cavaignac, président des Ministres, 2 September 1848.

[131] AN BB^{24} 348, Reg. S3 No. 8131. [132] Ibid.

Seeking compromise in a time of crisis, Boileau instructed Magny-lès-Jussey's mayor to divide the tops and branches of timber trees equally among residents but to reserve the tree trunks for landowners.[133] The order incensed the mayor and municipal council, who resigned en masse. Soon enough, however, the July elections would return these officials to their posts, whereupon they would see to it that the *affouage* protesters, all of humble profession and scant means, were duly punished.[134]

Magny-lès-Jussey's experience of concessions undone by revanchism was repeated across Franche-Comté throughout 1848 and 1849. In the Jura, the outlook for change initially looked bright. Responding in March 1848 to the clamor over *affouage*, commissioner Grévy authorized municipalities to divide both their firewood and timber equally *par feu*, provided there was no opposition within the community.[135] When the department's new Conseil général met in the fall, the prefect expressed support for extending these measures to the Jura as a whole. "In most of the communes of the Jura, timber trees are divided among the owners of buildings, following an ancient custom that was introduced to Franche-Comté, contrary to the most elementary principles of equity and humanity," noted a report presented in October 1848.[136] "The execution of this method has raised great difficulties for some time," it added, "and in the aim of ending them, the Conseil d'arrondissement of Dole asks that in the future the division of timber be carried out the same as it is with coppice (firewood), *per hearth* and *in equal amounts* among all members of the community, without the exclusion of anyone."[137] Though the attribution of the report is unclear – it is ascribed to Pagès, the third person to serve as prefect of the Jura in seven months, but he was not yet in office when it was presented – the author was unafraid to assert his views, however unpopular they might be with the Conseil général.

[133] Possibly Joseph Modeste Boileau, a member of the Légion d'honneur, born in 1811 in Cuve, Haute-Saône.

[134] AN BB^{24} 348, Reg. S3 No. 8131, Rapport à la Ministre de la Justice, 7 November 1848.

[135] A. Desaunais, "La révolution de 1848 dans le département du Jura (24 février–10 décembre 1848)," in *Mémoires de la société d'émulation du Jura pour 1946–1948, Volume du centenaire de la révolution de 1848 dans le Jura* (Lons-le-Saunier, 1948), 183–85.

[136] AN AD^{XIXi} 1 Jura 1848, Procès verbal des déliberations du Conseil général du Jura. Séances du 5–6 October 1848, Rapport du Préfet [Pagès] (Lons-le-Saunier: Imprimerie et lithographie de Frédéric Gauthier, 1848), 120–21. Pagès became prefect in November; thus the report should probably be attributed to Pagès's short-term predecessor, Jules Allin.

[137] AN AD^{XIXi} 1 Jura 1848, Rapport du Préfet [Pagès], 120–21 (emphasis in the original).

"I think, along with [the Conseil d'arrondissement of Dole]," the report asserted, "that the moment is come to eliminate from our legislation all that constitutes a privilege to the profit of whosoever."[138]

By January 1849, however, the position of prefect turned over again, and the prospects for systemic modifications to the Jura's *affouage* were dimming. The new prefect, Charles-Jean Besson, spoke firmly against change. "In the Jura, a region that was part of the former province of Franche-Comté, the custom was to distribute the *futaie* to landowners according to the size of their buildings[;] it is this custom that is followed," Besson noted in his report to the department's Conseil général in September 1849. "This method, regardless of the loud criticisms of it, must be respected," he concluded, adding, "The law is formal, it is everyone's duty to submit to it, whatever annoyance it may hold."[139]

The Jura's Conseil général wholeheartedly agreed. "Until the revolution of February, this custom [of allocating timber only to property owners] was respected, and if a few rare demands were raised against its implementation, the courts treated them as they deserved," the council declared, adding disapprovingly, "in 1848 some communes obtained permission from the departmental commissioner to divide their timber per hearth, like their coppice wood, and based on this they today demand that this derogation, required entirely by circumstance, be instituted as a general rule." Contending that "renters have no need" of timber for construction and repair, the council announced that it could not support the request of Dole's Conseil d'arrondissement. Instead, it recommended a minor modification allocating timber proportionally on a per-house rather than per-homeowner basis, and it clarified the policy regarding the eligibility of residents for firewood. Henceforth, the council ruled, individuals would have to live in the commune for one year before they could claim a portion of the communal coppice.[140]

In the Doubs, *affouage* struggles took a more radical course but ended in similar defeat. Of the six successive *commissaires du gouvernement* who governed the department in the spring of 1848, two, Charles Faivre and James Demontry, used their few weeks in power to install officials sympathetic to the Republican cause and issue far-reaching decrees

[138] Ibid.

[139] AN AD^{XIXi} 1 Jura 1849, Procès verbal des déliberations du Conseil général du Jura, Session de 2 September 1849, Rapport du Préfet, 7–8.

[140] AN AD^{XIXi} 1 Jura 1849, Procès verbal des déliberations du Conseil général du Jura, Séance du 2 September 1849, MM. Chevillard, Président, 37–43.

intended to succor the Doubs' poorest peasants. A Republican influenced by the communalist notions of Charles Fourier, Faivre decreed on 25 and 28 March – the critical period in late winter before the growing season began – that communal timber would no longer be delivered to homeowners but instead would be auctioned to the public, with proceeds from the sale going toward the taxes inhabitants paid for their communal firewood.[141] The rulings responded to grievances like that voiced in 1847 by the municipal council of Champagne that its residents were so impoverished they had to choose between buying bread or paying their *taxe affouagère*.[142] Nonetheless, the measures went only so far in appeasing the rural populace. Together with other well-intentioned but hamfisted decrees issued by Demontry, this effort at egalitarianism from above outraged landowners of all political stripes even as the landless complained that the fixes fell short.[143]

Faivre fled Besançon under a political cloud in April 1848, and his successors were in office too briefly to enact change.[144] By the time the Doubs Conseil general, reconstituted of landed elites, took up the *affouage* question in the fall, it had little interest in altering the status quo. "The Conseil général favors maintaining the ancient traditions for dividing the *futaie*, in light of the respect due to custom and the interests of agriculture, which has need of large buildings," ruled the Doubs' departmental council in December 1848.[145] Victor Joseph Loiseau, an official in Pontarlier, agreed. Belatedly responding to a nationwide agricultural and industrial survey that had been issued that spring, Loiseau argued that far from expanding the benefits of *affouage*, the department should impose greater restrictions on communes' wooded and pastoral resources so as to stem the exodus of rural inhabitants. Only descendants of long-standing local families and new settlers who purchased a formal right of residency should receive communal wood and timber or be allowed to participate in communal grazing, he insisted. All new residents, including renters and landless laborers, should be barred.[146]

141 Mayaud, *Les secondes républiques du Doubs*, 215–16, 222.

142 ADJ 11Qp215, Extrait du registre des deliberations du Conseil municipal de la commune de Champagne, 17 June 1847.

143 See discussion in Mayaud, *Les secondes républiques du Doubs*, 233 and 225–35.

144 On the expulsion of Faivre, see *Le Franc-Comtois*, 8 April 1848, pp. 1–2.

145 For deliberations of the council's 2 December 1848 session, see *L'ordre social, journal de Besançon et de la Franche-Comté*, 14 December 1848, p. 4, col. 1.

146 AN C 951 Doubs, Enquête sur le travail agricole et industriel (Loi du 25 mai 1848), Responses from Canton de Pontarlier, 20 December 1848.

In the Haute-Saône, the Conseil général proved slightly more amenable to modifying the firewood component of *affouage* distribution, apparently at least in part because of the sense of raised expectations that had followed the events of February. Hinting at wider conflict, the council explained in a December 1848 approval of a motion to revise article 105 of the Forest Code that it wished for "a legislative decision [to] bring an end to all uncertainty concerning the mode of dividing the *bois communaux* and that a policy be imposed that can satisfy the interests of everyone."[147] Half of coppice wood should be distributed per person and the other half per hearth, the council recommended. Timber was another matter, however. "In light of the respect due to custom and the interests of agriculture, which needs large buildings," observed the council, the *affouage* customs governing the distribution of timber should be maintained.[148] Emphasizing this point, the council added that whatever the solution might be, it should not "deviate from the respect that the law itself owes to established and legitimate rights."[149]

By including this caveat, the Haute-Saône Conseil général signaled to the prefect that it would not tolerate a radical overhaul in the allocation of communal forest resources and that any modification to *affouage* should be pursued with caution. The prefect upheld this instruction all too zealously, as the *procureur général* at Besançon, Gustave Loiseau, observed in a letter of February 1850.[150] Article 105 of the Forest Code was "an incessant cause of agitation and sometimes disorder" in the department, Loiseau noted, but municipal authorities had, of late, attempted to mitigate the unrest by bending the rules. "Last year, in the interest of the poor, branches from the timber felling were added to the coppice harvest to be distributed among the households. This mode of distribution," he added, "though contrary to ancient customs, was tolerated in light of the circumstances." However, Loiseau continued, "This year, the Prefect absolutely refused to tolerate it, even in the communes where everyone supports it; he insists that custom be absolutely respected." While this conduct was "perfectly legal," Loiseau fretted,

[147] Deliberations of the Haute-Saône Conseil général, printed in *L'ordre social, journal de Besançon et de la Franche-Comté.* 15 December 1848, p. 4, col. 2.

[148] Deliberations, in *L'ordre social, journal de Besançon et de la Franche-Comté.* 15 December 1848, p. 4, col. 2.

[149] Ibid.

[150] Gustave Loiseau, 1813–79. A collaborator and colleague of the legal compilationist and jurist Dalloz. See Max Roche and Michel Vernus, *Dictionnaire biographique du département du Doubs* (Lons-le-Saunier: Arts et Littérature S.A.R.I., 1996), 295.

"we fear that it will give rise to unrest." Using troops to carry out the *affouage* distribution was not an option, the beleaguered official warned, because their movement was "too difficult" in a woods that was being harvested.[151]

The shift in outlook on *affouage* from the reforming ambitions of February to the retrenchment of December can be explained in large part by the rural elites' electoral recapture of power from April 1848 forward. These elites – former seigneurs and large landowners, many of whom were closet legitimists – benefited from leaving *affouage* the way it was. Radical republicans, by contrast, sought to eradicate *affouage*'s vestiges of privilege and apply the same egalitarian principles as those underpinning universal suffrage. Recognizing *affouage*'s vital importance, candidates on the left highlighted it and other woodland grievances as they campaigned for the post of departmental representative in the spring of 1848.[152]

Among the candidates for the Haute-Saône's nine departmental representative seats, at least seven explicitly criticized the Forest Code and promised to seek revisions.[153] "There exists in the communities of the countryside, above all in those that own forests, a measure that is raising lively and extremely just complaints," announced one candidate in the *Journal de la Haute-Saône*, the department's main newspaper. The candidate continued,

> I mean to speak of the provisions of the Forest Code that, in the allocation of the *affouage*, award the timber to homeowners. ... It would seem more in keeping with the ideas of equality and brotherhood that henceforth are to govern us, if the timber as well as the coppice wood were divided up by hearth, without distinction between owners and nonowners of houses.[154]

Another candidate, from the Ognon River town of Voray, proposed that "communal woods be removed from the domination of forest agents" and left to the supervision of local citizens, who would "take turns" voluntarily guarding the forest as they had in the past.[155] Once again, the debates over woodland governance that had dominated the period from 1789 to 1799 revived, with much the same lines of analysis.

151 AN BB[30] 373, Rapport du procureur général près la Cour d'appel de Besançon, à M. le Garde des Sceaux, Ministre de la Justice, Paris, 5 February 1848.

152 The election was held on 23 April 1848.

153 Out of a total of forty-four campaign platforms printed in the *JHS*, 29 April 1848, p. 1.

154 J.-C. Placet, licencié en droit, Borey, 26 March 1848, in the *JHS*, 7 April 1848.

155 Bourdot de Voray, "Profession de foi," 27 March 1848, in the *JHS*, 7 April 1848.

Candidates in the Jurassien winemaking regions of Arbois and Poligny, pockets of radicalism where republican clubs had kept their hopes alive even in the darkest days of the Restoration, offered similarly pointed criticism of the forest situation. "The revision of the Forest Code regarding woodland pasturage and the gathering of grasses in appropriate periods, under the supervision of forest agents, [is] a vital matter for our district, where livestock and their yields form such a large component of prosperity," asserted Louis-Hyppolyte Bergère, a pharmacist and radical republican serving as the interim subcommissioner of Poligny.[156] Asserting that he wanted "liberty to be extended up to the point where anarchy begins," Bergère promised to work toward suppressing the 20 percent tax levied on the yield from communal woods.[157]

Bergère was supported in his views by Auguste Javel, an ardent Fourierist and printer of the left-leaning *Echo du Jura* as well as an influential member of Arbois' *comité démocratique*.[158] Presenting the *Echo*'s endorsements for the eight representative seats in the Jura, Javel extended the blame for the region's woodland woes beyond the Forest Code to the private landowners and investors who up to now had dominated local politics:

> The people know that the owners of forests and factories have abused their mandate for personal gain; that they have brought about and compelled the approval of laws of exception; that the general interest has been sacrificed to that of the influential deputies; [and that] the iron trade has been released from foreign competition by prohibitions that sever international relations; it has ruined the wine industry in removing its markets.[159]

"The people know all this," he added, "and a lot more, too."[160]

Not all republicans supported this emphasis on the issues. In the Doubs, the editors of the *Franc-Comtois*, mouthpiece of the Besançon left, observed, "In our opinion, the people of the countryside preoccupy themselves too exclusively with their material concerns."[161] Rebuking

[156] H. Bergère, sous-Commissaire du Gouvernement provisoire dans l'arrondissement de Poligny, in *L'echo du Jura*, 9 April 1848, p. 3, col. 1. On Bergère, see Roche and Vernus, *Dictionnaire biographique du département du Jura*, 19, 28.

[157] H. Bergère, sous-Commissaire du Gouvernement provisoire dans l'arrondissement de Poligny, in *L'echo du Jura*, 9 April 1848, p. 3, col. 1.

[158] See discussion in Michel Vernus, "La révolution de 1848 à Salins et Arbois: La présence du fouriérisme dans le mouvement démocratique," *Cahiers Charles Fourier* 10 (1999): 77–90.

[159] Philippe-Auguste Javel, editor, in *L'echo du Jura*, 9 April 1848.

[160] Ibid.

[161] "Besançon, 25 mars: Aux électeurs de nos campagnes" in *Le Franc-Comtois*, 23 March 1848.

candidates' tendency to emphasize specific grievances, the *Franc-Comtois* exhorted readers to focus on the big picture. "Scarcely had the decree of 5 March regarding the general elections appeared," despaired the journal's editors, "when a candidate was produced in every canton: candidates for the communal woods, candidates for the *affouages*, for the local pathways, the roads, the waters, the marshes or the woods, each of these candidates has promised, cajoled, [and] more or less intrigued."[162]

Focused on the issues or not, the left fared poorly in Franche-Comté's Easter 1848 elections.[163] Name recognition, journalistic endorsement, and pressure by local notables all played a role in the outcome, but so too did voters' reaction to the upheaval of the spring and the controversial egalitarian measures that interim authorities like Faivre had tried to apply – measures that smacked of the divisive reforms of 1793–94. In the Doubs, only one of the seven newly elected representatives was a long-term republican. The rest were republicans of recent vintage – so-called moderates who would soon co-opt the revolution in the name of stability and order. This same scenario played out in the Jura, where only three of the eight committed republicans endorsed by the *Echo du Jura* were elected.[164] The remaining five, a smattering of lawyers, landowners, a career military officer, and a forge master, professed moderate or even conservative views.[165]

Returns from the 30 July municipal elections were even worse. Though at least one candidate, a certain Jean-François from the Jurassien hamlet of Lemuy, hailed from humble roots ("the stance of our journal does not allow us to criticize [candidates'] statements of belief," sniffed the *Echo du Jura* in printing the aspirant's marginally literate platform, "but here is one that should not be considered seriously as a political piece"), the candidates who actually won were largely long-standing local notables, the very same persons who had been ousted in the upsurge of radicalism in March.[166]

As these elites resumed their posts and set about reestablishing their advantages, rural inhabitants' expectations declined.[167] So, too, did their enthusiasm for protest. An altercation in mid-July in the Haute-Saône

[162] Ibid. [163] See discussion in Mayaud, *Les secondes républiques du Doubs*, 266–68.

[164] The three included Jules Grévy. *L'echo du Jura*, 9 April 1848 and 30 April 1848.

[165] On the winners and losers in the Haute-Saône and their stance on woodland issues, see *JHS*, 29 April 1848.

[166] "Jean françois de Lemuy canton de Salins département du Jura," in *L'echo du Jura*, 5 May 1849.

[167] See discussion in Mayaud, *Les secondes républiques du Doubs*, 281–84.

village of Mailleroncourt-Saint-Pancras reflected the swing in public sentiment. According to the *Journal de la Haute-Saône*, the Mailleroncourtois had gathered at the town hall to pay their *affouage* tax when a woodcutter, "armed with a saber that he brandished, threatened to slit the belly of the first person who paid for their wood." Despite the woodsman's aggression, no one joined him in resisting the *affouage* payment or in denouncing the principles of inequality upon which it was based. Instead, the "citizens ... rendered him unable to bother them, and hastened to pay the amount of their contribution." For his trouble, the woodcutter spent the rest of the month in jail.[168]

Declining expectations were also evident in the peasantry's low voter turnout for the August elections for conseils généraux and conseils d'arrondissement. Their reduced participation not only reflected a loss of faith in the Republic but also reinforced its failure. By the year's end, these newly returned ruling bodies would vote to reject comprehensive changes to *affouage*.[169]

In this way, the drive to modify woodland custom in the interest of justice derailed in Franche-Comté. Skillfully invoking the hallowed yet hazy historical memory of the region's lost autonomy and asserting that "Francs-Comtois can only be judged well by Francs-Comtois," the region's reinstated elites drew on long-standing antipathy toward homogeneous solutions to local problems to defend their vested interests.[170] The same scenario had played out during the Federalist revolt and the "petite Vendée" of 1793; here it repeated in the context of the egalitarian reforms of the Second Republic. Yet even as they insisted on the value of custom, these same elites also benefited from their status as landowners in a post-Estates society founded on the legal, political, and moral superiority of private property. Beyond the Code's temporary and easily eluded restrictions on large-scale clearing and felling, no systematic oversight curbed private forest owners' exploitations.[171]

168 *JHS*, 26 July 1848.

169 Mayaud, *Les secondes républiques du Doubs*, 283–84; and Peter McPhee, *The Politics of Rural Life: Political Mobilization in the French Countryside, 1845–1852* (New York: Clarendon Press, 1992), 108–9.

170 Petition by the court of Besançon to the *Assemblée nationale*, 11 July 1848, protesting the government's plans to economize by transferring the court's jurisdiction to Dijon. The local judiciary was especially vital, it argued, in light of the complex and unwritten body of law governing Comtois forests and *fruitières*. Cited in Marlin, "Un problème d'affouage," 196.

171 AN AD^{XIXi} 1 Vosges 1844, Rapport de la Commission du Conseil général, 30 August 1844; also Société d'agriculture de la Haute-Saône, 8 March 1846, in *Recueil agronomique, industriel et scientifique*, 209–12.

Together with the forest administration's unremitting enforcement of the Forest Code and its divestment of use rights through *cantonnement*, Comtois landowners' unwillingness to relinquish their woodland advantages would contribute to the mass exodus of rural inhabitants from the middle of the nineteenth century forward. In villages like Magny-lès-Jussey, where unequal *affouage* persisted, emigration offered the landless their only option for improving their material conditions. By contrast, municipalities that did alter their mode of distribution – instituting instead a policy of auctioning off the *affouage* for the benefit of the community – were able to use the proceeds from the wood sales to offset the loss of agricultural earnings and concomitant decline of tax revenues, thereby sustaining those who remained.[172]

UPROOTING THE "GUILTY HOPES" OF LIBERTY

If the heady hopes of February ultimately fell short of expectations, it was not the first time the peasantry had been disappointed. Sixty years and three revolutions after rural inhabitants had expressed their hopes and grievances in the *cahiers de doléances* of 1789, their access to power remained circumscribed and the material conditions of the landless poor were as precarious as ever. Even universal suffrage did not bring improvement. Rather, the elections of 1848 put the nation on a new path toward repression that would be followed, from 1851 forward, by the revival of state authoritarianism.[173]

The collapse of popular democracy reinforced what was already well underway in national policy. Since 1789, the general trend in government, regardless of the reigning regime, had been the extension of state power, curtailment of customary rights, and elevation of individuated, private property ownership as the most vital element of civil society.

The Forest Code and its rigid enforcement from the July Monarchy forward embodied this trend. Conceived with an emphasis on the *bien public*, as defined in national strategic, economic, and hydrological terms, the Code promoted the interests of the state over the concerns of individual communities and instituted restrictions on collective usage so harsh that they criminalized basic activities like the gathering of deadwood and jeopardized rural inhabitants' very survival. At the same time, the Code

[172] See discussion in Jolas, "Bois communaux à Minot (Côte-d'Or)," 218–19 and 28–29.

[173] See discussion in Pierre Merlin, "Le coup d'état du 2 décembre 1851 dans le Jura, la répression, les hommes, l'exil," *Société d'émulation du Jura: Travaux* (1997): 281–313.

left private landowners, who together possessed more than half the nation's forests, largely to their own devices. This concession deeply undermined the Code's effectiveness. It also laid bare the extent to which its conservationist vision was shaped by political ideology rather than scientific principle. Despite admonitions by legitimists and liberals alike that the Code's restrictions on private forests were insufficient, and despite the hope that the 1848 revolution seemed to offer in this regard, the burdens of woodland conservation would continue to be shouldered most by the most marginal members of society.

A directive announced in January 1852, a month after Louis-Napoléon's successful coup to extend his presidency indefinitely, encapsulated this state of affairs. As implemented by the prefect of the Haute-Saône, the decree instructed the department's mayors to see to it that all liberty trees planted since the revolution of February be uprooted, and that all inscriptions on public edifices bearing the motto "Liberté, Egalité, Fraternité" be expunged. Though the majority of these trees had already been taken down in the years prior to the coup, the announcement underscored the new force of order in the French landscape. "These emblems, while containing nothing in themselves contrary to law or ethics, are nonetheless detrimental and dangerous by virtue of the memories that they retrace, the hateful divisions that they perpetuate, the hostile sentiment to which they could give rise, and the guilty hopes that they by nature foster," declared the prefect in his announcement of the decree. "They must disappear." In cases where a tree had become "an adornment for the commune," the prefect noted, a mayor could petition to preserve it, "not as a liberty tree but as decoration." As for the rest, they were to be hacked down and their wood "distributed to the needy."[174]

For all their demands for woodland liberty, for all their small gains through fire, firearms, lawsuits, petition, and open rebellion, six decades of popular political agitation and a quarter century of resistance to the dictates of the Forest Code had gained Franche-Comté's rural poor only the broken boughs of liberty. Their larger liberties – customary and collective access to the nation's forests and control over their own communal resources – had already been taken away.

[174] Announcement of *arrêté* of 6 January 1852 by Prefect Dieu, *JHS*, 10 January 1852.

Epilogue

"Homo is but Arbor Inversa"*

May 1856 was one of the rainiest months on record in France. Swollen by weeks of downpours, nearly every major river across the country burst its banks.[1] While the country had suffered an increasing number of violent floods in recent years, the inundations of 1856 set a new standard for destructiveness. The floods of the Rhône and Loire rivers were the most devastating of all. Ripping through Lyon and Avignon and engulfing surrounding villages, the Rhône rose as high as twenty-seven feet before pooling in the vast marshland of the Camargue, where it drowned hundreds of grazing cattle. The Loire likewise reached historic heights, submerging the entire middle valley of the river, ruining ninety-eight kilometers of railroad, and deluging thousands of hectares of cropland. In a village near Lyon, eighteen inhabitants drowned, and while there was comparatively little loss of life elsewhere, thousands lost their homes and had their livelihoods destroyed.[2]

Observers at the time noted that the floods were among the worst natural disasters France had ever faced.[3] Yet catastrophe also brought

* The quotation in the chapter title is from Evelyn, *Sylva, or A Discourse of Forest-Trees*, chap. 3, "Oak."

1 Maurice Champion, *Les inondations en France depuis le VIe siècle jusqu'à nos jours: Recherches et documents* (Paris: Dunod, 1861–62), vol. 3, pt. 2, cxxxvii–clxxx, and vol. 4, 157–97.

2 Champion, *Les inondations en France*, vol. 4, pt. 2, cxxv–cxxix and cxxxii–cxxxiii, and vol. 3, 144–79.

3 "Les inondations en 1856," in *L'année scientifique et industrielle*, ed. Louis Figuier (Paris: Hachette, 1857), 152–99; also Sara B. Pritchard, "Reconstructing the Rhône: The Cultural Politics of Nature and Nation in Contemporary France, 1945–1997," *French Historical Studies* 27, no. 4 (2004): 773.

opportunity. For Napoleon III, the floods offered a chance not only to burnish his public image but also to expand his program of national infrastructural improvement. Personally overseeing the relief effort on horseback and by boat, the Emperor garnered accolades as a man of action and compassion, a judgment that would serve his government well in the following year's elections.[4] In addition, he approved generous funding to repair the ravaged regions and stem future deluges.[5] In the same way that he simultaneously tamed and glorified Paris through the Haussmannization of its narrow streets and teeming quartiers, so too did Louis-Napoléon aim to master nature's volatility through the construction of dams, lakes, and reservoirs.[6] As he announced before the legislative chambers nine months after the floods, "By my honor I promise that rivers, like revolution, will return to their beds and remain unable to rise during my reign."[7]

Dams were not the only public works launched in the wake of the floods. Woodland administrators, local officials, and engineers similarly capitalized on the calamity to highlight the dangers of denuded slopes and insist on the urgency of upland reforestation. The links between deforestation, alpine pasturage, and intensified flooding and erosion had been widely discussed since the turn of the century in works like Jean-Antoine Fabre's *Essai sur la théorie des torrens et des rivières* (1797) and Etienne-François Dralet's *Description des Pyrénées* (1813).[8] The publication of Alexandre Surrell's *Étude sur les torrents des Hautes-Alpes* in 1841 incited further concern.[9] Nonetheless, it was not until the inundations of 1856 that calls for systematic, wide-scale reforestation began quite literally to gain ground. As one journalist noted, "It is not dikes that are the

[4] Annie Méjean, "Utilisation politique d'une catastrophe: Le voyage de Napoleon III en Provence durant la grande crue de 1856," *Revue historique* 295, no. 1 (1996): 133–51; also Champion, *Les inondations en France*, vol. 3, 186–87, and vol. 4, 69 and 77–78.

[5] "Lettre de l'Empereur sur les moyens de prévenir les inondations" in "Les inondations en 1856," 186–92.

[6] On public works as projects of national glorification, see Sara B. Pritchard, *Confluence: The Nature of Technology and the Remaking of the Rhône* (Cambridge, MA: Harvard University Press, 2011).

[7] *Le moniteur universel*, 15 May 1857, cited in Tamara Whited, "The Struggle for the Forest in the French Alps and Pyrenees, 1860–1940" (PhD dissertation, University of California, Berkeley, 1994), 58, n. 19.

[8] Fabre, *Essai sur la théorie des torrens et des rivières*; and Étienne-François Dralet, *Description des Pyrénées, considérées principalement sous les rapports de la géologie, de l'économie politique, rurale et forestière, de l'industrie et du commerce* (La Rochelle: Éditions La Découvrance, 2006).

[9] Alexandre Surrell, *Étude sur les torrents des Hautes-Alpes* (Paris: Dunod, 1870).

sole means of preventing [flooding] but rather the reforestation and resodding of the mountains."[10] Galvanizing even those who had not been previously convinced of alpine forests' importance, the floods helped assure strong legal and financial backing for what would come to be known as the *restauration des terrains en montagne* or RTM.[11]

Undergirded by scientific certainty and imperial ambition, alpine reforestation would enduringly transform France's upland landscapes and the communities that inhabited them. In the process, the forest administration would come to exercise a level of effectiveness and control of which Colbert could only have dreamed. Expanded in size and properly financed for the first time in its history, the Eaux et Forêts reached the apex of its power between 1860 and 1870.[12] Yet even at its moment of triumph, the contradictions and divisions that had driven two centuries of conflict persisted. Largely exempt from state oversight, France's privately owned woodlands, which constituted more than half the nation's forest cover, would continue their trend toward increased fragmentation and declining productivity. Likewise, *usagère* communities, dubious as ever of external schemes of improvement, would continue to defend their interests and exact concessions, however limited, as they had for two hundred years.

FROM LIBERTY TREE TO PRESIDENT PINE

By the time the roiling floodwaters of 1856 thrust the upland reforestation movement into being, the forest administration's influence had already been rising for three decades. The adoption of the Forest Code in 1827 and opening of an officers' training school at Nancy marked the first steps in this ascent. Better organized, more rigorously trained, and endowed with expanded powers of policing and coercion, the forest administration held the upper hand against woodland communities from this point forward.

Though some officials complained of the agency's excessive severity – the *procureur général* of the appeals court at Besançon, for example, protested that the administration "asks too often for incarceration, and

[10] M. J. Valserres, "Les inondations et le reboisement des montagnes," in *La nouvelle*, reprinted in *Annales agricoles et littéraires de la Dordogne* 21 (1860): 500. On scientific consensus regarding forest's value as an impediment to flooding and erosion, see also "Les inondations en 1856," 197–98.

[11] Whited, *Forests and Peasant Politics*, 58–63.

[12] *Les Eaux et Forêts*, 509–27.

for sums so minimal that this expedient becomes very onerous for the treasury" – few quibbled with the assumption that its oversight was needed.[13] The state's presence in the forest, as guiding hand and iron fist, was now a given.

Under Louis-Napoléon, who throughout his tenure showed exceptional interest in woodland matters, the forest administration became more confident still. The opening of two preparatory schools for guards and an across-the-board salary increase for woodland personnel boosted professional morale and reduced the incidence of corruption.[14] At the same time, it began to fulfill the role that had long been ascribed to it but seldom if ever achieved: that of ecological enforcer and guardian of the public interest. The forest administration did this by embarking on increasingly ambitious projects aimed at environmental amelioration rather simply accruing and exploiting commodities like naval timber and industrial fuel.

The first of these ecological initiatives involved the drainage and afforestation of the sandy bottomlands of the Sologne and the planting of pines in the *landes* of Gascony, which had long been besieged by sand dunes.[15] Both projects had a far-reaching impact. Even more ambitious was the work taken up after the 1856 floods, the *restauration des terrains en montagne*. Approved by the *Corps législatif* in July 1860 and initially focused on the Pyrenees and the Alps, the RTM project provided seeds, plants, and cash subsidies to private landowners, communes, and public establishments to replant voluntarily. Areas regarded as particularly hazardous were designated for mandatory reforestation and either expropriated by the state or seeded at the landowner's expense.[16]

From a hydrological standpoint, RTM was enormously effective. It transformed the face of the uplands, seeding some 350,000 acres (141,600 hectares) of uplands by the last decade of the century and mitigating the effects of future storms. It also changed the local economy, creating new employment in areas affected by emigration and the decline of charcoal-based metallurgy. In addition to work preparing and planting

[13] AN BB^{30} 373, Cabinet du Ministre de la Justice. Rapports politiques des procureurs généraux: Besançon, 8 November 1851.

[14] *Les Eaux et Forêts*, 515–16. [15] *Les Eaux et Forêts*, 511, 17–18, 80–81.

[16] Whited, *Forests and Peasant Politics*, 59; see also Gérard Brugnot and Yves Cassayre, "De la politique française de restauration des terrains en montagne à la prévention des risques naturels," in *Actes du colloque Les pouvoirs face aux risques naturels dans l'histoire, Grenoble, 22–23 mars 2001* (Grenoble: MSH Alpes, 2002), 2–3.

the terrain, RTM spurred the development of new jobs in alpine tourism, as visitors flocked to the region to view the mountains' scenic makeover.[17]

Socially, the project encountered greater problems. Many communities in the Savoie and the Ariège opposed reforestation on the grounds that it would deprive them of vital woodland access and interfere with pasturage rights. In response, a follow-up decree permitted sodding instead of tree planting where feasible.[18] Cultural distaste for conifers also played a role in local resistance to reforestation. "They clear to plant [the woods] with pines; they are not beautiful, pines," noted one Côte-d'Or old-timer, adding, "Nature is going to disappear."[19] Ecologically, too, the newly introduced plantations turned out to pose a challenge. Composed primarily of coniferous imports like Scots pine and Austrian black pine because of these species' ability to flourish even under poor conditions, the new forests' uniformity made them vulnerable to insect infestations and disease.[20]

At the time, however, most officials regarded the reforestation movement as a remarkable success.[21] Together with the powers attributed to the forest administration by the Forest Code, RTM's sheer visibility helped catapult woodland officers to new legitimacy and prominence in the latter part of the nineteenth century. Their feats drew appreciation not only within France but beyond its borders as well. Already by the mid-1860s, forest managers from across Europe were flocking to see the reforestation in progress.[22] Americans also took note, including George Perkins Marsh, the father of the modern American environmental movement. Admiring the "skill and perseverance of French foresters" in replanting steep alpine slopes, Marsh observed, "It is hardly hyperbolical to say that this is a case of impossibilities conquered."[23]

[17] On pine plantations and the development of a new woodland economy, see M. Becker and F. Le Tacon, "Santé de la forêt: Importance d'une sylviculture adaptée aux conditions de milieu," *Revue forestière française* 37, spec. issue (1985): 19–22. On forest tourism, see Fesquet, "L'arbre au secours des hommes," 165–66.

[18] Whited, *Forests and Peasant Politics*, 70–72.

[19] Cited in Jolas, "Bois communaux à Minot (Côte-d'Or)," 228.

[20] On the species composition of the plantations, see Husson, *Les forêts françaises*, 149–50. On cultural dislike of coniferous forests, see Jolas, "Bois communaux à Minot (Côte-d'Or)," 226. On cultural pride in conifers, see Klein, "La Franche-Comté touche toujours du bois," 61.

[21] See discussion in G.-A. Leroyer, "Notice sur le Congrès des délégués des Sociétés savantes, session de 1864," *Bulletin de la Société industrielle d'Angers et du département de Maine et Loire* 35 (1864): 187–88.

[22] *Les Eaux et Forêts*, 522.

[23] George Perkins Marsh, *The Earth as Transformed by Human Action: A Last Revision of Man and Nature* (New York: Charles Scribner's Sons, 1885), 309.

In this way, the forest administration overhauled its image during the Second Empire and shook off the stigma of the old, reviled Eaux et Forêts for good. Armed with sword, shotgun, and rifle, and attired in natty uniforms of green and gray, jaunty *képi* on their pates, forest officers were now no longer merely sentries against woodland encroachment, but respected stewards of the national patrimony, the first line of defense against allegedly abusive custom and retrograde routine.[24]

Fortuitously for the forest administration, factors that had nothing to do with their regulations eased woodland pressure during this period. The resulting recovery of the forest further burnished foresters and the Forest Code's reputations. Foremost among these factors was the abandonment of charcoal and adoption of coal as France's primary source of industrial and domestic energy. This shift was itself facilitated by the construction of railroad and canal networks from 1850 forward, which enabled raw materials and finished goods to be transported faster and more cheaply over long distances and gave rise to the concentration of manufacturing in large urban centers.[25] As industrial mechanization accelerated, creating abundant low-skill jobs, increasing numbers of peasants whose rural livelihoods had been restricted by the Forest Code poured out of the countryside to fill them. Together, these trends reduced local competition for fuelwood and eased reliance on woodland pasturage and other communal practices, thereby aiding the forest administration's efforts to supplant customary usage with scientific management.[26]

The emergence of new and expanded domestic markets for forest resources also augmented the forest administration's influence. With the invention of grinding machinery that allowed paper to be manufactured more cheaply from wood fiber than from rags, the state began adapting some forests to pulp production. As urban growth and rising prosperity accelerated the pace of construction, and as demand for telegraph poles and railroad ties flourished, the forest administration hastened to convert coppices to slower-growing but more lucrative timber stands. Here again, the development of canal and rail networks played a vital role in the development of these markets by making previously remote woodlands accessible. In this way, even as iron hulls replaced oak and naval timber

[24] *Les Eaux et Forêts*, 520–22.

[25] Xavier de Planhol, *An Historical Geography of France*, Janet Lloyd, trans. (Cambridge: Cambridge University Press, 1994), 269–72, 365–71, and 405; and Hugh D. Clout, *Agriculture in France on the Eve of the Railway Age*, Croom Helm Historical Geography Series (London: Croom Helm, 1979), 156.

[26] Planhol, *An Historical Geography of France*, 365–67.

ceased to be a critical component of France's national defense, hardwood timber continued to be a focus of woodland administrators' work.[27]

By the end of the Second Empire, woodland infractions were on the wane, the state's forest revenues were up, and the extent of customary usage in national woodlands had declined because of the state's aggressive enactment of *cantonnement*.[28] Not only had foresters achieved new potency in the nation's perennially unruly hinterlands, but they were also expanding their reach beyond France's borders. Having already imposed the Forest Code in Algeria following France's 1830 conquest, the forest administration now looked to apply the same notions of value, commodification, and development on the rich hardwoods of Southeast Asia, west Africa, and the Pacific.[29] The exploitation and import of these overseas resources further reduced pressure on forests within the hexagon and facilitated their reorientation toward recreation.[30]

France's woodland influence also extended beyond the lands under its dominion. Gifford Pinchot, the first head of the U.S. Forest Service, had no prior forestry schooling when, fresh out of Yale College, he enrolled as a student at the École nationale forestière at Nancy in November of 1889.[31] His six months there would comprise the entirety of his formal silvicultural training throughout his career. Returning to the United States and proclaiming himself the nation's "first scientifically trained forester," Pinchot set about creating the country's first managed forest, basing his methods in large part on what he had learned in France.[32] He would take control of U.S. national forests nine years later. Through him, the ideas and emphasies of conservation and management that were developed and elaborated in the French revolutionary era continue to resonate in the United States today.[33]

[27] On declining demand for naval timber, see Evans, "Wood since the Industrial Revolution," 46–48; also Glete, *Navies and Nations*, 2:421–29.

[28] Goujon, "Legislation et droits d'usage," 15–25.

[29] Whited, *Forests and Peasant Politics*, 199–208.

[30] C.A.R.A.N., *Histoire de forêts*, 122–30; and Jean Gadant, *L'atlas des forêts de France* (Paris: J.-P. de Monza, 1991), 205–13.

[31] de Steiguer, "The French National Forestry School," 18–20.

[32] At the Biltmore estate in North Carolina. Char Miller, "Sawdust Memories: Pinchot and the Making of Forestry History," *Journal of Forest History* 92, no. 2 (1994): 8. On Pinchot's regard for French forestry, see Gifford Pinchot, *Breaking New Ground*, commemorative ed. (Washington, DC: Island Press, 1998), 24 and 27.

[33] With troubling results. See discussion in Richard P. Tucker, *Insatiable Appetite: The United States and the Ecological Degradation of the Tropical World* (Berkeley: University of California Press, 2000).

Two hundred years after the 1669 Ordinance, Colbert's vision of a commanding state presence in the nation's forests, geared to assuring the country's broader economic and strategic well-being, would appear to have finally been realized not only in France but increasingly around the world as well. During the Revolution, innumerable efforts to revamp France's forest administration had been attempted and abandoned, thwarted by the irreconcilable impulses of liberalization and control, as well as by unremitting factionalism, popular unrest, economic crisis, and war. Though many theorists at the time believed that reorganizing and augmenting the state's authority was the best way to improve woodland management and safeguard the forests for the future, the *cahiers de doléances* made clear that rural inhabitants would not tolerate the persistence of the corrupt and ineffectual *maîtrises* of the *Eaux et Forêts*. The *cahiers*' demands for greater communal control were initially granted, but the oblique wording of the law and revolutionary legislators' failure to replace the *maîtrises* with a clear system of oversight rendered France's forests vulnerable to opportunism by private landowners and rural inhabitants alike.

The precariousness and uncertainty of the woodland situation continued well beyond the Revolution and into the nineteenth century. Though Napoleon took steps toward reviving and reinforcing the state's control, including setting up a military-based model of administration and issuing laws to clamp down on customary rights and private clearing, these efforts languished without an effective apparatus to implement them. It was not until the Restoration that forest reforms finally bore fruit in the form of the 1827 Forest Code. Representing the culmination of all the proposals, initiatives, and restructuring that had preceded it, as well as the state's growing intolerance for dissent, the Forest Code marked a turning point for the forest administration as well as for the forests themselves. From this moment forward, the scientific, orderly, hierarchical approach long advocated by theorists from Réaumur to Dralet took hold, carried out by a systematically trained corps of officers and guards. Communal usage, long impugned by these thinkers as benighted, abusive, and reactionary, was likewise singled out for repression.

With the advent of Louis-Napoléon and the revival of an unapologetically authoritarian state, the ideas of exploitation and management encapsulated in the Forest Code achieved their fullest realization. Further buttressed and expanded by the upland reforestation campaign, the forest administration completed the renovation of its public image from venal egotists and heedless oppressors to altruistic agents of watershed protection and the defenders of France's fertile plains.

There is perhaps no better symbol today of this metamorphosis than the "president pines" that dot the state and communal forests of Franche-Comté. Chosen by foresters and woodcutters for their exceptional height, width, and utility (based on their estimated board-foot volume), as well as their ease of access for visitors, presidents are meant to convey the magnificence of a carefully managed forest as well as the rewards of an orderly society.[34] Among the most well-known is the *sapin président* of the state-owned Forêt de la Joux, in the Jura, which bears a sign explaining that it was "elected" in 1964, is 45 meters high and 385 centimeters around, and is expected to produce 600 planks. This president has already "served" an estimated 230 years, but like all trees selected for the honor, its term will come to an abrupt end when foresters determine that its timber value has peaked.[35] At that point, this elder statesman, like the hoi polloi pines surrounding it, will be cut down, disarticulated, and sold to the highest bidder.[36]

With its overwhelming emphasis on human utility and quantifiable value, the president pine tradition embodies not only the aims and interests of foresters in the latter nineteenth century, the period when the tradition began, but also the victory of the centralized, bureaucratic, modern state over the diffuse and unpredictable pathways of popular democracy. In nearly every way, it contrasts sharply with the custom of the liberty tree, which was abolished by official decree during the same period that the concept of president pines was being devised. Whereas liberty trees were erected, often spontaneously, by a passionate and engaged collective, president pines were and still are selected by a panel of experts. Whereas liberty trees' meanings varied widely according to the beholder, president pines bear a placard itemizing their significance. Liberty trees varied in type – oaks were most common, but hornbeam, poplars, and even pines were also planted according to what was locally available. President pines are what their name implies: pines. And finally, whereas liberty trees were positioned prominently in public spaces and thereby offered other social benefits that are difficult to quantify, like aesthetic enhancement, shade, and shelter to citizens gathered beneath its canopy, president pines were, and continue to be, monuments in isolation, special only by comparison to other conifers and serving primarily as propaganda for the administration that controls them.

[34] See discussion in Klein, "La Franche-Comté touche toujours du bois," 57.

[35] On the age of the president pine in the Forêt de la Joux, see Agnès Kielbasa, "Promenons-nous dans les bois en Franche-Comté," *La Terre de Chez Nous*, 6 August 2005.

[36] On president pines' destiny, see Klein, "La Franche-Comté touche toujours du bois," 57.

In this way, the president pines, for all their individual beauty, also stand for an ugly outcome: the estrangement of *usagère* communities from the forests on which they relied, and the forest administration's replacement of a diverse range of benefits with a narrowly delineated set of values.

CONSERVATION'S ACHILLES' HEEL

By the latter half of the nineteenth century, the forest administration had emerged as a potent force of statebuilding. Yet profound limitations on its power also remained. As decreed by the Forest Code, the state had virtually no control over private woodlands, even though they accounted for more than two-thirds of France's forests. Compared with the dozens of regulations and policing efforts carried out in state and communal holdings, private owners were subject to only one significant legal constraint: they must provide four months' advance notice before clearing or face heavy fines and the potential obligation of reforesting the area at their own expense. The 1860 law on upland reforestation introduced further requirements, but for the most part private landowners were simply encouraged, rather than ordered, to replant their parcels.

This lenience had its origins in the radical liberalization of woodland oversight at the start of the Revolution. Reacting against the strict controls of the 1669 Ordinance, which subjected all forests, regardless of owner, to state management and in particular obliged private landowners to obtain advance approval for fellings and clearings, revolutionary legislators vested proprietors with the power to "administer and dispose of" their woods in whatever way they saw fit.[37] At the same time, legislators nationalized and auctioned off thousands of hectares of boskets and other isolated woods that had previously been part of ecclesiastical holdings. With few exceptions, these woods were cleared and divided.[38]

Both acts – the easing of woodland restrictions and the sale of ecclesiastical property – had their roots in physiocratic theory, which held that private tenure was superior to shared possession because individual proprietors were more likely to invest in improvements and thereby increase

[37] *Décret sur l'administration forestière*, 29 September 1791, title I, art. 6, in Duvergier, *Collection complète des lois*, 3:272.

[38] As per a 23 August 1790 law permitting the sale of *bois nationaux* smaller than fifty hectares and more than two kilometers from the *grandes masses*. See Bourgin, *La Révolution, l'agriculture, la forêt*, 194; and *Les Eaux et Forêts*, 271.

the productivity of the soil. Although contemporaries like Barère de Vieuzac denounced this view, arguing that "personal interest, even well intentioned, is only a selfish and rapacious usufructuary," the privatization and clearing of France's forests accelerated rapidly from 1791 forward, hastened significantly by the Directory's sales of vast areas of national woodlands to raise cash.[39]

Not until the Empire was well underway did the state take stock of the loss of the country's loss of thousands of hectares of forest and take steps to revive limits on clearing. The law authorized by Napoleon in 1803 – one of several that year related to woodlands – imposed a twenty-five-year moratorium on unauthorized deforestation.[40] Nonetheless, without an effective agency, enforcement remained spotty and abuses continued apace. In addition, the promulgation the following year of the Civil Code, with its firm insistence on the absolute rights of property, worked both to undermine the state's control over privately owned woodlands, no matter how vital these resources might be to society as a whole, and to increase the fragmentation of forest inheritances among family members.[41]

By the time legislative debates over the development of a new Forest Code began in December 1826, the precedent set in 1791 of unfettered private forests was virtually unassailable. Martignac and his fellow backers of the bill fought mightily simply to retain, on a limited and temporary basis, the clearing restrictions and timber protections established by Napoleon in 1803.

At the same time, every successive regime – royal, republican, and imperial – continued to sell immense swaths of national woodland to raise urgently needed funds. These sales persisted through the Second Empire, when the state used the deeply paradoxical approach of auctioning lowland forests to finance its mountain reforestation project.[42] Indeed, much of the benefit of the state's *cantonnement* of use rights in lowland forests in the nineteenth century lay in rendering its woodlands more marketable through the elimination of customary claims. More than 64,500 hectares of national woodland were sold during this period, most

[39] Barrere [Bertrand Barère de Vieuzac], *Rapport des comités réunis des Domaines, des Finances, de l'Aliénation des biens nationaux, de la Marine, du Commerce & d'Agriculture. Sur les bois & forêts nationales* (Paris: Imprimerie nationale, 1790), 1–29.

[40] *Loi relative aux bois des particuliers*, in Baudrillart, *Recueil chronologique*, tome 1, 636–37.

[41] Code Civil, 1804, book 2, chap. 2, art. 537. On the Civil Code and woodland fragmentation, see Badré, *Histoire de la forêt française*, 126.

[42] *Les Eaux et Forêts*, 522–23.

of which was subsequently cleared.[43] Overall, from the start of the Restoration in 1814 to the end of the Empire in 1870, the state alienated more than a quarter of its forests, a total of roughly 353,000 hectares.[44] Under these circumstances, one can imagine that the acclaim that observers showered upon Napoleon III's reforestation projects was fueled in part by relief that there would at last be compensation for the thousands of hectares the state had alienated and allowed to be cleared since the start of the Revolution.

As the demand for fuelwood declined and the extent of land under cultivation diminished in the later nineteenth century, France's total forest cover began rebounding significantly from its low point during the Revolution and the Directory. Nonetheless, with few controls on their owners, private woods remained vulnerable to subdivision, sale, and clearing. The partition of great woodland estates, in particular, accelerated in the first decade of the twentieth century, causing concerned legislators and local administrators to push through an amendment to the Forest Code that allowed for privately owned forests to be managed by the state in exchange for an annual fee.[45] In practice, however, the amount of private woods protected in this manner was minimal, and the vast majority of forests remained outside the authority of the Code.[46]

By the 1950s, the fragmentation of private forests was so extensive that only about 6 percent of individual forest owners owned more than ten hectares.[47] Though overall forest cover was on the rise in France and it remained the fourth most heavily wooded nation in Western Europe, the proliferation of small parcels impeded systematic timber production and led to yields of lesser quality and lower quantity, placing the country at an economic disadvantage relative to its neighbors.[48] With a level of private forest ownership greater than every other European country except the Netherlands and the overall productivity of these holdings comparatively

[43] Husson, *Les forêts françaises*, 161.

[44] *Les Eaux et Forêts*, 523; also Husson, *Les forêts françaises*, 161.

[45] S. V. Ciriacy-Wantrup and R. Bishop, "'Common Property' as a Concept in Natural Resources Policy," *Natural Resource Journal* 15 (1975): 327–28.

[46] *Les Eaux et Forêts*, 574–75.

[47] Stephen Haden-Guest et al., *A World Geography of Forest Resources* (New York: Ronald Press, 1956), 276.

[48] Only the Netherlands exceeded France's level of private forest ownership. Only 30 percent of French woodlands were publicly held, in comparison to 56 percent in West Germany, 44 percent in Belgium, 40 percent in Great Britain, and 70 percent in Switzerland. Haden-Guest et al., *World Geography of Forest Resources*, 276 and 364; also Gadant, *L'atlas des forêts de France*, 15, 17.

low, France was forced, as it had been for centuries, to augment its domestic deficit with imports.[49]

Not until 1963 was a law finally enacted to place France's private woodlands under closer scrutiny. In contrast to the broad autonomy that the Forest Code had granted landowners in 1827, the 1963 statute insisted on proprietors' responsibility for the "biological equilibrium" and "wise economic management" of their woods. In particular, it required individual forest owners to devise management plans for woods larger than 25 hectares and register them with a Centre régionaux de la propriété forestière, district offices that the 1963 law created to provide technical guidance and foster collaboration among proprietors.[50]

The 1963 measures soon bore fruit in the form of a rapid increase in the extent of private forests. Nonetheless, France continues to struggle with the long-term effects of woodland fragmentation. Although private forests produce nearly three-quarters of the wood harvested in France every year – an amount roughly proportional to the territory they cover – national shortfalls and foreign competition cause France to import more than ten million cubic meters of wood annually, primarily in the form of tropical hardwood logs, sawn wood, and wood-based products like furniture, pulp, and paper.[51]

This state of affairs reveals how much the ideological interventions of private property have continued to shape – or, as it turns out, misshape – the aims expressed at the time the 1827 Forest Code was created. The Code's authors repeatedly emphasized that they wished to conserve the country's vital forest resources for the public good – meaning, for the most part, perpetuating and expanding the production of timber and industrial fuelwood.[52] Yet by reducing the forest administration's oversight of the great majority of the nation's woodlands to an apologetically worded, temporary limit on clearing and felling, legislators rendered this goal nearly unattainable. Advocates of proprietary liberalism insisted that private ownership made for more efficient management of fields as well as

[49] On the deficit between wood supply and demand in 1960–63, see Claude Durand-Prinborgne, "La propriété forestière moderne," in *Actes du colloque sur la forêt, Besançon, 21–22 oct. 1966*, Cahiers d'Études Comtoises 12 (Besançon: Les Belles Lettres, 1967), 309.

[50] Durand-Prinborgne, "La propriété forestière moderne," 327.

[51] Institut national de l'information géographique et forestière, "La forêt française: Les résultats issus des campagnes d'inventaire 2006 à 2010 – Décembre 2011. Chiffres clés," http://inventaire-forestier.ign.fr/spip/spip.php?article709.

[52] See discussion in Fesquet, "L'arbre au secours des hommes," 164–65.

forests, but this clearly turned out not to be the case. Indeed, while the declining demand for wood-based charcoal and the growing market for construction wood in the second half of the nineteenth century prompted the forest administration to begin converting most state and communal woodlands from *taillis sous futaie* (coppice with standards) to timber-oriented *futaie*, private forest owners took much longer to follow suit, if ever. Woodland advocates' fears that individual proprietors would incline more toward the short-term gains of sales and clearing continue to prove accurate into the present. As one observer recently remarked, the management of private forests is "very inferior to that of state and communal woods as a result of [their] disintegration, excessive costs, and the carelessness of too many landowners."[53]

Not all of the consequences of fragmentation and neglect are bad: from an ecological standpoint, the existence of thousands of small, relatively abandoned woodlands provides habitat and migration corridors for insects and amphibians, birds, and mammals. Intensively managed monocrop woodlands rarely have the species diversity of these pocket forests. Nonetheless, as in the United States, without protection, these little oases are vulnerable to market forces and are easily swallowed when development promises greater returns.

If the state's lenience toward private property undermined its conservationist objectives, its limitations on customary usage went too far in the opposite direction. For more than half a century, woodland theorists from Duhamel du Monceau to Dralet had criticized *usagère* communities for everything from woodland grazing to their *jardinage* approach to harvesting timber.[54] These criticisms became the guiding principles of the Forest Code. Despite copious warnings by these same theorists that individual landowners needed to be reined in too, the Code's authors overwhelmingly targeted peasant practices, holding peasants responsible for all the forest's ills.

Protests from some legislators helped prevent an outright ban of certain usages, particularly woodland pasturage, which the contingent from the Doubs vehemently defended on the grounds that it was vital to the survival

[53] Georges Plaisance, *Guide des forêts de France* (Paris: Pierre Horay Éditeur, 1997), 118.

[54] On Dralet's criticisms of peasants' *jardinage*, the method of selective felling that he also embraced in certain highly regulated circumstances, see Whited, *Forests and Peasant Politics*, 29–30; and Curtis Sarles, "The Instatement of Order: State Initiatives and Hegemony in the Modernization of French Forest Policy," *Theory and Society* 35, no. 5–6 (2006): 577.

of the department's dairying cooperatives. Nonetheless, the exemptions were worded in a narrow manner that allowed the rights to be easily overridden should the forest administration deem them a threat to other, more sanctioned woodland benefits. Adding to the Code's potency was the increasingly technocratic authority of woodland officers, whose training at Nancy endowed them with a self-assurance and sense of duty lacking among most of their predecessors. Armed with the tenets of scientific forestry and the conviction that their mission was righteous, these woodsmen of the highest rank deployed *cantonnement* to isolate use rights and conversion – the transformation of coppices to timber groves – to eliminate them.[55] The substitution of clearly delineated property boundaries for the fluidity of customary usage was not merely a means of increasing woodland revenues; it was also an unambiguous assertion of authority, a forceful bid for exclusivity and control over a resource that had previously been administered collectively, on the basis of multiple values and shared responsibility.

In this respect, France's woodland policy and increasingly empowered forest administration succeeded only too well. Faced with the impossibility of continuing their traditional agricultural and silvopastoral practices and burdened by rural debt and mounting inflation, rural inhabitants, not forests, were now the ones in retreat. Though the mass migrations of the latter nineteenth century eventually affected the whole nation, they began earlier in Franche-Comté in large part because the hardscrabble rural economy, always tenuous to begin with, succumbed more rapidly to the Forest Code's constraints. Emigrating in droves from the 1830s forward, the Comtois peasantry in many cases departed for the very places where the Empire was now seeking to impose a new era of timber exploitation and natural resource control. As the prefect of the Doubs reported in July 1856, "The population of the department has shrunk by 10,000 souls; this fact must be attributed to the double movement of emigration toward the industrial centers and toward the new countries like America and Algeria."[56]

For those who remained behind, the outlook was bleak. While labor shortages drove up agricultural wages in some areas, the overall decline of

[55] Jean-Luc Mayaud, "Équilibre, tensions et déprise rurale aux XIXe et XXe siècles," in *Connaissance de la Franche-Comté: Régards sur les Vosges comtoises: À la découverte des terres comtoises du Parc naturel régional des Ballons des Vosges*, 269–87 (Besançon: Centre universitaire d'études régionales, Université de Franche-Comté, 1991), 281; and Pierre Pagney, "La région comprise entre forêt de Chaux et forêt de la Serre: Étude rurale," *Revue géographique de l'Est* 8, nos. 3–4, 1–2 (1968, 1969): 28.

[56] AN F^{1cIII} Doubs 8, dossier 3, July 1856. See also McPhee, *Social History of France*, 158 and 225–26.

farming intensified regional unemployment and caused food prices to rise. The decimation of vine and grain crops due to insect and fungal infestations further depressed the rural economy. At the same time, revenues from the sale of communal firewood and timber, a source of income on which communities relied, waned dramatically as a result of the formation of alliances between forge masters and railroad investors, which drove down wood prices and lowered property values.[57] "The countryside is depleted and it is urgent to come to their aid," the *procureur général* at Besançon testified in 1855.[58]

Yet even as some officials called on the state for help, others continued to agitate against the forest administration's management directives. Resistance was especially pronounced in the Ariège, where inhabitants vigorously protested the state's upland reforestation projects from 1860 forward.[59] Conflict also persisted in Franche-Comté, where communities rejected the uniform controls of the Forest Code in favor of their own, locally devised arrangements. In one especially long-running confrontation, the mayor of the metallurgical village of Syam, in the Jura, waged an aggressive crusade in the press from 1863 to 1887, urging communities everywhere to remove their woods from the state's oversight. The affair was eventually settled through compromise, but only after the mayor, Alphonse Jobez, had pushed it to the attention of the National Assembly.[60]

As in Jobez's case, the indignation of local leaders chafing under the interference of the forest administration was often the spark to rural discontent's tinder. A confrontation over grazing rights in the Forêt de Chaux in 1858 illustrates how complex – and personal – these disputes could be. According to correspondence concerning the incident, which eventually involved officials at the highest levels of the department, a handful of residents from the hamlets of Chissey and Germigney on the southern edge of Chaux had worked out an agreement whereby they pastured their horses and cows together with the communal herd of neighboring Chatelay because their farms were so far distant from their own village centers. This had been the case "since time immemorial," insisted the mayor of Chatelay, who noted that ushering the beasts – twenty-four animals in all – to join their official communes' herds would entail "considerable damage to the crops along the length of the

[57] See discussion in AN BB[30] 373. [58] AN BB[30] 373, 27 January 1855.

[59] Whited, *Forests and Peasant Politics*, 67–82.

[60] On the conflict, see Vion-Delphin et al., *Les hommes et la forêt en Franche-Comté*, 133.

road."[61] When the district forest officer flatly rejected this arrangement, citing the clause in the Forest Code that disallowed such practices, the mayor complained to the subprefect, noting that the forest officer was "motivated only by hatred and by vengeance" because of a prior complaint Chatelay's inhabitants had filed against him. Informed by his understanding of the local landscape and his personal relationship with the individuals involved, the mayor's account contrasts strikingly with the forest administration's response, which notes only the "irregularity" of the mingled herd and returns time and again to the Forest Code to emphasize why this "abuse" cannot be tolerated.[62] Rejecting the logic of geographic proximity, as well as the implication of personal animosity, the regional forest inspector repeated that the officer had "done his duty in carrying out the law," and it was on this basis that the villagers' unauthorized cooperation was ultimately rejected.[63]

That the forest administration increasingly prevailed in such conflicts by the end of the nineteenth century does not detract from the fact that local resistance, both indirect and explicit, powerfully shaped the state's policies in practice from the very first woodland *réformations* of Franche-Comté forward. From the months-long uprising by the demoiselles of Chaux in 1765 to the rash of sabotages and attacks on the forest administration in the 1840s, the Francs-Comtois time and again demonstrated their willingness to take up arms, risk their lives, and even kill in defense of their woodland interests. Rejecting the imposition of externally conceived and to their minds arbitrary forms of exploitation and management, the region's inhabitants forced the state to adapt its directives, or at least its enforcement of them, in ways that accommodated compound demands, not simply timber or fuelwood production.

Because of this tenacity, villages across France still to this day possess customary rights in state woodlands, ranging from fuelwood and timber allowances to animal grazing and mushroom gathering.[64] Timber and firewood allocations from communal woods also continue. Though the annual *affouage* apportionment of communal firewood is no longer as crucial as it once was in Franche-Comté – nowadays most households rely on oil, natural gas, or nuclear energy for their heat and power – the sale of communal timber has proved extremely lucrative. Some Comtois villages raise as much as 60 percent of their yearly income through communal

[61] ADJ 11Qp215, December 1856–January 1857. [62] Ibid. [63] Ibid.
[64] Per the 1827 Forest Code, no new rights have been granted since 1827.

timber sales.[65] These revenues reflect villages' success at taking care of and capitalizing upon their holdings, as well as the unexpected benefit of the state's aggressive *cantonnement* of customary usages after the Forest Code.[66] At the time, the loss of woodland access in state forests in exchange for a much reduced parcel of land was a source of bitter controversy and hardship. In the long run, however, the increase in communal forest holdings, augmented by the natural expansion of trees into former fields and pasture, has been a boon to smaller villages. Had rural communities not insisted on the legitimacy and preservation of their customary rights in the national forests at the time of the Revolution and in the years leading up through the Forest Code, they would not have obtained concessions on nearly the same scale that they did. The resulting legacy of their intransigence is visible today in the Forêt de Chaux, where the area owned by the state, some 13,060 hectares, is ringed by nearly 9,000 hectares of erratically shaped communal holdings.[67]

Not only have communes managed to uphold their forest benefits, but they have also seen a revolution in attitudes toward peasant practice. The shift began in the 1960s, with the creation of France's first "natural regional park," a designation intended to promote economic development and tourism and protect areas imperiled by demographic decline and the abandonment of forestry and farming.[68] Unlike France's national parks and natural reserves, which were also established from the 1960s forward, natural regional parks support traditional pastoralism and other activities in the interest of maintaining ecosystems and preserving landscape aesthetics central to French identity. Thus, cheesemaking, woodworking, clock manufacture, and *jardinage*-style logging all form a part of the Parc naturel régional du Haut-Jura and the Ballons de Vosges in Franche-Comté.

Amendments to the Forest Code since the 1970s have similarly emphasized forests' broader utility and the importance of customary practices in

[65] This figure is down from the prosperity of a generation ago, when the profits from "green gold" afforded some Comtois villages to waive municipal fees and taxes. Klein, "La Franche-Comté touche toujours du bois," 56–57.

[66] See discussion in Ciriacy-Wantrup and Bishop, "'Common Property' as a Concept," 713–27.

[67] Plaisance, *Guide des forêts de France*, 272. On the first step toward the *cantonnement*, see AD Doubs P 579, November–December 1829.

[68] The first Parc naturel régional was Scarpe-Escaut, created 13 September 1968. See discussion in Pierre Darmangeat, *Parcs et réserves de France*, 1991 ed. (Paris: Arthaud, 1991), 36.

maintaining their diverse economic, social, and environmental benefits. In 2005, the state launched a comprehensive rural development campaign that included decentralizing land management and encouraging local decision making of the sort that the Forest Code once pushed to the brink of obsolescence. The campaign reconciled activities like truffle and animal hunting with agricultural and silvicultural concerns. Most strikingly, it encouraged woodland pasturing in mountain regions as a means of restoring forest habitats and reducing the underbrush that fuels wildfires.[69] As a bulletin to departmental administrators explained, pastoralism "created these spaces, maintaining the quality of the scenery and biological diversity, preserving major equilibrium and contributing to the prevention of risks."[70] The revival of silvopastoralism is not unique to France; since at least 2008 the Ministerial Conference on the Protection of Forest in Europe has advocated it and other traditional practices, like coppicing, pollarding, and charcoal- and tar-making, as key strategies in the economic "maximization" and sustainable management of European forests in areas where "fully mechanised timber production" is not possible.[71] Once reviled, now vindicated, goats and sheep, charcoal burners and shepherds, have come to be viewed as indispensable agents of ecological repair.

In the realm of private property, French forest authorities have also worked to recreate collective modes of management. One of the key elements of the 1963 law on improving private forests' productivity was the establishment of Regional Forest Property Centers, eventually numbering eighteen in all, to foster collaboration among landowners, develop regional production goals, and provide technical advice and training.[72] A law of December 1985 similarly empowered regional committees with developing woodland management goals for private and

[69] *Projet de loi relatif au DATAR: Développement des territoires ruraux*, rapport 251 (2003–4), 8 April 2004.

[70] Ministère de l'Agriculture et de la Pêche, Direction générale de la forêt et des affaires rurales, sous-direction de l'environnement et de la ruralité, *Circulaire 2007–5006 rélatif à la protection et la mise en valeur des espaces pastoraux de la loi relative au développement des territories ruraux du 23 February 2005 et de l'ordonnance du 1er July 2004 relative aux associations syndicales des propriétaires*, http://agriculture.gouv.fr/spip/IMG/pdf/dgfarc20075006z.pdf, p. 3.

[71] Mauro Agnoletti et al., "The Introduction of Historical and Cultural Values in the Sustainable Management of European Forests: Document produced for the Ministerial Conference on the Protection of Forest in Europe (MCPFE)," *Global Environment* 2 (2008): 182.

[72] Durand-Prinborgne, "La propriété forestière moderne," 327–29.

public forests as a whole.[73] More recently still, the forest division of the Ministry of Agriculture informed departmental officials that in the interest of "encouraging pastoral activities," woodland managers and proprietors can negotiate multiyear agreements directly with flock owners, rather than go through the onerous approval process required by the Forest Code.[74]

All of these changes, presented today as conservation strategies, economic advocacy, and cost-saving simplifications, emulate the decision making regarding the forest that two and a half centuries ago was worked out at the level of the community. By encouraging such arrangements, the French government has in essence affirmed the virtues of custom, collectivity and cooperation over isolation and individuation and has validated their importance in rural livelihoods and landscapes.

Although this acknowledgment came two centuries too late for the Comtois communities that repeatedly struggled to defend their ancient firewood, timber, and grazing rights – from their occupation of the Forêt de Chaux in 1765, to their denunciations of the forest administration in their *cahiers* of 1789, to their bloody assaults on forest guards in the years after the 1827 Forest Code – it nonetheless signals that French policymakers have begun to reappraise the singular, centralized, commercially oriented conservation that has dominated woodland management since the nineteenth century. Pushed by contemporary pressures that emphasize sustainable solutions to problems of declining resources both in the hexagon and beyond, the French state is increasingly working to craft more effective and enduring approaches to allocating resources in ways that integrate stakeholders in all their variety while still seeing to the forest's long-term protection.[75] In this way, the Franche-Comté peasantry may claim to be the masters of their woods once again.

[73] *Loi forestière*, 4 December 1985.

[74] Ministère de l'Agriculture et de la Pêche, Direction générale de la forêt et des affaires rurales, sous-direction de l'environnement et de la ruralité, *Circulaire 2007–5006 rélatif à la protection et la mise en valeur des espaces pastoraux*.

[75] On the evolution of French environmental politics, see Michael Bess, *The Light-Green Society: Ecology and Technological Modernity in France, 1960–2000* (Chicago: University of Chicago Press, 2003).

Bibliography

Manuscript Sources

Archives Nationales

Série AD – Archives imprimées

AD^{XIXi} 1 Jura	Procès-verbaux du Conseil général du Jura, 1840, 1843, 1848–49, 1850–55
AD^{XIXi} 1 Vosges	Rapport de la Commission du Conseil général, 1844
AD^{XIXi} 1 Vosges	Rapport du Préfet au Conseil général, session du 1844–47

Série BB – Ministère de la Justice

BB^{3} 194	Affaires criminelles, 1792–93: Délits forestiers
BB^{3} 195	Affaires criminelles, 1828–29: Etats des délinquans forestiers
BB^{18} 267	Correspondance générale de la division criminelle, 1795
BB^{18} 945	Correspondance générale de la division criminelle, 1816: Instructions sur l'application de l'amnistie aux délits forestiers
BB^{18} 999	Correspondance générale de la division criminelle, 1822: Gardes forestiers
BB^{18} 1112	Correspondance générale de la division criminelle, 1824: Vols et rapines
BB^{18} 1144–45	Correspondance générale de la division criminelle, 1826–27: Propos séditieux
BB^{18} 1159	Correspondance générale de la division criminelle, 1828: Délits forestiers (Gers)
BB^{18} 1163–67	Correspondance générale de la division criminelle, 1828–29: Propos séditieux

BB^{18} 1186–87 Correspondance générale de la division criminelle, 1830: Désordres forestiers

BB^{18} 1188 Correspondance générale de la division criminelle, 1830–31: Désordres contre l'impôt sur les vins

BB^{18} 1192–95 Correspondance générale de la division criminelle, 1830: Désordres forestiers

BB^{18} 1209 Correspondance générale de la division criminelle, 1832: Propos séditieux

BB^{18} 1270 Correspondance générale de la division criminelle, 1823: Projet de code forestier

BB^{18} 1460–61 Correspondance générale de la division criminelle, 1848: Dévastations commises dans les bois et forêts dans divers départements à la suite de la révolution de février

BB^{24} 348 Affaires criminelles et des grâces, 1848

BB^{30} 274 Cour des comptes, 1827: Droits et indemnités des greffiers dans les affaires forestières

BB^{30} 358–66 Rapports politiques des Procureurs généraux, 1848

BB^{30} 373 Rapports politiques des Procureurs généraux, Besançon, 1849–59

Série C – Chambre des députés

C 951 Doubs Enquête sur le travail agricole et industriel (Loi du 25 mai 1848)

Série D – Missions des représentants du peuple et Comités des Assemblées

D^{XVIII} 1 Comité des domaines: Observations sur l'administration forestière [dossiers 29, 37], 1790–91, an II

Série E – Conseil du Roi

E564A, fol. 58 Arrêt du Conseil du 1er octobre 1686

E2055, fol. 18 & 115 Arrêts du Conseil des 18 janvier et 2 mai 1724

Série F – Versements des ministères et des administrations qui en dépendent

F^{1c}I 103 Esprit public, 1801

F^{1c}III Doubs 6 Esprit public et élections: Comptes rendus administr., 1790–an IX

F^{1c}III Doubs 7 Esprit public et élections: Comptes rendus administr., 1807–11

F^{1c}III Doubs 8 Esprit public et élections: Compte administratif et moral, comptes annuel, 1811–70

F^{1c}III Doubs 11 Esprit public et élections [dossier 5]: Comptes de tournées, 1799, 1809

F^{1c}III Doubs 12	Esprit public et élections: Corresp. et divers 1806–70
F^{1c}III Haute-Saône 5	Esprit public et élections: Extrait des registres des déliberations du Conseil général, 1790; Comptes rendus administratifs, 1792
F^{1c}III Haute-Saône 6	Esprit public et élections: Comptes rendus administratifs, 1793–an XII
F^{1c}III Jura 8	Esprit public et élections: Comptes analytiques des arrêtés, decisions et travaux, 1813; Comptes rendus administr., 1831, 1855–58
F^{2}I 136^{1}	Administration départementale [dossier 66]: Martelage et exploitation des bois destiné à la Marine, 1816–21
F^{3}I 2	Ministère de l'Intérieur, Administration communale [dossier 12]: Gardes forestiers, 1808–22
F^{3}I 3	Ministère de l'Intérieur, Administration communale [dossiers 1, 6]: Biens communaux, 1808, 1836–46
F^{7}3035	Police générale: Dilapidations des forêts, 1801–14
F^{7}9336	Police générale, 1830
F^{10}264	Agriculture: correspondance, an II
F^{10}328	Agriculture: correspondance concernant les communaux, an II–an IV
F^{10}329	Agriculture: Partage des communaux, 1792–an III
F^{10}403, 405–6	Agriculture: Bois et forêts, correspondance, 1778–1816
F^{10}459–89	Agriculture: Déstruction des animaux nuisibles, 1766–1834
F^{11}496	Subsistances: La statistique agricole de 1814 (circulaire)
F^{12}680	Commerce et industrie: Enquête sur les bouches à feu, 1788

Série Q – Domaines (Ancien régime)

Q^{1}181	Doubs: Eaux et forêts, 1727–83
Q^{1}182	Doubs: Eaux et forêts, 2e arrondissement (Beaume), 1739–60
Q^{1}183	Doubs: Eaux et forêts, 3e arrondissement (St. Hippolyte), 1749–84
Q^{1}416	Jura: 1er arrondissement (Dole), 1717–84
Q^{1}1001	Haute-Saône: 1er arrondissement (Gray), 1727–83

Archives départementales du Doubs, Besançon

Série C – Administrations provinciales avant 1790, sous-série 1C: Intendance de Franche-Comté

1C 1365	État général des bois et usines, 1783–84

Série L – Fonds de toutes origines de l'époque révolutionnaire (1790–1800)

L529 Salins. Forêts: Instructions, correspondance générale
L1380 Pontarlier. Forêts: Instructions, correspondance générale, délits. 1791–an IV
L1381 Pontarlier. Etats des forêts du Jura.

Série P – Fiscalité, finances publiques

P 420 Eaux et forêts depuis 1800. Cueillettes, parcours, amendes pour délits; Projet du Code forestier, 1818–29
P 425 État général des bois communaux et d'établissements publics soumis au régime forestier, 1828; délibérations des Conseils municipaux touchant les articles du Code forestier, 1828–29
P 579 Droits d'usage dans les forêts royales. Communes qui réclament des droits d'usage devant les tribunaux conformément à l'article 61 du Code forestier, 1829

Archives départementales de la Haute-Saône, Vesoul

Série B – Cours et juridictions d'ancien régime

B 255 Bailliage de Faucogney. Partage des biens communaux, 1790
B 290 Bailliage de Faucogney. Procédures criminelles, 1789
B 334 Bailliage de Faucogney. Procès verbal de rebellion au sujet de glandage, 1746
B 400 Bailliage de Faucogney. Rebellion au sujet de glandage, 1733
B 3187 Terre de Lure. Rapports: délits commis dans les bois et la plaine, 1788–91
B 3233 Cours extraordinaire, 1790
B 4209–28 Grand bailliage d'Amont, bailliage de Vesoul et les bailliages secondaires. Cahiers de doléances, 1789
B 9255 Grand bailliage d'Amont, Bailliage secondaire de Gray. Délits et dépradations, 1789–90

Série P – Finances, Cadastre, Postes, Eaux et forêts depuis 1800

17 P 23 Tribunal de Gray. Registre des jugements forestiers, 1848
17 P 25 Tribunal de Gray. Registre des jugements forestiers, 1851–52
17 P 352 Tribunal de Lure. Registre d'un garde forestier, Servance, 1806–12
17 P 358 Tribunal de Vesoul. Dossiers d'instruction pour les délits forestiers (an II–an V)

Archives départementales du Jura, Montmorot

Série C – Administrations avant 1790

C 128 [dossier 15] Intendance de Franche-Comté. Mémoire sur les difficultés qu'éprouve la ville de Dole de s'approvisionner de bois de chauffage, 1768

Série L - Epoque révolutionnaire, sous-série Lp

Lp 1618 Bailliage de Salins et Arbois. Cahiers de doléances, 1789
Lp 1618 Bailliage de Dole. Cahiers de doléances, 1789
Lp 2579 Statistique des bois et forêts, 1791–an VIII

Série M – Administration générale et économie – police, agriculture, forêts

M 25 Police générale: incendies, menaces, meurtres des gardes forestiers, Lons-le-Saunier potato riot, 1839–40
M 26 Police générale: Mutilation d'arbres dans la forêt royale de Chaux, June 1841
M 27 Police générale: Menaces envers les gardes forestiers et mutilations de baliveaux, Forêt de Chaux, 1842
M 28 Police générale: Incendies dans les forêts domainiales de Chaux et de Serre; meurtre et menaces envers les brigadiers forestiers, 1843–46
M 30 Police générale: Rapports mensuels, troubles, dévastations dans les forêts de Poligny, Arbois (Gen. Bachelu), Chaux, 1848
M 69 Suréte générale: Dole. Meurtre commis par un garde forestier, 1851
M 3257 Renseignements statistiques 1810–29: forges, haut-fourneaux, papeteries, verreries, tanneries
M 3259 Rapports périodiques des sous-préfets (Dole et Poligny): industrie et subsistances, 1832–35; Situation industrielle, 1856
M 3340 Statistique générale: statistique des bois et forêts, an VIII–1824

Série Q – Biens nationaux, domaines

11 Qp 215 Affouage et pâturage dans la Forêt de Chaux. Réclamations et contestations, 1844, 1847, 1856–57

Newspapers

L'echo du Jura, 1848–49
L'echo des Vosges, 1848
Le Franc-Comtois, journal de Besançon et des trois départements, 1848–49
Le journal de la Haute-Saône, 1848, 1852
Le moniteur universel, year IV–VII
L'ordre social, journal de Besançon et de la Franche-Comté, 1848
La sentinelle du Jura, 1832

Printed Primary Sources

Annales de statistique, ou Journal général d'économie politique, industrielle et commerciale; de géographie, d'histoire naturelle, d'agriculture, de physique, d'hygiene et de littérature. 3 vols. Paris: Imprimerie de Valade, 1802 (year IX).

Archives parlementaires de 1787 à 1860: Recueil complet des débats législatifs et politiques des chambres françaises, imprimé par ordre du corps législatif sous la direction de mm. J. Mavidal et E. Laurent. Première série (1787 à 1799), 82 vols.; *Deuxième série (1800 à 1860)*, 127 vols. Paris, 1867–1913.

Ardant, Isaac-Philibert. *Projet de code rural et de code forestier*. Paris: Imprimerie Testu, 1819.

Arrêt de la Cour royale de Besançon, pour les communes riveraines de la forêt de Chaux, contre l'Etat: 23 décembre 1841. Besançon: Imprimerie de Bintot, 1842.

Aveline, Jean-Baptiste. *Rapport et projet de décret sur le partage & le défrichement des biens communaux, faits et présentés, au nom du Comité d'Agriculture, par M. Aveline, Député du Département du Calvados, Président du Comité d'Agriculture*. 1792.

Balland, Cl. P. Jos. *Observations sur l'administration des forêts*. Paris: Imprimerie du Cercle Social, 1791.

Baron, François Louis Jérôme. *Projet de l'organisation de l'administration des Eaux et Forêts, proposé au Comité des Domaines de l'Assemblée Nationale, par M. Baron, Membre de ce Comité, & Député du département de la Marne*. Paris: Imprimerie nationale, 1790.

Barrere [M. Barère de Vieuzac]. *Rapport des comités réunis des Domaines, des Finances, de l'Aliénation des biens nationaux, de la Marine, du Commerce & d'Agriculture: Sur les bois & forêts nationales*. Paris: Imprimerie nationale, 1790.

Baudrillart, Jacques-Joseph. *Instruction sur la culture du bois, à l'usage des forestiers: Ouvrage traduit de l'allemand de G.I. Hartig, Maître des Forêts de la principauté de Solms, et Membre honoraire de la Société de physique de Berlin*. Paris: Levrault, 1805.

Nouveau manuel forestier, à l'usage des agens forestiers de tous grades, des arpenteurs, des Gardes des Bois Impériaux et communaux, des Préposés de la Marine pour la recherche des bois propres aux constructions navales. ... Traduit sur la 4e édition de l'ouvrage allemande de M. de Burgsdorf, Grand maître des Forêts de la Prusse ... et adapté à notre système d'administration d'après l'ordre du Gouvernement. 2 vols. Paris: Arthus-Bertrand, 1808.

Traité général des eaux et forêts, chasses et pêches. Part 1, tome 1, *Recueil chronologique des réglemens forestiers: contenant les ordonnances, édits et déclarations des rois de France; les arrêts du conseil et des cours souveraines; les lois, arrêtés du gouvernement, décrets, ordonnances du roi, arrêts de la Cour de Cassation, décisions ministérielles, circulaires et instructions administratives*. Paris: Imprimerie de Madame Huzard, 1821. (Note: Baudrillart initially intended the *Recueil* as one volume of the planned four-volume *Traité général des eaux et forêts, chasses et pêches*, but the work evolved into a five-volume work unto itself. To avoid confusion with the volume numbering of the Traité, I use "tome" to refer to the *Recueil* volumes.)

Traité général des eaux et forêts, chasses et pêches. Part 1, tome 3, *Recueil chronologique des réglemens sur les forêts, chasses et pêches, contenant les lois, ordonnances royales, arrêts de la Cour de Cassation, décisions ministérielles, et les circulaires et instructions administratives*. Paris: Imprimerie de Madame Huzard, 1824. (See note on preceding item.)

Baudrillart, Jacques-Joseph, and P. E. Herbin de Halle, eds. *Traité général des eaux et forêts, chasses et pêches*. 11 vols., 3 atlases. Paris: Imprimerie de Madame Huzard, 1821–48.

Boislisle, A. M., ed. *Correspondance des Contrôleurs généraux des Finances avec les Intendants des Provinces, publiée par ordre du ministre des finances, d'après les documents conservés aux archives nationales*. Vol. 1, *1683 à 1699*. Paris: Imprimerie nationale, 1874.

ed. *Correspondance des Contrôleurs généraux des Finances avec les Intendants des Provinces, publiée par ordre du ministre des finances, d'après les documents conservés aux archives nationales*. Vol. 2, *1699 à 1708*. Paris: Imprimerie nationale, 1883.

Boislisle, A. M., and P. de Brotonne, eds. *Correspondance des Contrôleurs généraux des Finances avec les Intendants des Provinces, publiée par ordre du ministre des finances, d'après les documents conservés aux archives nationales*. Vol. 3, *1708 à 1715*. Paris: Imprimerie nationale, 1897.

Bonguiot [Bonguyod], Marc-François. *Opinion de Marc-François Bonguiot (du Jura) sur l'administration forestière*. Paris: Imprimerie nationale, 1795 (year III).

Bourgin, Georges, ed. *Le partage des biens communaux: Documents sur la préparation de la loi du 10 juin 1793*. Collection de documents inédits sur l'histoire économique de la Révolution française. Paris: Imprimerie nationale, 1908.

ed. *La Révolution, l'agriculture, la forêt: Lois et règlements*. Paris: Éditions du Comité des Travaux historiques et scientifiques, 1989.

Bourgin, H. and G. *L'industrie sidérurgique en France au début de la Révolution*. Paris, 1920.

Brown, John Croumbie, ed. *French forest ordinance of 1669; with historical sketch of previous treatment of forests in France*. Edinburgh: Oliver and Boyd, 1883.

Buffon, Georges Louis Leclerc, comte de. *Mémoire sur la conservation et le rétablissement des forêts*. In *Mémoires de l'académie royale des sciences*, 140–56. Paris: Académie Royale des Sciences, 1739.

Oeuvres d'histoire naturelle. Vol. 8, *Époques de la nature*. Berne: Chez la Nouvelle Société Typographique, 1792. (Originally published in 1788 as volume 5 of the seven-volume supplement to Buffon's fifteen-volume *Histoire naturelle, générale et particulière*.)

Bulletin des lois [title variants: *Bulletin des lois de la République française; Bulletin des lois de l'Empire français; Bulletin des lois du Royaume de France*]. Paris: Imprimerie nationale des lois, 1794–1931.

Champion, Maurice. *Les inondations en France depuis le VIe siècle jusqu'à nos jours: Recherches et documents*. 6 vols. Vols. 3 and 4, Paris: Dunod, 1861–62.

Charnock, John. *A History of Marine Architecture*. 3 vols. London: R. Faulden, 1800–1802.

Clément, Pierre, ed. *Jean-Baptiste Colbert: Lettres, instructions et mémoires*. 8 vols. Vol. 4, *Administration provinciale, agriculture, forêts, haras, Canal du Languedoc, routes, canaux et mines*. Paris: Imprimerie impériale, 1867.

Code Napoleon; or, The French Civil Code. Translated by George Spence. London: William Benning, 1827.

Comte, Charles. *Traité de la propriété*. 2 vols. Vol. 1. Paris: Chamerot / Ducollet, 1834.

Cotheret. "Sylviculture: Discours et considérations générales sur les forêts." In *Travaux de la Société d'Émulation du département du Jura, 1836*, 26–35. Lons-le-Saunier, 1837.

"Considérations générales sur les forêts." *L'annuaire du département du Jura* (1842): 292–302.

Cousin, Gilbert. *La Franche-Comté au milieu du XVIe siècle; ou, Description de la Haute-Bourgogne connue sous le nom de Comté, par Gilbert Cousin, de Nozeroy (1552 et 1562)*. Edited by Émile Monot. Publications de la Société d'émulation du Jura. Lons-le-Saunier: Lucien Declume, 1907.

Couturier, Jean-Pierre. *Project de décret sur les véritables principes d'une administration forestière, constitutionnelle et paternelle, par Couturier, Député à la Convention nationale, par le Département de la Moselle*. Paris: Imprimerie nationale, n.d. [1794/5?].

Cunin, Étienne. *Observations sur le projet d'aliénation des forêts nationales; par Étienne Cunin, Membre du Comité des domaines, député du département de la Meurthe*. Paris: Imprimerie nationale, n.d. [1790?].

Dalloz, Victor Alexis Désiré, Amand Dalloz, and Henri Thiercelin. *Répertoire méthodique et alphabétique de législation, de doctrine et de jurisprudence en matière de droit civil, commercial, criminel, administratif, de droit de gens et de droit public*. New ed. Vol. 25, Paris: Bureau de la Jurisprudence générale du royaume, 1849.

De Bry, Jean Antoine Joseph Baron. *Mémoire statistique du département du Doubs*. Paris: Imprimerie impériale, year XII (1804).

Delisle de Moncel. *Mémoire sur le repeuplement, l'augmentation & la conservation à venir des bois dans les départemens de la Meurthe, Moselle, Aisne, Meuse, Marne, &c*. Nancy: Chez H. Haener, Imprimeur du Roi, 1791.

"Des villages de Cassini aux communes d'aujourd'hui." http://cassini.ehess.fr/cassini/fr/html/index.htm.

Dralet, Étienne-François. *Traité du régime forestier, ou Analyse méthodologique et raisonnée des arrêts, règlements, décisions, instructions et circulaires concernant l'organisation des officiers et employés forestiers, et la partie administrative de leurs fonctions*. 2 vols. Paris: Arthus-Bertrand, 1812.

Description des Pyrénées, considérées principalement sous les rapports de la géologie, de l'économie politique, rurale et forestière, de l'industrie et du commerce. La Rochelle: Éditions La Découvrance, 2006.

Droz, François Nicolas Eugène. *Réflexions sur les inconvéniens & les dangers des nouveaux systêmes d'administration, relativement à la Franche-Comté*. Besançon, 1789.

Dubois-Bellegarde [Dubois de Bellegarde], Antoine. *Quelques idées sur les forêts et bois nationaux, par le Citoyen Antoine Dubois-Bellegarde, ci-devant Député du département de la Charente au Corps législatif*. Paris: Imprimerie de Lemaire, year VIII (1800).

Duhamel du Monceau, Henri-Louis. *Élements de l'architecture navale, ou traité pratique de la construction des vaisseaux*. Paris, 1752.

Des semis et plantations des arbres et de leur culture, ou, Méthodes pour multiplier et élever les arbres, les planter en massifs & en avenues: former les forêts & les bois: les entretenir & rétablir ceux qui sont dégradés: faisant partie du Traité complet des bois & des forêts. Paris: H. L. Guerin & L. F. Delatour, 1760.

"Art du charbonnier ou manière de faire le charbon de bois." In *Descriptions des Arts et Métiers faites ou approuvées par Messieurs de l'Académie royale des Sciences*, 1761.

De l'exploitation des bois; ou, Moyens de tirer un parti avantageux des taillis, demi-futaies et hautes-futaies, et d'en faire une juste estimation: avec la description des arts qui se pratiquent dans les forêts: faisant partie du Traité complet des bois & des forests. 2 vols. Paris: H. L. Guerin & L. F. Delatour, 1764.

Du transport, de la conservation et de la force des bois; ou l'on trouvera des moyens d'attendrir les bois, de leur donner diverses courbures, sur-tout pour la construction des vaisseaux; faisant la conclusion du Traité complet des bois & des forests. Paris: L. F. Delatour, 1767.

Duvergier, J. B., ed. *Collection complète des lois, décrets, ordonnances, réglemens, avis du Conseil-d'Etat, publiée sur les éditions officielles du Louvre, de l'Imprimerie nationale, par Baudouin; et du Bulletin des Lois; de 1788 à 1830 inclusivement, par ordre chronologique.* 2nd ed. 30 vols. Paris: Chez A. Guyot et Scribe, 1834–38.

Édit portant règlement général pour les eaux et forêts, Saint-Germain-en-Laye, 31 août 1669. Edict no. 571; reprinted in *Recueil général des anciennes lois françaises, depuis l'an 420, jusqu'à la révolution de 1789*, 29 vols., edited by François-André Isambert, Athanase-Jean-Léger Jourdan (Decrusy), and Alphonse-Honoré Taillandier, 18:219–311. Paris: Berlin-Leprieur, 1821–33.

Essuile, J. F., "Observations sur la consommation des bois en France et sur leur reproduction." In *Observations sur divers objets importants.* Berlin: Société typographique, 1787.

Ethis de Novéau, "Essai sur cette question: S'il seroit plus utile en Franche-Comté de donner à chacun la liberté de clorre ses héritages pour les cultiver à son gré, que de les laisser ouverts, pour le vain [*sic*] pâturage, après la récolte des premiers fruits?" In *Mémoires de l'Académie de Besançon*, 2. Besançon: Imprimerie de Cl. Jos. Daclin, 1767.

Evelyn, John. *Sylva, or a Discourse of Forest-Trees, and the Propagation of Timber in His Majesties Dominions: To which Is Annexed POMONA; Or an Appendix Concerning Fruit-Trees in Relation to Cider, the Making and Several Ways of Ordering It.* London: Royal Society of London for Improving of Natural Knowledge, 1664. In *The Writings of John Evelyn*, edited by Guy de la Bédoyère. Woodbridge, UK: Boydell Press, 1995.

Fabre, Jean-Antoine. *Essai sur la théorie des torrens et des rivières.* Paris: Bidault, 1797 (year V).

Festy, Octave, ed. *Les animaux ruraux en l'an III: Dossier de l'enquête de la commission de l'agriculture et des arts.* 2 vols. Paris: vol. 1, P. Hartmann; vol. 2, Tépac, 1941–46.

Froidour, Louis de. *Nouvelle instruction pour les gardes des eaux & forêts, pesches & chasses: avec une maniere tres-facile pour dresser leurs procez verbaux & raports, conformément à l'ordonnance du mois d'aoust 1669*. Paris: Chez Henry Charpentier, 1692.

Gerbaux, Fernand, and Charles Schmidt, eds. *Procès-verbaux des Comités d'agriculture et de commerce de la Constituante, de la Législative et de la Convention*. 4 vols. Collection de documents inédits sur l'histoire économique de la Révolution Française, Ministère de l'Instruction Publique. Paris: Imprimerie nationale, 1906–10.

Girod-Chantrans, Justin. "Quelques réflexions politiques sur l'accroissement des forges et des usines de l'ancienne province." *Mémoires de la Société d'agriculture, sciences, commerce et arts de la Haute-Saône* 2 (1808): 6.

Godard, Charles, and Léon Abensour, eds. *Cahiers de doléances du bailliage d'Amont*. 2 vols. Vol. 1, Besançon, 1918; vol. 2, Auxerre, 1927.

Gollut, Loys. *Les mémoires historiques de la République séquanoise et des princes de la Franche-Comté de Bourgogne*. 2 vols. 1592; new ed., Roanne: Editions Horvath / Bibliothèque municipale de Besançon, 1978.

Guyot, M., ed. *Répertoire universel et raisonné de jurisprudence civile, criminelle, canonique et bénéficiale; ouvrage de plusieurs jurisconsultes*. 17 vols. Paris: Visse, 1784.

Harvey, William. *De circulatione sanguinis. The circulation of the blood. Two anatomical essays by William Harvey, together with nine letters written by him*. Translated by Kenneth J. Franklin. Oxford: Blackwell Scientific Publications, 1958.

Herbin de Halle, P. E., and J. Peuchet, eds. *Statistique générale et particulière de la France et ses Colonies: avec une Nouvelle Description topographique, physique, agricole, politique, industrielle et commerciale de cet état*. 7 vols. Paris: F. Buisson, 1803 (year XII).

Houry. "Mémoire sur cette question: Quels sont les moyens d'augmenter la production de Bois, et de diminuer leur consommation sans nuire a l'agriculture?" *Mémoires de la Société d'agriculture, sciences, commerce et arts de la Haute-Saône* 1 (1806): 61–82.

Houry and Descolins. "Rapport sur l'usage de la Houille et de son Escarbille, fait à la séance publique du 3 frimaire an 14." *Mémoires de la Société d'agriculture, sciences, commerce et arts de la Haute-Saône* 1 (1806): 139–42.

Institut national de l'information géographique et forestière, Inventaire forestier. *Le mémento: La forêt en chiffres et en carte*. 2013. http://inventaire-forestier.ign.fr/spip/spip.php?article583.

Institut national de l'information géographique et forestière, Inventaire forestier national. *La forêt française: Les résultats issus des campagnes d'inventaire 2005 à 2009: Les résultats pour la region Franche-Comté*. http://inventaire-forestier.ign.fr/spip/IMG/pdf/IFN_Publi_2010_Franche-Comte.pdf.

Isambert, François-André, Athanase-Jean-Léger Jourdan (Decrusy), and Alphonse-Honoré Taillandier, eds. *Recueil général des anciennes lois françaises, depuis l'an 420 jusqu'à la Révolution de 1789*. 29 vols. Paris: Belin-Leprieur, 1821–33.

Jouvenot, Robert, ed. *Le bailliage de Baume-les-Dames en 1789: Les cahiers de doléances*. Paris: Les Belles Lettres, Annales littéraires de l'université de Besançon, 1985.

Jugement du Tribune de première instance de Dole (Jura) pour les communes riveraines de la forêt de Chaux [les maintenant dans l'exercise de leurs droits de panage, pacage et affouage dont elles jouissaient avant et lors de la promulgation du Code forestier de 1827] contre l'Etat (Du 28 août 1839). Dole: Imprimerie Pillot, 1842.

Latruffe-Montmeylian. *Des droits des communes sur les biens communaux, ou examen historique et critique des démemembremens des Usages communaux, opérés autrefois sous les noms de Réserves et de Triages, et aujourd'hui sous celui de Cantonnement.* 2 vols. Paris: Libraire de la Forest, 1825.

Lemarchant-Gomicourt [Lemarchant de Gomicourt], A.-J. *Rapport fait … sur la destruction des loups.* Paris: Conseil des Cinq-Cents, year V (1797).

Lequinio de Kerblay, Joseph Marie. *Voyage dans le Jura.* Vol. 2. Paris: Caillot, 1800 (year IX).

Leroyer, G.-A. "Notice sur le Congrès des délégués des sociétés savantes, session de 1864." *Bulletin de la Société industrielle d'Angers et du département de Maine et Loire* 35 (1864): 187–88.

"Les inondations en 1856." In *L'Année scientifique et industrielle*, edited by Louis Figuier, 152–99. Paris: Hachette, 1857.

Loi forestière, 4 December 1985, and Loi d'orientation sur la forêt, 9 July 2001, no. 2001-602. www.legifrance.gouv.fr/.

Lorenz, J. B. *Manuel du Forestier ou Recueil des Lois, Réglemens et Arrêtés relatifs aux Forêts, rendus depuis l'Ordonnance de 1669. Suivi de l'instruction de l'Administration générale actuelle et d'un aperçu général sur la culture, conservation, usage et aménagement des forêts et de la chasse, enfin ce que les Forestiers doivent observer pendant les différentes saisons de l'année.* 2 vols. Sarrebourg: Imprimerie de Jean-Simon Iarreis, an X (1802).

Marc, J. A. "Quelques vues qui pourront servira à la solution de cette question: Quels sont les moyens d'augmenter la production de Bois, et de diminuer leur consommation sans nuire à l'Agriculture ni aux Arts?" *Mémoires de la Société d'agriculture, sciences, commerce et arts de la Haute-Saône* 1 (1806): 29–57.

Marquiset, Armand. *Manuel de l'usager dans les bois communaux.* Paris: Gauthier frères, 1834.

Mémoire de l'Intendant de Franche-Comté, publié d'après un manuscrit de la Bibliothèque de Besançon. Edited by Maurice Pigallet. Paris: Librarie ancienne Honoré Champion, 1914.

Mémoire pour le Prince Pierre d'Arenberg, demandeur en cantonnement et défendeur en révendication; contre les communes de Remoray et de Froidefontaine, défenderesses en révendication; contre les communes de Boujon, de Rondefontaine, … [toutes voisines de la forêt de Haute-Joux, dans le val de Mièges (Jura), soit demanderesses en révendication ou en supplément de cantonnement, soit défenderesses en cantonnement, et divers particuliers aussi défendeurs en cantonnement. Signé: Bryon, avoué; Clerc de Landresse, conseil], 1833.

Mémoires et documents inédits, pour servir à l'histoire de la Franche-Comté. Vols. 1, 4, 9. Besançon: Académie des sciences, belle-lettres & arts de Besançon, 1838, 1867, 1900.

Ministère de l'Agriculture et de la Pêche, Direction générale de la forêt et des affaires rurales, sous-direction de l'environnement et de la ruralité. *Circulaire 2007–5006 rélatif à la protection et la mise en valeur des espaces pastoraux de la loi relatif au développement des territories ruraux (DATAR) du 23 February 2005 et de l'ordonnance du 1er July 2004 relatif aux associations syndicales des propriétaires.* http://agriculture.gouv.fr/IMG/pdf/dgfarc20075006z.pdf.

Mirabeau, Victor de Riqueti, marquis de. *L'ami des hommes, ou Traité de la population.* Avignon, 1756.

Mirabeau, Victor de Riquetti, marquis de, and François Quesnay. *Philosophie rurale, ou, Économie générale et politique de l'agriculture réduite à l'ordre immuable des loix physiques & morales, qui assurent la prospérité des empires.* Vol. 2, Amsterdam: Chez les Libraires associés, 1763.

Nodier, Charles, J. Taylor, and Alph. de Cailleux. *Voyages pittoresques et romantiques dans l'ancienne France, la Franche-Comté.* Vol. 3. Paris: Imprimerie de J. Didot l'Ainé, 1825; Reprint, Éditions Jeanne Laffitte, Marseille, 1977.

Observations sur le régime des forêts, et sur la nécessité de conserver la forme actuelle de leur administration: Arrêt à l'assemblée générale des grands-maîtres des Eaux et Forêts de France, le 29 janvier 1790. Paris: Imprimerie L. Jorry, 1790.

Ordonnance de Louis XIV pour les armées navales et arcenaux de marine. 15 April 1689. Paris: Chez Estienne Michallet, 1689.

Ordonnances forestières de 1619. In *Recueil des arrêts, ordonnances, et lois de la Chambre des comptes du Parlement de Dole.*

Perrod, M. *Table générale récapitulative des travaux et mémoires de la Société d'émulation du Jura depuis sa fondation, 1818–1917.* Lons-le-Saunier, 1918.

Poncelin. "Réponse à la question: 'Quel est le meilleur aménagement possible des bois de la province de Franche-Comté?'" Mémoires de l'Académie de Besançon. Besançon: Académie de Besançon, 1771.

Poullain-Granprey [Poullain de Grandprey], Joseph Clement. *Projet d'un code forestier, présenté au nom des Comités des Domaines, d'Aliénation, d'Agriculture, de Commerce, des Finances & de la Guerre; par Poullain-Granprey, Député par le département des Vosges.* Paris, an III (1793).

"Protestation d'un serf du Mont-Jura. Contre l'Assemblée des Notables, le Mémoire des princes du sang, le clergé, la noblesse & le Tiers-Etat. Au roi." 1789. French Revolutionary Research Collection, Fiche B2313 9.2/29 Fiche 2; also available on the Bibliothèque nationale website at http://gallica.bnf.fr/ark:/12148/bpt6k67044.

Quesnay, François. *Oeuvres économiques et philosophiques.* Edited by Auguste Oncken. New York: Burt Franklin, 1969.

Rauch, François Antoine. *Harmonie hydro-végétale et météorologique.* 2 vols. Paris: Levrault, 1802.

Rougier de la Bergerie, J. B. *Mémoire au Roi et aux Chambres législatives sur la déstruction des bois et sur les graves conséquences qui peuvent en résulter.* Paris: G.-A. Dentu, 1831.

Traité d'agriculture pratique, ou Annuaire des cultivateurs du département de la Creuse et pays circonvoisins, avec des vues générales sur l'économie rurale... Paris,year III [1795].

Rougier de la Bergerie, Jean-Baptiste [Baron]. *Mémoire et observations sur les abus des défrichemens et la destruction des bois et forêts; avec un projet d'organisation forestière.* Auxerre: Imprimerie de Laurent Fournier; Paris: Chez Madame Huzard, Imprimerie-Librairie, year IX (1800–1801).

Rousset, Alphonse. *Dictionnaire géographique, historique et statistique des communes de la Franche-Comté, Département du Jura.* 6 vols. Besançon: Bintot, 1853–57.

Schéma Régional de Gestion Sylvicole. "Forêt et sylviculture en Franche-Comté." April 2006. www.foretpriveefrancaise.com/data/info/737643-Franche_Comte.pdf.

Scott, M. le Baron. *Essai patriotique, ou Mémoire pour servir à prouver l'inutilité des communaux, l'avantage qu'il y auroit de les défricher, ainsi que toutes les terres incultes; celui que l'état retireroit de la protection accordée à l'agriculture, & les causes qui en empêchent les progrès.* Geneva, 1775.

Service régional de l'information statistique et économique, Direction Régionale de l'Alimentation, de l'Agriculture et de la Forêt, Franche-Comté. "Fiche thématique – forêt." May 2010.

Société d'Agriculture de la Haute-Saône. "Modes d'exploitation des bois" and "Deboisements." In *Recueil agronomique, industriel et scientifique*, 8 March 1846, vol. 5, 209–12. Vesoul: Imprimerie de L. Suchaux, 1849.

Stendhal. *Le rouge et le noir.* Paris: Gallimard Folio Classique, 2000 (1830).

Telles d'Acosta, D. A. *Instruction sur les bois de marine et autres. Contenant des détails relatifs à la physique & à l'analyse du chêne.* Paris: Duchesne [&] Clousier, 1782; Supplément, 1784; Second supplément, 1786.

Les Bois de Marine, 1770.

Projet de réforme et plan d'une nouvelle administration des forêts et bois du Roi. 1789.

Projet d'une nouvelle administration pour les forêts de France et servir de supplément à notre "Instruction sur les bois." 1791.

Tessier, Alexandre-Henri, l'abbé, André Thouin, and Louis-Augustin-Guillaume Bosc, eds. *Encyclopédie méthodique: Agriculture.* 7 vols. Paris: Panckoucke (vols. 1–3); Agasse (vols. 4–7), 1787–1821.

Tourney, E. *Nouvelle méthode d'aménagement et d'exploitation des forêts; suivie de la 3e édition de la Nouvelle méthode de semis, plantation et d'aménagement.* Paris: M. E. Tournay, 1832.

Varenne de Fenille, Philibert Charles. *Observations sur l'aménagement des forêts, et particulièrement des forêts nationales: Présentées à l'Assemblée Nationale, par la Société Royale d'Agriculture, le 9 juin 1791.* Imprimerie de la Feuille du Cultivateur, 1791.

Vautrin, M., Chevalier de Saint-Louis. *Mémoire sur la nécessité d'étendre aux bois des particuliers, notamment des grands propriéetaires, les lois établies pour l'administration des forêts royales, des communes et des établissemens publics, et pour la répression des délits qui s'y commettent.* Paris, 1816.

Vergnes, Jacques Paul de, Général. *Mémoire sur la statistique du département de la Haute-Saône.* Paris: Imprimerie des Sourds-Muets, 1801 (year IX).

Vion-Delphin, François, and François Lassus, eds. *Le bailliage de Quingey en 1789: Les cahiers de doléances.* Paris: Les Belles-Lettres, 1989.

Vuillier. *Mémoire concernant les droits d'usage indéfini de la ville de Dole, dans la forêt impériale de Chaux*. Dole: Imprimerie de Joly, 1812.

Young, Arthur. *Travels in France during the Years 1787, 1788, and 1789*. Garden City, NY: Doubleday, 1969.

Secondary Sources

Books

Acerra, Martine. *Rochefort et la construction navale française: 1661–1815*. 4 vols. Paris: Libraire de l'Inde, 1993.

Acerra, Martine, Jose Merino, and Jean Meyer, eds. *Les marines de guerres européenes, XVIIe–XVIIIe siècles*. Paris: Presses de l'Université de Paris-Sorbonne, 1985.

Ado, Anatoli. *Paysans en révolution: Terre, pouvoir et jacquerie, 1789–1794 (Traduction revue et complétée d'après la seconde édition russe, 1987)*. Translated by Serge Aberdam. Paris: Société des Études Robespierristes, 1997.

Agulhon, Maurice. *Apogée et crise de la civilisation paysanne, 1789–1914*, Paris: Éditions du Seuil, 1976.

Allen, Robert C. *Enclosure and the Yeoman: Agricultural Development of the South Midlands, 1450–1850*. Oxford: Clarendon Press, 1992.

Arnold, Ellen. *Negotiating the Landscape: Environment and Monastic Identity in the Medieval Ardennes*. Philadelphia: University of Pennsylvania Press, 2012.

Arnould, Paul, Micheline Hotyat, and Laurent Simon. *Les forêts d'Europe*. Paris: Nathan, 1997.

Badré, Louis. *Histoire de la forêt française*. Paris: Éditions Arthaud, 1983.

Bamford, Paul Walden. *Forests and French Sea Power, 1660–1789*. Toronto: University of Toronto Press, 1956.

Privilege and Profit: A Business Family in Eighteenth-Century France. Philadelphia: University of Pennsylvania Press, 1989.

Bertaud, Jean-Paul. *The Army of the French Revolution: From Citizen-Soldiers to Instrument of Power*. Translated by R. R. Palmer. Princeton, NJ: Princeton University Press, 1988.

Bess, Michael. *The Light-Green Society: Ecology and Technological Modernity in France, 1960–2000*. Chicago: University of Chicago Press, 2003.

Biet, Christian, ed. *Les droits de l'homme*. Paris: Imprimerie nationale, 1989.

Boichard, Jean, ed. *Le Jura*. Toulouse: Privat, 1986.

Braudel, Fernand, and E. Labrousse. *Histoire économique et sociale de la France*. Vol. 2, *Dès dernier temps de l'âge seigneurial aux préludes de l'âge industriel, 1660–1789*. Paris: Presses Universitaires de France, 1970.

Briselance, Marie-France. *Voyage en Franche-Comté littéraire*. Baume-les-Dames: Éditions Cêtre, 1991.

Brossault, Colette. *Les intendants de Franche-Comté 1674–1790*. Paris: Boutique de l'Histoire, 1999.

Brown, Howard G. *Ending the French Revolution: Violence, Justice, and Repression from the Terror to Napoleon*. Charlottesville: University of Virginia Press, 2006.

C.A.R.A.N., ed. *Histoire de forêts: La forêt française du XIIIe au XXe siècle*. Paris: Centre historique des Archives nationales, 1997.
Clade, Jean-Louis. *Mutations, permanences, ruptures dans le canton de Rougemont, Doubs: 1789–1815*. Collection du bicentenaire de la Révolution française 4. Paris: Les Belles Lettres, 1986.
Clout, Hugh D. *Agriculture in France on the Eve of the Railway Age*. Croom Helm Historical Geography Series. London: Croom Helm, 1979.
The Land of France, 1815–1914. London Research Series in Geography 1. London: Allen & Unwin, 1983.
Cole, Charles Woolsey. *Colbert and a Century of French Mercantilism*. 2 vols. Reprint, Hamden, CT: Archon Books, 1964.
Corvol, Andrée. *L'homme et l'arbre sous l'Ancien Régime*. Paris: Economica, 1984.
L'homme aux bois: Histoire des relations de l'homme et de la forêt XVIIe–XXe siècle. Paris: Fayard, 1987.
Darmangeat, Pierre. *Parcs et réserves de France*. 1991 ed. Paris: Arthaud, 1991.
Davis, Diana K. *Resurrecting the Granary of Rome: Environmental History and French Colonial Expansion in North Africa*. Athens: Ohio University Press, 2007.
Dee, Darryl. *Expansion and Crisis in Louis XIV's France: Franche-Comté and Absolute Monarchy, 1674–1715*. Rochester, NY: University of Rochester Press, 2009.
Devèze, Michel. *La vie de la forêt française au XIVe siècle*. 2 vols Paris: S.E.V.-P.E.N., 1961.
Dey, Aristide, and Jules Finot. *Histoire de la sorcellerie au comté de Bourgogn:, Procès de sorcellerie au bailliage de Vesoul de 1606 à 1636*, edited by Michel Collée. Marseille: Lafitte Reprints, 1983. Reprint of 1861 (Dey) and 1875 (Finot).
Doyle, William. *The Oxford History of the French Revolution*. Oxford: Oxford University Press, 1989.
Origins of the French Revolution. 3rd ed. Oxford: Oxford University Press, 1999.
Dull, Jonathan R. *The Age of the Ship of the Line: The British and French Navies, 1650–1815*. Studies in War, Society, and the Military. Lincoln: University of Nebraska Press, 2009.
Dupâquier, Jacques, ed. *Histoire de la population française*. 4 vols. Paris: Presses Universitaires de France, 1988.
Faber, Karl Georg. *Andreas van Recum, 1765–1828; ein rheinischer Kosmopolit*. Bonn: Röhrscheid, 1969.
Fagan, Brian M. *The Little Ice Age: How Climate Made History 1300–1850*. New York: Basic Books, 2000.
Febvre, Lucien. *Philippe II et la Franche-Comté, la crise de 1567, ses origines et ses conséquences: Étude d'histoire politique, religieuse et sociale*. Abridged ed. Paris: Flammarion, 1970.
Fiétier, Roland, ed. *Histoire de la Franche-Comté*. Toulouse: Edouard Privat, 1977.
Forrest, Alan. *Paris, the Provinces and the French Revolution*. London: Arnold, 2004.

Gadant, Jean. *L'atlas des forêts de France*. Paris: J.-P. de Monza, 1991.

Gille, Bertrand. *Les sources statistiques de l'histoire de France: Des enquêtes au XVIIe siècle à 1870*. Geneva: Libraire Droz, 1964.

Girardot, Jean. *Le département de la Haute-Saône pendant la Révolution*. 3 vols. Vesoul: Société d'agriculture, lettres, sciences et arts de la Haute-Saône, 1973–74.

Glete, Jan. *Navies and Nations: Warships, Navies and State Building in Europe and America, 1500–1860*. 2 vols. Stockholm: Almqvist & Wiksell International, 1993.

Gonzales, Antonio, and Pierre Gresser. *Nouvelle histoire de la Franche-Comté et des francs-comtois*. Pontarlier: Éditions du Belvédère, 2014.

Goubert, Pierre. *The Ancien Régime: French Society, 1600–1750*. Translated by Steve Cox. New York: Harper & Row, 1973.

The French Peasantry in the Seventeenth Century. Cambridge: Cambridge University Press, 1986.

Gravier, Gabriel. *Franche-Comté: Pays des légendes*. 4 vols. Lons-le-Saunier: Marque-Maillard, 1980.

Greenbaum, Louis S. *Talleyrand, Statesman-Priest: The Agent-General of the Clergy and the Church of France at the End of the Old Regime*. Washington, DC: Catholic University of America Press, 1970.

Gresser, P., ed. *Franche Comté*. Le Puy-en-Velay: C. Bonneton, 1983.

Gresser, Pierre. *Le crépuscule du Moyen Âge en Franche-Comté*. Besançon: Éditions Cêtre, 1992.

Gresset, Maurice. *La Franche-Comté à la veille de la Révolution*. Collection du bicentenaire de la Révolution française 8. Paris: Les Belles Lettres, 1988.

L'introduction de la vénalité des offices en Franche-Comté: 1692–1704. Annales littéraires de l'université de Besançon 394. Paris: Les Belles Lettres, 1989.

Grossi, Paolo. *An Alternative to Private Property: Collective Property in the Juridical Consciousness of the Nineteenth Century*. Translated by Lydia Cochrane. Chicago: University of Chicago Press, 1981.

Grove, Richard H. *Green Imperialism: Colonial Expansion, Tropical Island Edens, and the Origins of Environmentalism, 1600–1860*. New York: Cambridge University Press, 1995.

Guérin, Mathieu. *Paysans de la forêt à l'époque coloniale: La pacification des aborigènes des hautes terres du Cambodge, 1863–1940*. Bibliothèque d'histoire rurale. Caen: Association d'histoire des sociétés rurales, 2008.

Guha, Ramachandra. *The Unquiet Woods: Ecological Change and Peasant Resistance in the Himalaya*. Berkeley: University of California Press, 1990.

Guyot, M., ed. *Répertoire universel et raisonné de jurisprudence civile, criminelle, canonique et bénéficiale; ouvrage de plusieurs jurisconsultes*. 17 vols. Paris: Visse, 1784.

Haden-Guest, Stephen, Eileen M. Teclaff, and John K. Wright. *A World Geography of Forest Resources*. New York: Ronald Press, 1956.

Hanson, Paul R. *The Jacobin Republic under Fire: The Federalist Revolt in the French Revolution*. University Park: Pennsylvania State University Press, 2003.

Hayhoe, Jeremy. *Enlightened Feudalism: Seigneurial Justice and Village Society in Eighteenth-Century Northern Burgundy*. Rochester, NY: University of Rochester Press, 2008.

Huffel, Gustave. *Économie forestière*. 2nd ed. 3 vols. Paris: Librairie Agricole de la Maison Rustique, 1919–26.

Hufton, Olwen. *The Poor of Eighteenth-Century France, 1750–1789*. Oxford: Clarendon Press, 1974.

Husson, J. P. *Les forêts françaises*. Nancy: Presses universitaires, 1995.

Jones, Colin. *The Cambridge Illustrated History of France*. Cambridge: Cambridge University Press, 1994.

Jones, Peter. *The Peasantry in the French Revolution*. Cambridge: Cambridge University Press, 1988.

Layton, Robert. *Anthropology and History in Franche-Comté: A Critique of Social Theory*. Oxford: Oxford University Press, 2000.

Les Eaux et Forêts du 12e au 20e siècle. Histoire de l'administration française. Paris: Éditions du Centre national de la recherche scientifique, 1987.

Locatelli, R., Claude-Isabelle Brélot, J. -M. Debard, Maurice Gresset, and J. -F. Solnon. *La Franche-Comté: À la recherche de son histoire (1800–1914)*. Cahiers d'études comtoises 31. Paris: Les Belles Lettres, 1982.

Locatelli, René, Denis Brun, and Henri Dubois, eds. *Les salines de Salins au XIIIème siècle: Cartulaires et livre des rentiers*. Annales littéraires de l'université de Besançon 448; Cahiers d'études comtoises 47. Paris: Les Belles Lettres, 1991.

Lower, A. R. M. *Great Britain's Woodyard: British America and the Timber Trade*. Montreal: McGill-Queens University Press, 1973.

Lyons, Martyn. *Napoleon Bonaparte and the Legacy of the French Revolution*. New York: St. Martin's Press, 1994.

Margadant, Ted W. *French Peasants in Revolt: The Insurrection of 1851*. Princeton: Princeton University Press, 1979.

Urban Rivalries in the French Revolution. Princeton, NJ: Princeton University Press, 1992.

Markoff, John. *The Abolition of Feudalism: Peasants, Lords, and Legislators in the French Revolution*. University Park: Pennsylvania State University Press, 1996.

Marsh, George Perkins. *The Earth as Transformed by Human Action: A Last Revision of Man and Nature*. New York: Charles Scribner's Sons, 1885.

Mayaud, Jean-Luc. *La Franche-Comté au XIX siècle*. L'histoire de la Franche-Comté 7. Wettolsheim: Mars et Mercure, 1979.

Les secondes républiques du Doubs. Annales littéraires de l'université de Besançon 338. Paris: Les Belles Lettres, 1986.

McPhee, Peter. *The Politics of Rural Life: Political Mobilization in the French Countryside, 1845–1852*. New York: Clarendon Press, 1992.

A Social History of France 1780–1880. London: Routledge, 1992.

Revolution and Environment in Southern France, 1780–1830: Peasants, Lords, and Murder in the Corbières. Oxford: Oxford University Press, 1999.

The French Revolution, 1789–1799. Oxford: Clarendon Press, 2002.

Merlin, Colette. *Ceux des villages ... La société rurale dans la "Petite Montagne" jurassienne à la veille de la Révolution*. Cahiers d'études comtoises. 52. Besançon: Annales littéraires de l'université de Besançon, 1994.

Le pain de melée: Campagnes jurassiennes dans l'ancienne France. Bourg-en-Bresse: Taillanderie, 1994.
Merriman, John M. *The Agony of the Republic: The Repression of the Left in Revolutionary France, 1848–1851.* New Haven, CT: Yale University Press, 1978.
Police Stories: Building the French State, 1815–1851. Oxford: Oxford University Press, 2006.
Michel, Guy J. *La Franche-Comté sous les Habsbourg (1493–1678).* Wettolsheim: Mars et Mercure, 1978.
Michelet, Jules. *Histoire de France.* 2nd ed. 17 vols. Vol. 2, Paris: Hachette, 1835.
Mingay, G. E., ed. *The Unquiet Countryside.* London: Routledge, 1989.
Moreau, Richard, and René-André Schaeffer. *La forêt comtoise.* Besançon: Centenaire de la Société forestière de Franche-Comté et des Provinces de l'Est, 1990.
Ørebech, Peter. *The Role of Customary Law in Sustainable Development.* Cambridge: Cambridge University Press, 2005.
Pinchot, Gifford. *Breaking New Ground.* Commemorative ed. Washington, DC: Island Press, 1998.
Plaisance, Georges. *Guide des forêts de France.* Paris: Pierre Horay Éditeur, 1997.
Planhol, Xavier de. *An Historical Geography of France.* Translated by Janet Lloyd. Cambridge: Cambridge University Press, 1994.
Pritchard, Sara B. *Confluence: The Nature of Technology and the Remaking of the Rhône.* Cambridge, MA: Harvard University Press, 2011.
Rackham, Oliver. *Trees and Woodland in the British Landscape.* London: J. M. Dent, 1989.
Radkau, Joachim. *Wood: A History.* Translated by Patrick Camiller. Malden, MA: Polity Press, 2012.
Ragache, Gilles. *Le retour des loups.* Paris: Éditions Ramsay, 1990.
Rajan, S. Ravi. *Modernizing Nature: Forestry and Imperial Eco-development 1800–1950.* Oxford: Clarendon Press, 2006.
Riley, James C. *The Seven Years War and the Old Regime in France: The Economic and Financial Toll.* Princeton, NJ: Princeton University Press, 1986.
Roche, Daniel. *France in the Enlightenment.* Translated by Arthur Goldhammer. Cambridge, MA: Harvard University Press, 1999.
Roche, Max, and Michel Vernus. *Dictionnaire biographique du département du Doubs.* Lons-le-Saunier: Arts et Littérature S.A.R.I., 1996.
Dictionnaire biographique du département du Jura. Lons-le-Saunier: Arts et Littérature S.A.R.I., 1996.
Rousseau, Jean, and Michel Bouvier. *La grande forêt de Chaux: Ses riverains, ses usagers, ses gardiens.* Dole: Les Presses Jurassiennes, 1980.
Sahlins, Peter. *Forest Rites: The War of the Demoiselles in Nineteenth-Century France.* Cambridge, MA: Harvard University Press, 1994.
Saint Jacob, Pierre de. *Les paysans de la Bourgogne du nord au dernier siècle de l'Ancien Régime.* Publications de l'université de Dijon 21. Paris: Société les belles lettres, 1960.
Sclafert, Thérèse. *Cultures en Haute-Provence: Déboisements et pâturages au Moyen Age.* Les hommes et la terre 4. Paris: S.E.V.P.E.N., 1959.

Scott, James C. *Weapons of the Weak: Everyday Forms of Peasant Resistance.* New Haven: Yale University Press, 1985.

Domination and the Arts of Resistance: Hidden Transcripts. New Haven, CT: Yale University Press, 1990.

Shapiro, Gilbert, and John Markoff. *Revolutionary Demands: A Content Analysis of the Cahiers de Doléances of 1789.* Stanford, CA: Stanford University Press, 1998.

Sivaramakrishnan, K. *Modern Forests: Statemaking and Environmental Change in Colonial Eastern India.* Stanford, CA: Stanford University Press, 1999.

Smith, David. *The Practice of Silviculture.* 8th ed. New York: Wiley, 1986.

Soll, Jacob. *The Information Master: Jean-Baptiste Colbert's Secret State Intelligence System.* Ann Arbor: University of Michigan Press, 2009.

Solnon, J. -F. *Quand la Franche-Comté était espagnole.* Paris: Arthème, Fayard, 1983.

Surrell, Alexandre. *Étude sur les torrents des Hautes-Alpes.* 2 vols. Paris: Carilian-Goeury et Victor Dalmont, 1841.

Sutherland, D. G. *France, 1789–1815: Revolution and Counterrevolution.* New York: Oxford University Press, 1986.

Tackett, Timothy. *Religion, Revolution, and Regional Culture in Eighteenth-Century France: The Ecclesiastical Oath of 1791.* Princeton, NJ: Princeton University Press, 1986.

Tucker, Richard P. *Insatiable Appetite: The United States and the Ecological Degradation of the Tropical World.* Berkeley: University of California Press, 2000.

Vardi, Liana. *The Physiocrats and the World of the Enlightenment.* Cambridge: Cambridge University Press, 2012.

Vernus, Michel. *La vie comtoise au temps de l'Ancien Régime.* 2 vols. Vol. 1, Lons-le-Saunier: Éditions Marque-Maillard, 1988.

Vidler, Anthony. *Claude-Nicholas Ledoux: Architecture and Social Reform at the End of the Ancien Regime.* Cambridge, MA: MIT Press, 1990.

Vion-Delphin, François, Pierre Gresser, André Robert, and Claude Royer, eds. *Les hommes et la forêt en Franche-Comté.* Paris: Editions Bonneton, 1990.

Vivier, Nadine. *Propriété collective et identité communale: Les biens communaux en France 1750–1914.* Paris: Publications de la Sorbonne, 1998.

Waquet, Jean-Claude. *Les grands maîtres des Eaux et Forêts de France de 1689 à la Révolution, suivi d'un dictionnaire des grands maîtres.* Geneva: Droz, 1978.

Warde, Paul. *Ecology, Economy, and State Formation in Early Modern Germany.* Cambridge: Cambridge University Press, 2006.

Weulersse, Georges. *La physiocratie à l'aube de la Révolution, 1781–1792.* Edited by Corinne Beutler. 4 vols. Vol. 4, Paris: Éditions de l'École des Hautes Études en Sciences Sociales, 1985.

Whited, Tamara. "The Struggle for the Forest in the French Alps and Pyrenees, 1860–1940." PhD dissertation, University of California, Berkeley, 1994.

Forests and Peasant Politics in Modern France. New Haven, CT: Yale University Press, 2000.

Williams, Michael. *Deforesting the Earth: From Prehistory to Global Crisis.* Chicago: University of Chicago Press, 2003.

Zupko, Ronald Edward. *French Weights and Measures before the Revolution: A Dictionary of Provincial and Local Units*. Bloomington: Indiana University Press, 1978.

Articles and Book Chapters

Acerra, Martine. "Marine militaire et bois de construction: Essai d'évaluation (1779–1789)." In *Révolution et espaces forestiers: Colloque des 3 et 4 juin 1987*, edited by Denis Woronoff, 113–18. Paris: L'Harmattan, 1988.

Agnoletti, Mauro, coord.; Steven Anderson, Elisabeth Johann, Mart Kulvik, Andrey V. Kushlin, Peter Mayer, Cristina Montiel Molina, John Parrotta, Ian D. Rotherham, and Eirini Saratsi. "The Introduction of Historical and Cultural Values in the Sustainable Management of European Forests: Document produced for the Ministerial Conference on the Protection of Forest in Europe (MCPFE)." *Global Environment* 2 (2008): 172–99.

Allen, Edward A. "Deforestation and Fuel Crisis in Pre-Revolutionary Languedoc, 1720–1789." *French Historical Studies* 13 (1984): 455–73.

Antonetti, G. "Le partage des forêts usagères ou communales entre les seigneurs et les communautés d'habitants." *Revue historique de droit français et étranger* 41 (4th ser.), no. 2 (1963): 238–442.

Appleby, Andrew B. "Epidemics and Famine in the Little Ice Age." In *Climate and History: Studies in Interdisciplinary History*, edited by Robert I. Rotberg and Theodore K. Rabb, 63–84. Princeton, NJ: Princeton University Press, 1981.

Augustin, Jean-Marie. "La propriété et les droits d'usage des habitants de la terre de Mouthe dans la forêt du Noirmont." In *Mélanges offerts à la mémoire de Roland Fiétier par ses collègues de Besançon*, edited by François Lassus. Annales litteraires de l'Université de Besançon. Paris: Les Belles Lettres, 1984.

Bamford, Paul Walden. "French Forest Legislation and Administration, 1660–1789." *Agricultural History* 29 (1955): 97–107.

Becker, M., and F. Le Tacon. "Santé de la forêt: Importance d'une sylviculture adaptée aux conditions de milieu." *Revue forestière française* 37, no. spécial (1985): 7–28.

Belhoste, Jean-Francois. "La sylviculture du XVIe au XIXe siècle." *Sciences et techniques en perspective* 36 (1996): 53–59.

"Forges et forêts en révolution." In *Révolution et espaces forestiers: Colloque des 3 et 4 juin 1987*, edited by Denis Woronoff, 127–36. Collection alternative[s] rurales. Paris: L'Harmattan, 1988.

"Une sylviculture pour les forges, XVIe–XIXe siècles." In *Forges et forêts: Recherches sur la consommation proto-industrielle de bois*, edited by Denis Woronoff, 219–61. Paris: Éditions EHESS, 1990.

Benoît, Bruno. "Le Jura, les jurassiens et Lyon à l'époque du soulèvement lyonnais (mai–octobre 1793)." In *Le Jura contre Paris: Le mouvement fédéraliste jurassien de 1793. Actes de la table-ronde du 3 avril 1993*, edited by Henri Hours, 143–58. Lons-le-Saunier: Société d'émulation du Jura, 1994.

Benoit, Serge. "La consommation de combustible végétal et l'évolution des systèmes techniques." In *Forges et forets*, edited by Denis Woronoff, 87–150. Paris: Éditions EHESS, 1990.

Bertrand, Christelle. "Les hommes et la forêt dans les communautés villageoises de la maîtrise particulière des Eaux et Forêts de Vesoul, au XVIIIe siècle." *Haute-Saône S.A.L.S.A.* 26 (1994): 151–222.

Besson, Louis abbé., "La Petite Vendée." *Mémoires de l'Académie des sciences, belle-lettres et arts de Besançon* (1861): 52–72.

Blais, R. "Contribution à une histoire des gardes forestiers au XVIIIe siècle." *Revue forestière française* 38, no. 1 (1986): 17–26.

Blin, L. "Sur trois maîtrises des eaux et forêts et la navigation de la Saône." *Annales de Bourgogne* 42 (1970): 170–90.

Boichard, Jean. "L'espace agricole comtois à travers le temps." In *Provinces et états dans la France de l'Est: Le rattachement de la Franche-Comté à la France*, edited by Maurice Gresset, 317–49. Cahier d'études comtoises. Besançon: Annales litteraires de l'université de Besançon, 1977.

Bourgenot, Louis. "L'histoire des forêts feuillues en France." In "Eléments d'histoire forestière," special issue of *Revue forestière française* (1977): 7–26.

Brélot, Claude-Isabelle. "Le sentiment provincial en Franche-Comté pendant la première moitié du XIXe siècle: Persistances et sociologie." In *Provinces et états dans la France de l'Est: Le rattachement de la Franche-Comté à la France*, edited by Maurice Gresset, 109–26. Cahier d'études comtoises. Besançon: Annales litteraires de l'université de Besançon, 1977.

"Pour une histoire des forêts comtoises dans la première moitié du XIXe siècle: Le procès de la Haute-Joux." *Société d'émulation du Jura: Travaux* (1979): 181–225.

"Les 'fédéralistes' jurassiens: Républicanisme, contre-révolutionnaires et réseaux d'amitié." In *Le Jura contre Paris: Le mouvement fédéraliste jurassien de 1793. Actes de la table-ronde du 3 avril 1993*, edited by Henri Hours, 61–74. Lons-le-Saunier: Société d'émulation du Jura, 1994.

Brosselin, Arlette, Andrée Corvol, and François Vion-Delphin. "Les doléances contre l'industrie." In *Forges et forêts: Recherches sur la consommation proto-industrielle de bois*, edited by Denis Woronoff, 11–28. Paris: Éditions EHESS, 1990.

Brugnot, Gérard, and Yves Cassayre. "De la politique française de restauration des terrains en montagne à la prévention des risques naturels." In *Actes du colloque Les pouvoirs face aux risques naturels dans l'histoire, Grenoble, 22–23 mars 2001*. Grenoble: MSH Alpes, 2002.

Buttoud, G. "Les propriétaires forestiers privés et l'État." In "Société et forêts," special issue, *Revue forestière française* (1980): 96–101.

"Les projets forestiers de la Révolution (1789–1796)." *Revue forestière française* 1 (1983): 9–20.

Cameron, Iain. "The Policing of Forests in Eighteenth-Century France." In *Police and Policing: Past and Present Society Colloquium*, 1–27. Oxford: Oxford University Press, 1983.

Carrez, Maurice. "L'opinion publique aux environs de Sancey et Belvoir (Doubs) sous la Révolution française: Essai d'interprétation." *Annales historiques de la Révolution française* (October–December 1992): 539–70.

Chauvaud, Frédéric. "Les répresentations morbides de la forêt au XIX siècle." In *La forêt: Perceptions et représentations*, edited by Andrée Corvol, Micheline Hoytat, and Paul Arnould, 367–74. Paris: L'Harmattan, 1997.

Ciriacy-Wantrup, S. V., and R. Bishop. "'Common Property' as a Concept in Natural Resources Policy." *Natural Resource Journal* 15 (1975): 713–27.

Clout, Hugh D. "Industrial Development in the Eighteenth and Nineteenth Centuries." In *Themes in the Historical Geography of France*, edited by Hugh D. Clout, 447–82. London: Academic Press, 1977.

Coquerelle, Suzanne. "L'armée et la répression dans les campagnes (1848)." *Société d'histoire de la Révolution de 1848: Études* 18 (1955): 121–59.

Corvol, Andrée. "L'affouage au XVIII siècle: Intégration et exclusion dans les communautés d'Ancien Régime." *Annales E.S.C.* 36, no. 3 (1981): 390–407.

"Un nouveau Colbert." In *Actes du colloque pour le tricentenaire de la mort de Colbert*, edited by Roland Mousnier, 230–39. Paris: C.D.U. / S.E.D.E.S., 1983.

"La privatisation des forêts nationales aux XVIIIe et XIXe siècles." *Histoire économique et financière de la France: Études et documents* 2 (1990): 211–22.

de Colyar, H. A. "Jean-Baptiste Colbert and the Codifying Ordinances of Louis XIV." *Journal of the Society of Comparative Legislation* 13, no. 1 (1912): 56–86.

de Steiguer, J. E. "The French National Forestry School One Hundred Years after Pinchot." *Journal of Forestry* 92, no. 2 (1994): 18–20.

Demard, J.-C. "L'utilisation du bois dans la tradition paysanne et artisinale comtoise." In "Société et forêts," special issue, *Revue forestière française* (1980): 281–300.

Desaunais, A. "La révolution de 1848 dans le département du Jura (24 février–10 décembre 1848)." In *Mémoires de la société d'émulation du Jura pour 1946–1948, Volume du centenaire de la révolution de 1848 dans le Jura*, 49–359. Lons-le-Saunier, 1948.

Devèze, Michel. "Superficie et propriété des forêts du Nord et de l'Est de la France vers la fin du règne de François 1er (1540–1547)." *Annales E.S.C.* 3 (1960): 485–92.

"La crise forestière en France dans la première moitié du XVIIIe siècle et les suggestions de Vauban, Réaumur, Buffon." In *La forêt et les communautés rurales-XVIe-XVIIIe siècles: Recueil d'articles parus en revue*, 63–84. Paris: Publications de la Sorbonne, 1982.

"Les forêts françaises à la veille de la Révolution de 1789." In *La forêt et les communautés rurales-XVIe-XVIIIe siècles: Recueil d'articles parus en revue*, 179–210. Paris: Publications de la Sorbonne, 1982.

Duhem, Gustave. "La révolution des pommes de terre à Lons-le-Saunier en 1840." *Société d'émulation du Jura: Travaux* (1948): 17–32.

Dumoulin, Jacqueline. "Poursuites en réparation des délits et contraventions commis dans les bois communaux soumis au regime forestier au XIXe siècle." *Revue historique de droit français et étranger* 80, no. 2 (2002): 147–77.

Durand-Prinborgne, Claude. "La propriété forestière moderne." In *Actes du colloque sur la forêt, Besançon, 21–22 oct. 1966*, 305–32. Cahiers d'études comtoises 12. Besançon: Les Belles Lettres, 1967.

Durand-Vaugaron, L. "Le Loup en Bretagne pendant cent ans, 1773–1872, d'après les documents inédits." *Annales de Bretagne* 70, no. 3 (1963): 291–338.

Egret, Jean. "La Révolution aristocratique en Franche-Comté et son échec, 1788–1789." *Revue d'histoire moderne et contemporaine* 1, no. 4 (1954): 245–71.

Evans, F. T. "Wood since the Industrial Revolution: A Strategic Retreat?" In *History of Technology*, edited by A. Rupert Hall and Norman Smith, 37–55. London: Mansell, 1982.

Evans, Julian. "Coppice Forestry: An Overview." In *Ecology and Management of Coppice Woodlands*, edited by G. P. Buckley, 18–27. London: Chapman and Hall, 1992.

Febvre, Lucien. "Un sécretaire d'Erasme: Gilbert Cousin et la réforme en Franche-Comté." *Bulletin de la Société de l'histoire du protestantisme français* 56 (1907): 97–148.

Feeny, David, Firket Berkes, Bonnie J. McCay, and James M. Acheson. "The Tragedy of the Commons: Twenty-Two Years Later." *Human Ecology* 18, no. 1 (1990): 1–19.

Fel, André. "Petite Culture 1750–1850." In *Themes in the Historical Geography of France*, edited by Hugh D. Clout, 215–45. London: Academic Press, 1977.

Ferrez, Jean. "La Convention face au mouvement Jurassien, janvier–août 1793." In *Le Jura contre Paris: Le mouvement fédéraliste jurassien de 1793. Actes de la table-ronde du 3 avril 1993*, edited by Henri Hours, 77–101. Lons-le-Saunier: Société d'émulation du Jura, 1994.

"Les jurassiens devant le Tribunal révolutionnaire." In *Le Jura contre Paris: Le mouvement fédéraliste jurassien de 1793. Actes de la table-ronde du 3 avril 1993*, edited by Henri Hours, 173–210. Lons-le-Saunier: Société d'émulation du Jura, 1994.

Fesquet, Frédéric. "L'arbre au secours des hommes: Les bienfaits de la forêt dans le discours forestier au XIX et XXe siècles." In *La forêt: Perceptions et représentations*, edited by Andrée Corvol, Micheline Hoytat, and Paul Arnould, 163–72. Paris: L'Harmattan, 1997.

Fière, Remy. "Pour aller à Mouthe, descendre sous le zero." *Libération*, 3 February 2001. www.liberation.fr/guide/0101362860-pour-aller-a-mouthe-descendre-sous-le-zero.

Ford, Caroline. "Nature, Culture and Conservation in France and Her Colonies 1840–1940." *Past and Present* 183, no. 1 (2004): 173–98.

Fortmann, Louise, and John W. Bruce. "Why Tenure Matters." In *Whose Trees? Proprietary Dimensions of Forestry*, edited by Louise Fortmann and John W. Bruce, 1–9. Boulder, CO: Westview Press, 1988.

Gadant, J. "Le chêne aux XVIIe siècle: Bois de marine et arbre à fruits." *Revue forestière française* 29, no. 1 (1977): 57–59.

Gauthier, Florence. "De Mably à Robespierre: De la critique de l'économique à la critique du politique, 1775–1793." In *La guerre du blé au XVIIIe siècle*, edited by Florence Gauthier and Guy-Robert Ikni, 111–44. Paris: Les Editions de la Passion, 1988.

Gay, Annie. "Les acquéreurs de biens nationaux à Dole." *Société d'émulation du Jura. Travaux* (1990): 229–42.

Geneau, G. "La législation forestière sous l'Ancien Régime." *Revue des Eaux et Forêts* 174 (1940): 32–41, 171–85, 248–63.

Gibson, Clark C., Elinor Ostrom, and Margaret A. McKean. "Forests, People, and Governance: Some Initial Theoretical Lessons." In *People and Forests: Communities, Institutions, and Governance*, edited by Clark C. Gibson, Margaret A. McKean, and Elinor Ostrom, 227–42. Cambridge, Mass.: MIT Press, 2000.

Gille, Bertrand. "Un recensement des usines comtoise en 1744." *Revue d'histoire de la sidérurgie* 2 (1961): 257–76.

"L'enquête sur les bois de 1783." Paper presented at the Actes du 88ème congrès national des sociétés savantes, section d'histoire moderne et contemporaine, Clermont-Ferrand, 1963.

Goujon, Pierre. "Legislation et droits d'usage en France dans la premiere moitié du XIXe siècle." In *La Coutume et la lois: Etudes d'un conflit*, edited by Claude Journès et al., 15–25. Lyon: Presses universitaires de Lyon, 1986.

Graham, Hamish. "Profits and Privileges: Forest and Commercial Interests in Ancien Regime France." *French History* 16, no. 4 (2002): 381–401.

"Policing the Forests of Pre-industrial France: Round Up the Usual Suspects." *European History Quarterly* 33, no. 2 (2003): 157–82.

"Greedy or Needy? Forest Administration and Landowners' Attitudes in South-western France during the Eighteenth Century." *Rural History* 16, no. 1 (2005): 1–20.

"Fleurs-de-lis in the Forest: 'Absolute' Monarchy and Attempts at Resource Management in Eighteenth-Century France." *French History* 23, no. 3 (2009): 311–35.

Gresset, Maurice. "Les débuts du regime français en Franche-Comté (1674–1675)." In *Provinces et états dans la France de l'Est: Le rattachement de la Franche-Comté à la France*, edited by Maurice Gresset, 19–38. Cahier d'études comtoises. Besançon: Annales litteraires de l'université de Besançon, 1977.

"Les Francs-Comtois entre la France et l'Empire." In *Régions et régionalisme en France du XVIIIe siècle à nos jours*, 106–13. Paris: Presses Universitaires de France, 1977.

Hardin, Garrett. "The Tragedy of the Commons." *Science* 162 (1968): 1243–48.

Hours, Henri. "L'administration du Jura pendant le mouvement fédéraliste: Rôle et enjeu des institutions." In *Le Jura contre Paris: Le mouvement fédéraliste jurassien de 1793. Actes de la table-ronde du 3 avril 1793*, 21–40. Lons-le-Saunier: Société d'émulation du Jura, 1994.

Hunt, David. "Peasant Movements and Communal Property during the French Revolution." *Theory and Society* 17 (1988): 255–83.

Husson, Jean-Pierre. "L'histoire du paysage forestier français: Les trois ages de la forêt." *Gavroche* 62 (1992): 8–10.

Izibert, Pierre. "La forêt de Chaux." *Pays comtois: Le magazine du patrimoine, de l'histoire et de l'art de vivre*, March–April 1998, 28–39.

Jennings, Jeremy. "The Debate about Luxury in Eighteenth- and Nineteenth-Century French Political Thought." *Journal of the History of Ideas* 68, no. 1 (2007): 79–105.

Jolas, T. "Bois communaux à Minot (Côte-d'Or)." In "Société et forêts," special issue, *Revue forestière française* (1980): 218–29.

Jones, Colin. "The Politics and Personnel of Social Welfare." In *Beyond the Terror: Essays in French Regional and Social History, 1794–1815*, edited by Gwynne Lewis and Colin Lucas, 53–91. Cambridge: Cambridge University Press, 1983.

Jones, Peter. "Agrarian Radicalism during the French Revolution." In *Reshaping France: Town, Country, and Region during the French Revolution*, edited by Peter Jones and Alan Forrest, 137–51. Manchester: Manchester University Press, 1991.

Journès, Claude. "L'utilisation d'une forêt dans l'Ain du début du dix-neuvième siècle: Enjeux et conflits." In *La coutume et la loi: Études d'un conflit*, edited by Claude Journès, Andrée Corvol, Luc Ferry, Pierre Goujon, Françoise Fortunet, J. Poumarède, Jacques Michel, and Évelyne Serverin, 39–60. Lyon: Presses universitaires de Lyon, 1986.

Jouvenot, Robert. "Les cahiers de doléances en Franche-Comté." In *Questions d'histoire sur la Franche-Comté, à l'occasion du 99e congrès national des Sociétés savantes*, edited by Comité des travaux historiques et scientifiques (CTHS) Orientations de recherche, 238–42. Besançon: CTHS Bulletin de la section d'histoire moderne et contemporaine, 1974.

Keyser, Richard. "The Transformation of Traditional Woodland Management: Commercial Sylviculture in Medieval Champagne." *French Historical Studies* 32, no. 3 (2009): 353–84.

Kielbasa, Agnès. "Promenons-nous dans les bois en Franche-Comté." *La Terre de chez nous*, 6 August 2005, 7.

Klein, Philippe. "La Franche-Comté touche toujours du bois." *Pays comtois*, March–April 1997, 50–69.

Lachaussée, E. "L'évolution de la sylviculture en France depuis 100 ans, ou des principes de Lorentz et Parade à la sylviculture à but industriel et commercial." *Bulletin de la Société Forestière de Franche-Comté et des Provinces de l'Est* 33, no. 1 (1966): 5–10.

LaFargue, Jérôme. "Entre la ruse et l'impuissance: Les paysans face à la règle de droit au XIXe siècle." In *Sociabilité et politique en milieu rural: Actes du colloque organisé à l'Université Rennes 2, les 6, 7 et 8 juin 2005*, edited by Annie Antoine and Julian Mischi, 87–100. Rennes: Presses Universitaires de Rennes, 2008.

Larrère, Raphaël. "Rauch ou Rougier de la Bergerie: Utopie ou Réforme?" In *Révolution et espaces forestiers: Colloque des 3 et 4 juin 1987*, edited by Denis Woronoff. Collection Alternative[s] rurales, 247–56. Paris: L'Harmattan, 1988.

Lassus, François. "La sidérurgie comtoise d'Ancien Régime (XVIe–XVIIIe siècles)." In *De la mine à la forge en Franche-Comté: Des origines au XIXe siècle*, edited by Jean-Paul Jacob and Michel Mangin, 139–51. Paris: Annales littéraires de l'université de Besançon (Belle Lettres), 1990.

Lewis, Gwynne. "Political Brigandage and Popular Disaffection in the South-east of France 1795–1804." In *Beyond the Terror: Essays in French Regional and Social History, 1794–1815*, edited by Gwynne Lewis and Colin Lucas, 195–231. Cambridge: Cambridge University Press, 1983.

Liston, Andrew, and Arne Rumbold. "A Forest Study Tour in North East France." *Quarterly Journal of Forestry* 83, no. 2 (1989): 107–13.

Lorion, André. "Les inondations de la Seine sous le Consulat et l'Empire." *Revue de l'Institut Napoléon*, no. 127 (1973): 59–62.

Lutun, Bernard. “Les approvisionnements de la Marine (1661–1761): De la règlementation. Règles générales (1674).” *La revue administrative* 55, no. 327 (2002): 292–311.

Marlin, Roger. “Un problème d’affouage en Franche-Comté au milieu du XIX siècle.” *Revue d’histoire économique et sociale* 2 (1957): 167–97.

“Essai d’interprètation de l’insurrection paysanne du Haut-Doubs, dite Petite Vendée, en septembre 93.” *Mémoires de la Société d’émulation du Doubs*, new ser., no. 31 (1989): 33–52.

Matagne, Patrick. “The Politics of Conservation in France in the 19th Century.” *Environment and History* 4 (1998): 359–67.

Matteson, C. Kieko. “‘Bad citizens’ with ‘murderous teeth’: Goats into Frenchmen, 1789–1827.” *Proceedings of the Annual Meeting of the Western Society for French History* 34 (2006): 148–62.

Mayaud, Jean-Luc. “Équilibre, tensions et déprise rurale aux XIXe et XXe siècles.” In *Connaissance de la Franche-Comté: Régards sur les Vosges comtoises: À la découverte des terres comtoises du Parc naturel régional des Ballons des Vosges*, 269–87. Besançon: Centre universitaire d’études régionales, Université de Franche-Comté, 1991.

“Pour une généalogie catholique de la mémoire contre-révolutionnaire: La Petite Vendée du Doubs.” In *Religion-Révolution: Actes du colloque de Saint-Florent-Le-Vieil, 13–15 mai 1993*, edited by Jean-Clément Martin, 215–27. Paris: Anthropos/Economica, 1994.

McPhee, Peter. “‘The Misguided Greed of Peasants’? Popular Attitudes to the Environment in the Revolution of 1789.” *French Historical Studies* 24, no. 2 (2001): 247–69.

Méjean, Annie. “Utilisation politique d’une catastrophe: Le voyage de Napoleon III en Provence durant la grande crue de 1856.” *Revue historique* 295, no. 1 (1996): 133–51.

Merlin, Colette. “Impositions, charges et résistance paysanne dans la Petite Montagne jurassienne à la veille de la Révolution.” In *La Franche-Comté à la veille de la Révolution*, edited by M. Gresset, 13–27. Besançon: Annales littéraires de l’université de Besançon, 1988.

Merlin, Pierre. “Le coup d’état du 2 décembre 1851 dans le Jura, la répression, les hommes, l’exil.” *Société d’émulation du Jura: Travaux* 1995 (1997): 281–313.

“Poligny et l’idée démocratique sous la Seconde République, février 1848–2 décembre 1851.” *Patrimoine polinois* 15 (2000): 91–101.

Merriman, John M. “The Demoiselles of the Ariège, 1829–1831.” In *1830 in France*, edited by John M. Merriman, 87–118. New York: New Viewpoints, 1975.

Miller, Char. “Sawdust Memories: Pinchot and the Making of Forestry History.” *Journal of Forest History* 92, no. 2 (1994): 8–12.

Monnet, Robert. “Les voluntaires comtois, ‘soldats de l’an II.’” *Revue historique des armées* 5, no. 2 (1978): 158–66.

Monniot, Suzanne. “Le role de la forêt dans la vie des populations franc-comtoises de la conquête française à la révolution, 1674–1789.” *Revue d’histoire moderne* 12, nos. 29/30, n.s. 6 (1937): 449–68.

Mukerji, Chandra. “The Great Forestry Survey of 1669–1671: The Use of Archives for Political Reform.” *Social Studies of Science* 37, no. 2 (2007): 227–53.

Neumann, Roderick P. "Land, Justice, and the Politics of Conservation in Tanzania." In *People, Plants, and Justice: The Politics of Nature Conservation*, edited by Charles Zerner, 117–33. New York: Columbia University Press, 2000.

Ogé, Frédéric. "Appropriation communautaire et/ou appropriation étatique de la forêt sous la Révolution." In *Propriété et révolution: Actes du colloque de Toulouse, 12–14 octobre 1989*, edited by Geneviève Koubi, 127–33. Paris: Éditions du Centre national de la recherche scientifique, 1990.

Ostrom, Elinor. "How Types of Goods and Property Rights Jointly Affect Collective Action." *Journal of Theoretical Politics* 15, no. 3 (2003): 239–70.

Ostrom, Elinor, Joanna Burger, Christopher B. Field, Richard B. Norgaard, and David Policansky. "Revisiting the Commons: Local Lessons, Global Challenges." *Science* 284, no. 5412 (1999): 278–82.

Pagney, Pierre. "La région comprise entre forêt de Chaux et forêt de la Serre: Étude rurale." *Revue géographique de l'Est* 8, nos. 3–4, 1–2 (1968, 1969): 229–336 and 3–125.

Pardé, J. "Des temps gallo-romains aux temps contemporains: Premiers pas et progrès des aménagements." In "L'aménagement forestier hier, aujourd'hui, demain," special issue, *Revue forestière française* (1999) 23–44.

Petit, Vincent. "Étude sur l'identité comtoise au XIXe siècle, telle qu'elle est conçue par les milieux politiques et littéraires de l'époque.. . ." *Lettres comtoises, revue de l'Association du livre et des auteurs comtois*, no. 9 (October 2004): 17–35.

Pilbeam, Pamela. "Popular Violence in Provincial France after the 1830 Revolution." *English Historical Review* 91, no. 359 (1976): 278–97.

Pingué, Danièle. "Les Jacobins de province et Paris: Le cas franc-comtois." *Annales historiques de la Révolution française*, no. 330 (2002): 101–13.

Plack, Noelle L. "Agrarian Reform and Ecological Change during the Ancien Régime: Land Clearance, Peasants and Viticulture in the Province of Languedoc." *French History* 19, no. 2 (2005): 189–210.

Plaisance, Georges. "La révolte des Demoiselles." *Barbizier almanach populaire comtois* 5 (1951): 483–86.

"Les droits d'usage forestiers et leur vocabulaire." In *Colloque sur la forêt*, 209–18. Annales littéraires de l'Université de Besançon 88; Cahiers d'études comtoises 12. Besançon: Les Belles Lettres, 1966.

"Les droits d'usage en Forêt de Chaux et leur cantonnement." In *Actes du symposium international d'histoire forestière, 24–28 septembre 1979*. Nancy: ENGREF, 1982.

Price, Roger. "Popular Disturbances in the French Provinces after the July Revolution of 1830." *European Studies Review* 1, no. 4 (1971): 323–55.

Pritchard, Sara B. "Reconstructing the Rhône: The Cultural Politics of Nature and Nation in Contemporary France, 1945–1997." *French Historical Studies* 27, no. 4 (2004): 765–99.

Rey, Maurice. "La forêt de Chaux au moyen âge: Étude de contacts." Paper presented at the Colloque sur la forêt, Besançon, 21–22 October 1966.

Richard, Jean. "Les Loups et la communauté villageoise: Quelques documents." *Annales de Bourgogne* 21 (1949): 284–96.

Rousseau, Jean. "Les droits de la ville de Dole dans la forêt de Chaux: Procès contre l'État (1835–1843)." *Société d'émulation du Jura: Travaux*, no. 1980 (1981): 357–67.

Sarles, Curtis. "The Instatement of Order: State Initiatives and Hegemony in the Modernization of French Forest Policy." *Theory and Society* 35, no. 5–6 (2006): 565–85.

Schaeffer, René. "La forêt résineuse comtoise et la politique forestière au XVIIIème siècle." *Bulletin de la Société Forestière de Franche-Comté et des Provinces de l'Est* 34, no. 6 (1969): 161–67; 242–48.

Sée, Henri. "Les forêts et la question du déboisement en Bretagne à la fin de l'Ancien Régime." *Annales de Bretagne: Revue publiée par la faculté des lettres de Rennes* 36 (1924–25): 1–30.

Solakian, Daniel. "De la multiplication des chèvres sous la Révolution." In *Révolution et espaces forestiers: Colloque des 3 et 4 juin 1987*, edited by Denis Woronoff. Collection Alternative[s] rurales, 53–62. Paris: L'Harmattan, 1988.

"Territoires naturels et communautés: La loi du 14 août 1792–10 juin 1793." In *La nature en révolution 1750–1800*, edited by Andrée Corvol, 145–55. Paris: Éditions L'Harmattan, 1993.

Spitzer, Alan B. "The Elections of 1824 and 1827 in the Department of the Doubs." *French History* 3, no. 2 (1989): 153–76.

Tallett, Frank. "Religion and Revolution: The Rural Clergy and Parishioners of the Doubs, 1780–1797." PhD thesis, University of Reading, 1981.

Turc, Lucien. "L'aménagement des bois de sapin de Franche-Comté." *Revue forestière française* 9 (1950): 445–57.

Vandergeest, Peter, and Nancy Lee Peluso. "Empires of Forestry: Professional Forestry and State Power in Southeast Asia, Parts 1 and 2." *Environment and History* 12, nos. 1 and 4 (2006): 31–64, 359–93.

Vernus, Michel. "La révolution de 1848 à Salins et Arbois: La présence du fouriérisme dans le mouvement démocratique." *Cahiers Charles Fourier* 10 (1999): 77–90.

Vigier, Philippe. "Les troubles forestiers du première XIXe siècle français." In "Société et Forêts," special issue, *Revue forestière française* (1980): 128–35.

Vion-Delphin, François. "La fourniture des bois de marine en Franche-Comté de la conquête à la Révolution." Paper presented at the 103rd Congrès national des Sociétés savantes, Nancy-Metz, 1978.

"Forêts et cahiers de doléances: L'exemple de la Franche-Comté." In *Révolution et espaces forestiers: Colloque des 3 et 4 juin 1987*, edited by Denis Woronoff. Collection alternatives rurales, 11–22. Paris: L'Harmattan, 1988.

"Les forêts du Nord de la Franche-Comté à la veille de la Révolution d'après les cahiers de doléances des bailliages d'Amont et de Baume." In *La Franche-Comté à la veille de la Révolution*, edited by Maurice Gresset, 39–68. Paris: Les Belles Lettres, 1988.

"Forêt, frontière et commerce du bois: Noirmont et Val de Joux à la fin du XVIIIe siècle." *Société d'émulation du Jura: Travaux* 1990 (1989): 199–217.

"L'homme et la forêt dans le Haut-Doubs au XVIIIe siècle: L'exemple de la gruerie de Pontarlier." In *La Forêt, 113e congrès national des Sociétés savantes, Strasbourg, 1988*, edited by Andrée Corvol, 259–82. Paris: Éditions du C.T.H.S., 1991.

"L'homme et la forêt dans les Vosges saônoises au XVIII siècle: L'exemple de la gruerie de Faucogney." In *Connaissance de la Franche-Comté: Régards sur les Vosges comtoises: À la découverte des terres comtoises du Parc naturel régional des Ballons des Vosges*, 219–42. Besançon: Centre Universitaire d'Études Régionales (CUER), Université de Franche-Comté, 1991.

"Le braconnage en Franche-Comté: Une pratique populaire au XVIIIe siècle." In *Forêt et chasse*, edited by Andrée Corvol, 201–11. Paris: L'Harmattan, 2004.

Warde, Paul. "Fear of Wood Shortage and the Reality of the Woodland in Europe, c.1450–1850." *History Workshop Journal* 62, no. 1 (2006): 28–57.

Weipert, R. *Du mode de jouissance des bois communaux et de la distribution de l'affouage aux habitants des communes: Necessité de la révision de l'art. 105 du Code forestier*. Lure: Imprimerie de Bettend, 1852.

Woronoff, Denis. "La 'dévastation révolutionnaire' des forêts." In *Révolution et espaces forestiers: colloque des 3 et 4 juin 1987*, edited by Denis Woronoff. Collection Alternative[s] rurales, 44–52. Paris: L'Harmattan, 1988.

"La politique des autorisations d'usines et la question du bois." In *Forges et forêts*, edited by Denis Woronoff, 57–85. Paris: Éditions EHESS, 1990.

Young, David Bruce. "A Wood Famine? The Question of Deforestation in Old Regime France." *Forestry* 49, no. 1 (1976): 45–56.

Index